Biology of the Eucestoda

Volume 2

Biology of the Eucestoda

Volume 2

edited by

C. Arme

Parasitology Research Laboratory
Department of Biological Sciences
University of Keele
Staffordshire, UK

P. W. Pappas

Zoology Department
Ohio State University
Columbus, Ohio, USA

1983

ACADEMIC PRESS

A subsidiary of Harcourt Brace Jovanovich, Publishers

LONDON · NEW YORK
PARIS · SAN DIEGO · SAN FRANCISCO
SÃO PAULO · SYDNEY · TOKYO · TORONTO

ACADEMIC PRESS INC. (LONDON) LTD.
24/28 Oval Road
London NW1

United States edition published by
ACADEMIC PRESS INC.
111 Fifth Avenue
New York, New York 10003

British Library Cataloguing in Publication Data

Biology of the Eucestoda.
Vol. 2
1. Cestoda
I. Arme, C. II. Pappas, P. W.
595.1'21 QL391.C4

ISBN 0-12-062102-9

LCCCN 8370978

Phototypeset by Latimer Trend & Company Ltd, Plymouth, Devon
and printed in Great Britain by Galliard (Printers) Ltd, Great Yarmouth

Contributors

C. ARME *Parasitology Research Laboratory, Department of Biological Sciences, University of Keele, Staffordshire ST5 5BG, UK*

J. BARRETT *Department of Zoology, U.C.W., Aberystwyth, SY23 3DA, UK*

J. F. BRIDGES *Parasitology Research Laboratory, Department of Biological Sciences, University of Keele, Staffordshire ST5 5BG, UK*

K. S. CHEAH *ARC, Meat Research Institute, Bristol BS18 7DY, UK*

G. C. COLES *Department of Zoology, Morrill Science Centre, University of Massachusetts, Amherst, Mass. 01003, USA*

G. W. ESCH *Department of Biology, Wake Forest University, Winston-Salem, NC 27109, USA*

R. S. FREEMAN *Department of Microbiology and Parasitology, University of Toronto, Canada M5S 1A1*

B. G. HARRIS *Biology Department, North Texas State University, Denton, Tx. 76203, USA*

M. B. HILDRETH *Department of Biology, Tulane University, New Orleans, La. 70118, USA*

D. HOOLE *Parasitology Research Laboratory, Department of Biological Sciences, University of Keele, Staffordshire ST5 5BG, UK*

C. R. KENNEDY *Department of Biological Sciences, University of Exeter, EX4 4PS, UK*

R. D. LUMSDEN *Department of Biology, Tulane University, New Orleans, La. 70118, USA*

P. W. PAPPAS *Zoology Department, Ohio State University, Columbus, Ohio 43210, USA*

M. D. RICKARD *School of Veterinary Science, University of Melbourne, Princes Highway, Werribee, Victoria 3030, Australia*

L. S. ROBERTS *Department of Biology, Texas Tech. University, Lubbock, Tx. 79409, USA*

H. W. STUNKARD *American Museum of Natural History, Central Park West at 79th Street, New York, NY 10024, USA*

J. E. UBELAKER *Department of Biology, Southern Methodist University, Dallas, Tx. 75275, USA*

Preface

Historically the class Cestoidea of the phylum Platyhelminthes has been divided into two sub-classes: the Cestodaria and the Eucestoda. The former contains monozoic organisms that produce a ten-hooked (decacanth) larva, whereas the Eucestoda are mostly polyzoic and their life-cycles involve a six-hooked (hexacanth) larva. It is members of this latter group that are commonly referred to as "tapeworms".

There is an imbalance in the amount of published information concerning these two groups. On the one hand our knowledge of the Cestodaria is limited, many aspects of their biology being poorly understood, and there is even disagreement over whether they belong to the class Cestoidea at all! In contrast, there is a relatively large body of information on eucestodes. They parasitize a wide range of vertebrate and invertebrate hosts, and numerous aspects of their biology have been the subjects of extensive research. In fact it is virtually impossible to examine any major parasitological journal and not find articles on eucestodes. In addition, many papers dealing with certain aspects of eucestode biology appear in non-parasitological journals specializing in other zoological or biological disciplines.

Much information on the biology of eucestodes may also be found in general textbooks of parasitology. A few specialized texts are also available dealing with specific topics and entire volumes have appeared recently devoted to single species. There is, however, no single authoritative source that brings together a wide range of eucestode biology, and this work is an attempt to fill this obvious void in the parasitological literature. We are aware that certain areas have received only brief treatment and may only appear as part of a number of chapters, whereas some might consider they merited a separate section. The decisions on what areas to include was ours and ours alone. The areas that we have selected have been covered by individuals who are experts in their fields. The opinions of some of these authors are in conflict with one another; however we feel that the presentation of a variety of ideas and viewpoints will stimulate research and enhance, rather than detract from, the value of the work.

Keele and Columbus
June, 1983

C. Arme
P. W. Pappas

Contents of Volume 2

Contents of Volume 1

Chapter 7

Host–Parasite Interface

P. W. Pappas

I. INTRODUCTION

The term "host–parasite interface", as originally defined by Read *et al.* (1963), refers to the "region of chemical juxtaposition of regulatory mechanisms of both host and parasite". Such juxtaposition occurs where host and parasite are in intimate contact, and where biochemicals can move relatively freely between host and parasite. Therefore, the external membranes of parasites (those membranes in intimate contact with the host or the surrounding environment) constitute a primary component of the host–parasite interface, and the chemical and physical characteristics of these membranes are important parameters of the host–parasite relationship.

Historically, tapeworms have been a favourite model for the study of host–parasite interfaces; since they lack any remnant of a functional digestive tract, tapeworms have only one interface with the host, that being the tegument (or, more precisely, the distal plasma membrane of the tegument). However, the host–parasite interface cannot be viewed simply

BIOLOGY OF THE EUCESTODA Vol. 2
ISBN 0–12–062102–9

as a "membrane", for the interface includes other components such as the glycocalyx, an "unstirred" region within the glycocalyx, and molecular species adsorbed to the glycocalyx. Thus, these various components of the interface form a dynamic region composed primarily of material of worm origin. Processes such as absorption occur at (or in) the surface membrane; extracellular digestion may occur distally to the surface membrane through the action of membrane-bound enzymes, and possibly distally to the glycocalyx if adsorbed enzymes of host origin are considered (Pappas, 1980a, 1980b).

This chapter will review the biogenesis, composition and "functions" of the various components of host–parasite interface, but it will not include discussions of either the cestode tegument in general, or theoretical aspects of membrane structure and transport; for a general discussion of the tapeworm tegument readers are directed to the accounts of Lumsden (chapter 5, volume 1; 1975), Lumsden and Murphy (1980), Lumsden and Specian (1980) or Chappell (1980); an excellent and extensive review of the theoretical aspects of membrane biology in cestodes is also available (Podesta, 1980).

II. BIOGENESIS OF THE TEGUMENTARY MEMBRANE

The mechanisms of membrane formation in tapeworms are not fully understood. In larval and adult cestodes, radioactive precursors for membrane formation (i.e. monosaccharides and amino acids) are incorporated into macromolecules in cell bodies of the tegumental syncytium. The macromolecules are then transported through the distal cytoplasm to the membrane in the form of vesicles (Lumsden, 1966; Oaks and Lumsden, 1971; Trimble and Lumsden, 1975; Oaks *et al.*, 1981). Pulse-chase experiments (Lumsden, 1966; Oaks and Lumsden, 1971; Trimble and Lumsden, 1975) and biochemical analyses of the protein (Knowles and Oaks, 1979) and lipid fractions (Cain *et al.*, 1977) leave little doubt that the vesicles fuse with the membrane, but the biochemical events accompanying such fusion are unknown. Following incorporation into the membrane, some elements may be modified; e.g. proteins of the tegumental (brush border) membrane of *Hymenolepis diminuta* may be phosphorylated following assimilation (Knowles and Oaks, 1979). At least three different types of cytoplasmic vesicles are found in the tegument of *Spirometra mansonoides*, and these may represent the packaging of different types of material for membrane maintenance and/or synthesis (Oaks and Mueller, 1981).

Like the glycocalyces of other absorptive surfaces, the glycocalyx of the cestode tegument is a product of the underlying cells (or syncytium) (Lumsden, 1966; Oaks and Lumsden, 1971) and is not of host origin (Lumsden, 1974). As discussed in a following section (Section VI), molecules of host origin and cations are adsorbed to the glycocalyx subsequent to membrane synthesis.

III. COMPOSITION OF THE TEGUMENTARY MEMBRANE

Ultrastructural studies and preliminary chemical analyses indicate that the brush border plasma membrane of tapeworms is a "fluid-mosaic", consisting of proteins, and glyco- and lipoproteins, embedded in a lipid bilayer (Singer and Nicolson, 1972). Intramembraneous particles, thought to represent membrane proteins, are demonstrable in freeze-fracture and freeze-etch preparations of the larvae of *Taenia crassiceps* (Belton, 1977; Conder *et al.*, 1981), *T. taeniaeformis* and *Echinococcus granulosus* (Conder *et al.*, 1981), and in adult *H. diminuta* (Lumsden and Murphy, 1980).

The surface (brush border) plasma membrane must first be separated from the underlying cytoplasm for biochemical analysis, and a technique for such a separation has been developed by Oaks and his co-workers (Knowles and Oaks, 1979). Adult tapeworms are incubated in a dilute solution of non-ionic detergent (Triton X-100) to remove the tegument (cytoplasm and membrane). The isolated tegument is treated alternately with solutions of low and high osmolarity and the brush border membrane fraction collected by differential centrifugation. The purity of this brush border membrane fraction has been verified by electron microscopy, and lectin binding and enzymatic studies (Knowles and Oaks, 1979), and this membrane preparation appears superior to that obtained using saponin (Oaks *et al.*, 1977; Rahman *et al.*, 1981a); much higher specific enzyme activities are obtained, without any apparent loss of *membrane* protein, using Triton followed by osmotic disruption rather than saponin and no disruption. This technique was originally developed using *H. diminuta* (Knowles and Oaks, 1979), but it has also been used successfully to isolate brush border membrane from *H. microstoma* (Pappas and Narcisi, 1982) and *S. mansonoides* (Friedman *et al.*, 1980, 1981).

Numerous polypeptides and at least ten glycopolypeptides are demonstrable in the membrane fraction of *H. diminuta* following SDS-polyacrylamide electrophoresis (Knowles and Oaks, 1979), and 20 polypeptides have been identified in a similar preparation from *S. mansonoides* spargana (Friedman *et al.*, 1980). However, the number of demonstrable polypeptides in any species is probably limited only by the sensitivity of the analytical technique; for example, more than 30 polypeptides of molecular masses ranging from 300 000 to less than 14 000 daltons are demonstrable in the brush border of *H. diminuta* using SDS-polyacrylamide gradients (Fig. 1).

Very little data are available regarding the chemical nature and distribution of proteins and glycoproteins in the brush border membranes of cestodes. Phosphorylated proteins occur in the membrane of *H. diminuta* as well as glycoproteins (Knowles and Oaks, 1979), and membrane-bound enzymes are demonstrable in membrane preparations of *H. diminuta* (Knowles and Oaks, 1979; Gamble and Pappas, 1980, 1981a, 1981b; Pappas, 1980c, 1981, 1982a) and *H. microstoma* (Pappas and Narcisi, 1982) (see Section IV). A high affinity binding site, possibly a protein, for vitamin B_{12} is present in the brush border membrane of *S. mansonoides* (Friedman *et al.*, 1981). Based on preliminary radioiodination experiments, at least seven polypeptides are oriented externally in the membrane of *H. diminuta* (Knowles and Oaks, 1979). Additionally, membrane-bound enzymes of *H. diminuta* can be solubilized, without complete dissolution of the membrane, in non-ionic detergent (Pappas, 1980c), indicating that these proteins are not embedded deeply in the lipid bilayer (Fig. 1).

To obtain detailed knowledge regarding the chemical nature and spatial relationships of brush border membrane proteins requires solubilization of the proteins in a "native" state. The brush border membrane of *H. diminuta* is solubilized completely in 0·1% SDS (a strongly anionic detergent) or 1% Zwittergent 3–12 (an amphoteric detergent) (Knowles and Oaks, 1979; Gamble and Pappas, 1980), but it is unlikely that proteins solubilized with either detergent are in a native conformation for the following reasons: (1) both detergents are ionic and would be expected to alter the ionic characteristics of the solubilized proteins: (2) solubilization of the brush border membrane of *H. diminuta* in SDS causes an apparent dissociation of the membrane-bound alkaline phosphohydrolase enzyme(s) into subunits of unequal mass (Pappas, 1980c); and (3) there is a dramatic increase in the membrane-bound 5′-nucleotidase activity of *H. diminuta* upon solubilization of the membrane in SDS (Pappas, 1981), suggesting some type of conformation change in the enzyme(s).

Although SDS is an effective detergent of the membrane of *H. diminuta* at 0°C, non-ionic detergents are completely ineffective in solubilizing the membrane at this temperature (Gamble and Pappas, 1980). At 37°C, however, non-ionic detergents partially solubilize the membrane proteins of *H. diminuta* as evidenced by solubilization of membrane-bound enzymes. But, even 1% Triton X-100, which solubilizes nearly 100% of the membrane-bound alkaline phosphohydrolase, 5′-nucleotidase, ATPase and RNase activities of the membrane, solubilizes only 25–30% of the total protein (Papps, 1982a) (Fig. 1, Table I); a majority of the proteins solubilized by Triton X-100 have acidic isoelectric points (pI's) (Fig. 2).

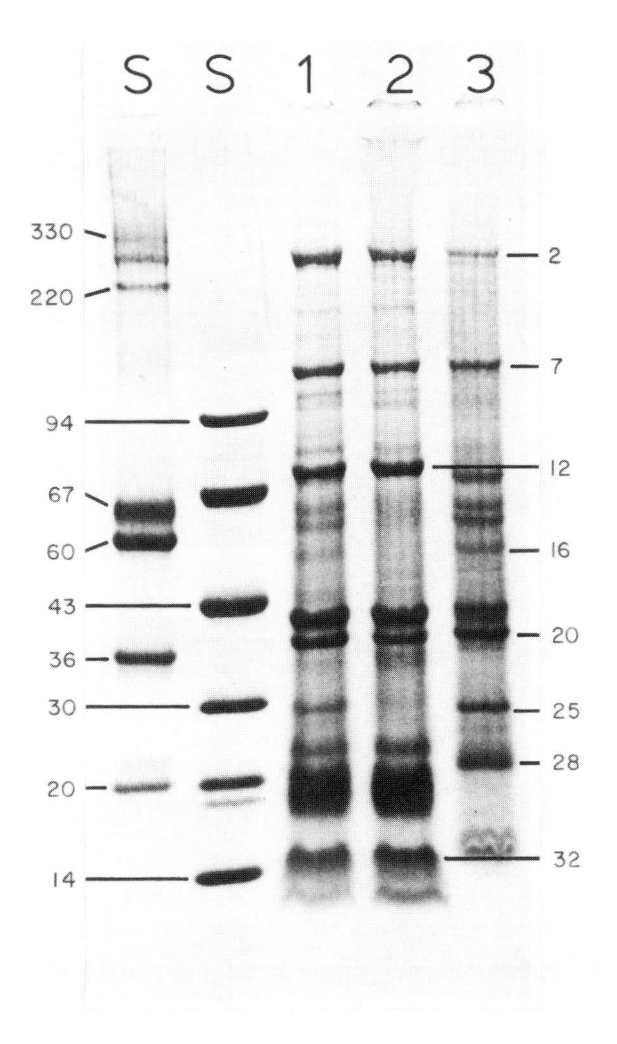

Fig. 1. Sodium dodecyl sulfate-polyacrylamide gradient (5–15% acrylamide) gel comparing standard proteins of known masses and various preparations of the brush border plasma membrane of *Hymenolepis diminuta.* Standard proteins are in the two left columns (S), and masses (in kilodaltons) are indicated on the left side of the figure. For the brush border membrane preparation, prepared according to Knowles and Oaks (1979) (Column 1), and the Triton-insoluble (Column 2) and Triton-soluble (Column 3) fractions, 25 μg of protein was used; since many of the less intensely staining polypeptides (using Coomassie blue R-250) are not evident in the photograph, the major polypeptides have been numbered on the right side of the figure. The selective solubilization of brush border polypeptides is evident upon comparison of Columns 1, 2 and 3. For example, polypeptide No. 2 is only partially solubilized; polypeptides 8–12 and 29–35 are not solubilized, while polypeptides 13–16, 25, 27 and 28 are solubilized.

Table I. Solubilization of membrane-bound enzymes of the brush border plasma membrane of *Hymenolepis diminuta* by various detergents. Data from Gamble and Pappas (1980) and Pappas (1982a).

Detergent and treatment	Percentage of enzyme activity solubilized		
	Alkaline phosphohydrolase activity	Phosphodiesterase activity	Ribonuclease activity
0.1% Sodium dodecyl sulfate (30 min @ 0°C)	98	98	97
1% β-octyl-D-glucoside (30 min @ 0°C)	48	N.D.[a]	43
30 mM β-octyl-D-glucoside (60 min @ 37°C)	59	100	N.D.
1% Dodecyltrimethylammonium bromide (30 min @ 0°C)	46	N.D.	38
1% Zwittergent 3–12 (30 min @ 0°C)	74	N.D.	65
1% Zwittergent 3–12 (30 min @ 25°C)	98	N.D.	98
1% Tween 80 (30 min @ 0°C)	0	N.D.	0
1% Tween 80 (60 min @ 37°C)	58	N.D.	N.D.
1% Brij 35 (30 min @ 0°C)	0	N.D.	0
1% Brij 35 (60 min @ 37°C)	0	N.D.	0
2% Lubrol PX (60 min @ 37°C)	57	41	N.D.
1% Triton X-100 (30 min @ 0°C)	0	N.D.	N.D.
1% Triton X-100[b] (60 min @ 37°C)	98	100	98

[a] N.D. = not determined.
[b] This combination is also effective in solubilizing the ATPase and 5′-nucleotidase activities.

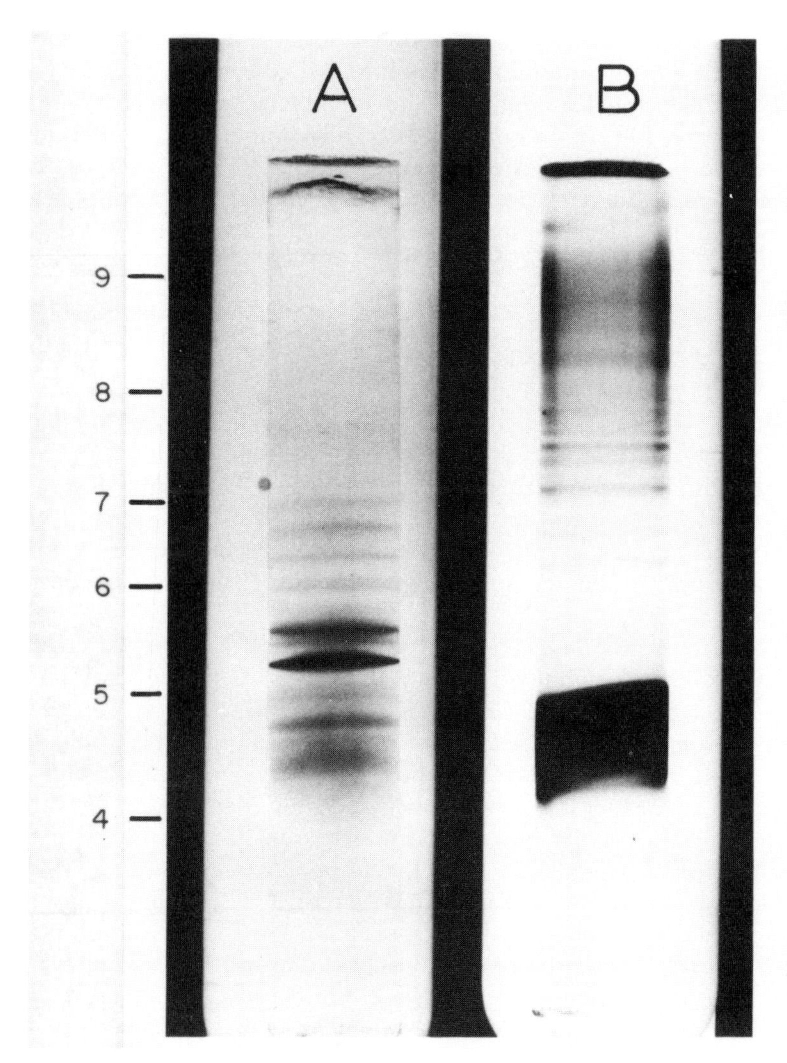

Fig. 2. Isoelectric focussing gels (5% acrylamide, 1% 3–10 ampholyte; 1% Triton X-100) of the Triton-soluble proteins from the brush border plasma membrane of *Hymenolepis diminuta.* Gel A = 50 μg of protein stained with a mixture of Coomassie blue and Crocein scarlet. Gel B = 5 μg of protein "stained" for alkaline phosphohydrolase activity according to Pappas (1980c); the pH gradient of the gels following focussing is indicated along the left side of the figure. Note that the major proteins solubilized by Triton X-100 have isoelectric points below pH 7 (i.e. they are anionic at physiological pH), and that the alkaline phosphohydrolase activity is represented by one or more proteins with a pH of approximately 4.5, and numerous minor bands of activity between pH 6 and 9.

Lipids compose 39% of the dry weight of the brush border membrane of *H. diminuta*. Neutral and polar lipids constitute 41% and 59% of the lipid fraction, respectively. Cholesterol is the primary neutral lipid, but small amounts of sterol esters, free fatty acids and mono-, di- and triglycerides occur as well. Phosphotidyl ethanolamines and cerebrosides-cardiolipins occur in nearly equal amounts in the membrane, and together they constitute nearly 50% of the polar lipids. Smaller amounts of phosphotidyl cholines, lyso-phosphotidyl cholines, phosphotidyl inositols, phosphotidyl serines and sphingomyelins are also present. A wide variety of fatty acids are present in the polar lipid fractions, but in each fraction 16:0, 18:0 and 18:1 fatty acids predominate (Cain *et al.*, 1977).

IV. MEMBRANE-BOUND ENZYMES

The ability of cestodes to hydrolyse substrates in the external medium appears to be due solely to the presence of membrane-bound enzymes associated with the brush border plasma membrane of the tegument. These enzymes may be liberated into the medium in small amounts (Kuo, 1979) during *in vitro* incubations, probably as a consequence of normal membrane turnover, but it is unknown whether this phenomenon is of any significance *in vivo*.

The following membrane-bound enzymes have been demonstrated in cestodes: alkaline phosphohydrolase (E.C. 3.1.3.1); 5′-nucleotidase (E.C. 3.1.3.5); monacyl hydrolase (E.C. 3.1.1.-); ribonuclease (E.C. 3.1.4.22); Type I phosphodiesterase (E.C. 3.1.4.1), and adenosine triphosphatase (E.C. 3.6.1.4). Adenyl cyclase (E.C. 4.6.1.1) activity also occurs in at least one species of cestode (Conway-Jones and Rothman, 1978), but this activity probably plays a more important role in "regulation" than in "digestion" (see Section VII). Type II phosphodiesterase (E.C. 3.1.4.18) (Gamble and Pappas, 1981a; Pappas and Narcisi, 1982), cyclic adenosine-3′,5′-monophosphate phosphodiesterase (E.C. 3.1.4.17) (Pappas and Narcisi, 1982), disaccharidase (maltase and lactase) (Pappas, 1980a; Pappas and Narcisi, 1982), and leucine aminopeptidase (Uglem *et al.*, 1973; Pappas and Narcisi, 1982) activities are not demonstrable in either *H. diminuta* or *H. microstoma*, but the data of Read and Rothman (1958) suggest a membrane-bound disaccharidase is present in *Cittotaenia*.

Monoacyl hydrolase hydrolyses monoglycerides to free fatty acids and glycerol. Such activity apparently occurs at the surface of *H. diminuta* as demonstrated by the observation that during *in vitro* incubations of living *H. diminuta* with mono-olein, free oleic acid is liberated into the medium (Bailey and Fairbairn, 1968). Since *H. diminuta* lacks demonstrable triacyl hydrolase (lipase) activity (Bailey and Fairbairn, 1968; Ruff and Read,

1973), this surface activity appears specific for monoglycerides. No further data are available on this enzyme activity.

Alkaline phosphohydrolase hydrolyses a wide variety of phosphoric monoesters, including various synthetic phosphoesters and phosphorylated monosaccharides (e.g. hexose, pentose and triose phosphates). Many techniques have been used to demonstrate this activity in the surface membranes of *H. diminuta*, including EM cytochemistry (see Lumsden *et al.*, 1968; Dike and Read, 1971a, for examples), end-product analysis of the incubation medium (see Arme and Read, 1970; Dike and Read, 1971a; Lumsden and Berger, 1974, for examples), and studies measuring the effects of hydrolysis products on mediated transport (see Dike and Read, 1971a, 1971b; Uglem *et al.*, 1974; Pappas and Read, 1974; Kuo, 1979, for examples); these studies have been reviewed previously (Pappas, 1980a). Additionally, alkaline phosphohydrolase activity is demonstrable in isolated brush border membrane preparations of *H. microstoma* (Pappas and Narcisi, 1982) and *H. diminuta* (Knowles and Oaks, 1979; Gamble and Pappas, 1980, 1981a; Pappas, 1980c, 1981). Based on various studies using intact worms, it was suggested by Dike and Read (1971a) that multiple phosphohydrolase activities occur in the surface membrane of *H. diminuta*, a hypothesis echoed by additional investigators (Lumsden and Berger, 1974; Pappas and Read, 1974; Kuo, 1979). Also, isoelectric focussing of the soluble membrane proteins of *H. diminuta*, followed by localization of enzyme activity, indicates multiple phosphohydrolase activities which can be differentiated on the basis of their pI's (Fig. 2). However, preliminary analyses of substrate and inhibitor specificities of the alkaline phosphohydrolase activity in isolated brush border membrane preparations of *H. diminuta* suggest that only a single enzyme is present, or that if multiple enzymes are present they have very broad and overlapping specificities and cannot be differentiated at present (Pappas, 1982b).

In isolated membrane preparations of *H. diminuta*, alkaline phosphohydrolase activity is maximal at pH 8.3; upon solubilization in either anionic or non-ionic detergent the pH optimum rises to 8.8–9.0 (Gamble and Pappas, 1981a; Pappas, 1982a). Activity increases dramatically in the presence of Mg^{++}, Mn^{++} or Ca^{++}, and decreases in the presence of *o*-phenanthroline (which chelates Zn^{++}), Cu^{++}, Fe^{++}, Pb^{++}, F^{-}, molybdate, 2-mercaptoethanol, cysteine, p-chloromercuribenzoate, or *excess* Mn^{++} or Mg^{++} (Pappas, 1981c). Treatment of the membrane preparation with either anionic or non-ionic detergent solubilizes the alkaline phosphohydrolase activity (Table I), and both soluble and membrane-bound activities are stable at elevated temperatures (Pappas, 1982b). SDS-polyacrylamide electrophoresis demonstrates that a majority of the alkaline phosphohydrolase activity is associated with polypeptides of molecular weights of 104 000, 172 000 and >340 000 (Knowles and Oaks, 1979; Pappas, 1980c).

Hydrolysis of adenosine-5′-monophosphate (AMP) by 5′-nucleotidase produces inorganic phosphate and adenosine. Data of several studies indicate that nucleotides, including AMP, are hydrolysed at the surface of living *H. diminuta* (Lumsden *et al.*, 1968; Dike and Read, 1971a; Pappas and Read, 1974; Kuo, 1979), and similar activity is demonstrable in brush order membrane preparations of *H. diminuta* (Knowles and Oaks, 1979; Pappas, 1981, 1982a) and *H. microstoma* (Pappas and Narcisi, 1981). Hydrolysis of AMP by membrane preparations of *H. diminuta* is optimal at pH 8.6 (Pappas, 1982a), and stimulated by Zn^{++}, Ca^{++} and Mg^{++}. Hydrolysis is inhibited by molybdate, while other potential inhibitors (F^-, *p*-chloromercuribenzoate and N-methymaleimide) have varying effects depending on the pH of the assay medium (Pappas, 1981). Studies using intact worms indicate that the alkaline phosphohydrolase and 5′-nucleotidase activities are distinct (reviewed by Pappas, 1980a), but analyses of kinetic parameters and substrate and inhibitor specificities are equivocal (Pappas, 1982b).

Phosphodiesterase activity hydrolyses internucleotide bonds (and synthetic nucleotide derivatives) and cyclic diester bonds (such as that found in cyclic-AMP), and various phosphodiesterases are differentiated on the basis of substrate specificity. Type I-phosphodiesterase is present in isolated membrane preparations of *H. diminuta* (Gamble and Pappas, 1981a) and *H. microstoma* (Pappas and Narcisi, 1982). It functions as an exonuclease producing 5′-nucleotides and requiring a free 3′-OH group to function. Neither Type II-phosphodiesterase, which produces 3′-nucleotides and requires a free 5′-OH group, nor cyclic-AMP phosphodiesterase, which hydrolyses cyclic-AMP to AMP, has been detected in membrane preparations of *H. diminuta* or *H. microstoma* (Gamble and Pappas, 1981a; Pappas and Narcisi, 1982).

Type I-phosphodiesterase activity in isolated and solubilized membrane preparations of *H. diminuta* is maximal at pH 8.6, and activity is stimulated by Ca^{++} or Mg^{++}. Several observations indicate that the alkaline phosphohydrolase and phosphodiesterase activities are separate enzymes: (1) potential inhibitors of enzyme activity have distinctly different effects on the two activities (Table II); (2) alkaline phosphohydrolase activity is inhibited 100% by EDTA, while phosphodiesterase is inhibited maximally only 65%; and (3) the two activities can be resolved following SDS-polyacrylamide gel electrophoresis and localization of enzyme activity (Gamble and Pappas, 1981a). Additionally, the membrane-bound RNase (endonuclease) and phosphodiesterase activities display different pH optima and specificities, indicating the endo- and exonuclease activities represent separate enzymes (Gamble and Pappas, 1981a, 1981b).

ATPase activity associated with the surface membranes has been demonstrated in several species of cestodes using various techniques; such activity occurs in *H. citelli* (Rothman, 1966), *H. microstoma* (Pappas and Narcisi, 1982), *H. diminuta* (Lumsden *et al.*, 1968; Pappas and Read, 1974; Kuo, 1979; Knowles and Oaks, 1979; Rahman *et al.*, 1981b; Pappas, 1981; plus others), and *Cysticercus cellulosae* (Sosa *et al.*, 1978), but only the ATPase associated with *H. diminuta* has been studied in any detail.

ATP hydrolysis by intact *H. diminuta*, or an isolated membrane preparation of this species, is inhibited by a wide variety of compounds including phosphorylated monosaccharides and molybdate; F^- inhibits ATP hydrolysis by intact worms, but not by isolated membrane preparations (Lumsden *et al.*, 1968; Pappas and Read, 1974; Kuo, 1979; Pappas, 1981). ATP hydrolysis by isolated membrane preparations is maximal at pH 7.4 (although near maximum hydrolysis rates are obtained at any pH between 7.2 and 8.2) and stimulated by divalent cations (Ca^{++}, Mg^{++} and Mn^{++}); activity is inhibited by molybdate and ethracrynic acid, but not by F^-, SCN^-, HCO_3^-, *p*-chloromercuribenzoate or N-methylmaleimide. Na^+ and K^+ do not stimulate membrane-bound ATPase activity significantly, and ouabain does not inhibit the activity (Pappas, 1981; Rahman *et al.*, 1981b).

It is not known whether the 5′-nucleotidase and ATPase activities of *H. diminuta* represent distinct enzymes, or whether the two activities simply represent the sequential hydrolysis of terminal phosphates from the nucleotide by a single enzyme. Analyses of 5′-nucleotidase and ATPase activities in intact worms (Pappas and Read, 1974; Kuo, 1979) and isolated membrane preparations (Pappas, 1981) are equivocal.

Membrane-bound RNase activity occurs in *H. diminuta* (Pappas *et al.*, 1973a; Gamble and Pappas, 1981b) and *H. microstoma* (Pappas and Narcisi,

Table II. The effects of selected inhibitors on the alkaline phosphohydrolase and Type I-phosphodiesterase activities of the solubilized brush border plasma membrane of *Hymenolepis diminuta*. Data from Gamble and Pappas (1981a).

	Percentage Inhibition of Activity	
Inhibitor	Alkaline phosphohydrolase activity	Phosphodiesterase activity
Glucose-1-phosphate	51	0
Glucose-6-phosphate	54	0
Fructose-1,6-diphosphate	63	0
Inorganic phosphate	93	0
Pyrophosphate	81	11
Yeast RNA	0	54
Polyadenylic acid (poly A)	0	46

1982), but only the enzyme of the former species has been studied in detail. RNase activity associated with *H. diminuta* is an enzyme of worm origin; it cannot be washed off the worms, it cannot be demonstrated in media in which worms have been incubated, and it displays kinetic parameters distinctly different from rat (host) RNase (Pappas *et al.*, 1973a; Gamble and Pappas, 1981b).

In intact worms and isolated brush border membrane preparations of *H. diminuta*, RNase displays typical saturation kinetics with a pH optimum of about 8.3 (Pappas *et al.*, 1973a; Gamble and Pappas, 1981b; Pappas, 1982a). Solubilized RNase activity is inhibited by F^-, cysteine, *p*-chloromercuribenzoate and N-ethylmaleimide; activity is also inhibited by DNA, but this is apparently due to non-productive binding of DNA to the RNase, since there is no demonstrable DNase activity in solubilized membrane preparations of *H. diminuta* (Gamble and Pappas, 1981b).

Hydrolysis of RNA, poly A, poly C, poly I and poly U by solubilized RNase is inhibited by EDTA, suggesting a requirement for divalent cations, yet addition of Ca^{++} or Mg^{++} to solubilized RNase preparations following EDTA treatment does not restore activity (Gamble and Pappas, 1981b). Additionally, RNase activity of intact worms and isolated membrane preparations is inhibited by several divalent cations (including Mg^{++}, Ca^{++}, Mn^{++}, Cu^{++} and Zn^{++}) (Pappas *et al.*, 1973a; Gamble and Pappas, 1981b). The inconclusive nature of these data may be due to several factors, including: (1) difficulties in removing cations from the solubilized membrane preparations (Pappas, 1980c, 1981, 1982b); (2) irreversible changes in the enzyme (possible conformational) due to chelation of cations; or (3) interactions of the cations with the anionic substrate (RNA) (Gamble and Pappas, 1981b). Cation requirements of the RNase need to be examined further.

Native (rat liver) and degraded (yeast) RNA and synthetic polynucleotides are hydrolysed by *H. diminuta* RNase. The action of RNase against native and degraded RNA produces a variety of ethanol-insoluble polynucleotides (i.e. it acts as an endonuclease), and the enzyme appears to hydrolyse preferentially higher molecular weight species of RNA (Pappas and Gamble, 1978). Poly A, poly C and poly U are hydrolysed by solubilized RNase, while poly G is not, but poly G inhibits hydrolysis of the other homopolymers. Apparently poly G interacts with the RNase, but the internucleotide bonds are resistant to hydrolysis (i.e. the binding is non-productive) (Gamble and Pappas, 1981b). The low rate of RNA hydrolysis, compared to hydrolysis rates of poly A, poly C and poly U (Table III), probably reflects the fact that RNA has a variety of linkages between heterologous and homologous nucleotide pairs, and that some of these linkages are resistant to hydrolysis. As noted above, the endonuclease (RNase) and exonuclease (phosphodiesterase) activities of the brush border of *H. diminuta* appear to be distinct enzymes.

Table III. Maximum hydrolysis rates for various substrates by the solubilized RNase of *Hymenolepis diminuta* brush border plasma membrane. Data from Gamble and Pappas (1981b).

Substrate	V_{max}[a]
Yeast RNA	2.52
Polyadenylic acid (poly A)	5.28
Polycytidylic acid (poly C)	5.35
Polyguanylic acid (poly G)	0
Polyuridylic acid (poly U)	11.56

[a] mg substrate hydrolyzed × mg protein^{-1} × hr^{-1}.

V. ABSORPTION

Of the functions commonly associated with the host–parasite interface, none has been studied in greater detail than that of absorption. Cestodes are permeable to a large number of solutes, and a variety of absorptive mechanisms (e.g. diffusion, "active transport", and pinocytosis) occur. Previous studies have dealt mainly with parameters relating to the mechanism (i.e. active or passive) and stereospecificity of solute movement across the external plasma membrane, rather than the underlying mechanisms of solute movement (and accumulation in the cases of active transport). It is now recognized that the syncytial nature of the cestode tegument, the compartmentalization of organic solutes and ions, and the "unstirred layer" at the surface of the tegument must be considered in the evaluation of transport mechanisms. These factors will not be considered in detail here, for they have been addressed elsewhere (Podesta, 1980). Likewise, descriptive kinetic parameters have been excluded from this text since they are of little value in differentiating the various mechanisms of membrane transport, and since they may vary depending on experimental conditions; for those interested in such parameters, a compilation of such values up to 1975 appears in Pappas and Read (1975).

A. Macromolecules

Adult cestodes are not readily permeable to colloidal particles and high molecular weight solutes. Rothman (1967) reported that *H. diminuta* adults are permeable to ferritin and colloidal thorium dioxide and carbon. However, using these same compounds and radioactively labelled

Chlorella-protein, Lumsden *et al.* (1970a) were unable to detect the movement of any colloidal particles or protein across the tegument of *H. diminuta.*

Proteins of host origin have been identified in the "bladder" fluids of larval cyclophyllidean cestodes, suggesting these larvae are permeable to high molecular weight solutes. Host proteins have been identified in hydatid cysts from sheep (Norman *et al.*, 1964; Chordi and Kagan, 1965; Coltorti and Varela-Díaz, 1972), rats (Hustead and Williams, 1977a), mice, humans and gerbils (Coltorti and Varela-Díaz, 1972), and in cysticerci of *T. taeniaeformis* and *T. crassiceps* from rats and mice, respectively (Hustead and Williams, 1977a). Also, hydatid cysts transplanted from one host to another and subsequently analysed for the presence of both hosts' immunoglobulins are found to contain proteins of both host species (Varela-Díaz and Coltorti, 1972; Coltorti and Varela-Díaz, 1975). Additionally, the absorption of radioactively labelled *Chlorella*-protein by *T. crassiceps* larvae (Esch and Kuhn, 1971), iodinated bovine serum albumin, ribonuclease and IgG by *T. taeniaeformis* and *T. crassiceps* cysticerci (Hustead and Williams, 1977a, 1977b), and peroxidase and IgG by hydatid cysts (Coltorti and Varela-Díaz, 1974) has been demonstrated experimentally.

The mechanism by which proteins enter cysticerci and hydatid cysts is not known, although diffusion (Esch and Kuhn, 1971) or fissures in the external membranes (Coltorti and Varela-Díaz, 1975) have been suggested. It is possible that the ionic nature of proteins plays some role in absorption, as suggested by Hustead and Williams (1977a), since cysticerci are permeable to protein (Esch and Kuhn, 1971) but not to inulin (Pappas and Read, 1973).

Larvae and adults of *Schistocephalus solidus* and larvae of *Ligula intestinalis* absorb proteins and colloidal particles. *S. solidus* absorbs peroxidase, lanthanum nitrate and ruthenium red by pinocytosis, with the absorbed material being sequestered in membrane-bound vesicles ("pinosomes") in the cytoplasm of the tegument (Hopkins *et al.*, 1978; Threadgold and Hopkins, 1981). The significance of this pinocytotic activity is unknown, but it has been suggested to function in either nutrition or protection from the host's immune response (Threadgold and Hopkins, 1981).

B. Carbohydrates

Glucose uptake by cestodes occurs by "active transport". Uptake of glucose *via* a saturable (mediated) system and/or accumulation of glucose against a concentration difference has been demonstrated in *H. diminuta* adults (Phifer, 1960a; Pappas *et al.*, 1974; Read *et al.*, 1974; Uglem, 1976) and larvae (Arme *et al.*, 1973), *H. microstoma* (Pappas and Freeman, 1975),

Calliobothrium verticillatum (Fisher and Read, 1971; Pappas and Read, 1972a) *T. crassiceps* larvae (Murrell, 1968; Pappas *et al.*, 1973b) and *T. taeniaeformis* von Brand *et al.*, 1964). Similarly, galactose uptake is mediated and/or accumulative in *H. diminuta* (Read, 1961, 1967), *C. verticillatum* (Fisher and Read, 1971) and *T. crassiceps* larvae (Pappas *et al.*, 1973b). Glucose absorption is inhibited by a variety of monosaccharides (Table IV); in *H. diminuta* (Read, 1961), *C. verticillatum* (Fisher and Read, 1971) and *T. crassiceps* larvae (Pappas *et al.*, 1973b), glucose and galactose are mutually competitive inhibitors indicating that only a single transport site is involved, and β-methylglucoside also enters *H. diminuta* through this same site (Uglem *et al.*, 1978).

Glucose and galactose absorption by *H. diminuta* adults (Read *et al.*, 1974; Starling, 1975; Uglem, 1976), and glucose absorption by *H. microstoma* (Pappas and Freeman, 1975), *C. verticillatum* (Fisher and Read, 1971; Pappas and Read, 1972a), *T. crassiceps* larvae (Pappas *et al.*, 1973b) and *T. taeniaeformis* (von Brand and Gibbs, 1966) are Na^+-dependent; in the absence of external Na^+, glucose enters worms only by diffusion and solute accumulation does not occur. Influxes of Na^+ and glucose are coupled in *C. verticillatum* (Pappas and Read, 1972a) and *H. diminuta* (Read *et al.*, 1974; Uglem, 1976; Love and Uglem, 1978) as well. Glucose uptake by *H. diminuta* is also Cl^--dependent and coupled (Pappas *et al.*, 1974; Pappas and Hansen, 1977; Uglem and Prior, 1980). However, in Cl^--free media glucose uptake by *H. diminuta* is only inhibited partially, and accumulation of glucose against a concentration difference occurs (Pappas *et al.*, 1974; Pappas and Hansen, 1977).

Glucose uptake by *H. diminuta* cystercoids is partially Na^+-dependent, and the effects of phlorizin and Na^+-deletion on glucose uptake are nearly identical. These data suggest two operational glucose transport systems in cysticercoids (Arme *et al.*, 1973), one of which is not Na^+-dependent. A similar situation has also been hypothesized for *H. diminuta* adults (i.e. two glucose transport systems), but various kinetic analyses and inhibitor studies have been unable to demonstrate unequivocally the presence of two systems (reviewed by Roberts, 1980).

The mechanism of ion-coupled glucose transport and accumulation in cestodes is uncertain. It is known that the effects of Na^+ (Pappas *et al.*, 1974) and Cl^--deletion (Pappas and Hansen, 1977) on glucose uptake by *H. diminuta* are readily reversible, and that the coupling coefficients for Na^+ and glucose in both *H. diminuta* (Love and Uglem, 1978) and *C. verticillatum* (Pappas and Read, 1972a) and for Cl^- and glucose in *H. diminuta* (Pappas and Hansen, 1977) are two (2) (i.e. two Na^+ and Cl^- are transported with each glucose). Deletion of Na^+ or Cl^- from the incubation medium decreases the rate of uptake and the affinity of the transport site for glucose, and other cations or anions are poor replacements

Table IV. The actions of various monosaccharides and other compounds as inhibitors of glucose uptake in various species of cestodes (modified after Pappas and Read, 1975).

Inhibitor[a]	*Hymenolepis diminuta*	*Hymenolepis microstoma*	*Taenia crassiceps* larvae	*Calliobothrium verticillatum*
Glucose	+[b]	+	+	+
α-Methylglucoside	+	+	+	+
β-Methylglucoside	+	+	+	N.D.
1-Deoxyglucose	+	N.D.	N.D.	N.D.
2-Deoxyglucose	N.D.	—	N.D.	N.D.
3-O-Methylglucose	+	—	—	—
6-Deoxyglucose	+	N.D.	N.D.	N.D.
Glucosamine	N.D.	N.D.	—	—
N-Acetylglucosamine	N.D.	—	—	—
Galactose	+	+	+	+
2-Deoxygalactose	N.D.	—	N.D.	N.D.
6-Deoxygalactose	—	—	+	N.D.
Galacitol	N.D.	—	—	N.D.
Allose	+	N.D.	N.D.	N.D.
Mannose	N.D.	—	—	—
1,5-Anhydro-D-Mannitol	—	N.D.	—	N.D.
Fructose	—	—	N.D.	—
Phlorizin	+	+	+	+
Phloretin	—	+	N.D.	—
Ouabain	—	—	—	+

[a] Data for table taken from following sources: *Hymenolepis diminuta*—Read, 1961; Phifer, 1960b; McCracken and Lumsden, 1975a; Dike and Read, 1971b; Uglem and Love, 1977; Uglem *et al.*, 1978. *Hymenolepis microstoma*—Pappas and Freeman, 1975. *Taenia crassiceps* larvae—Pappas *et al.*, 1973b. *Calliobothrium verticillatum*—Fisher and Read, 1971.

[b] A (+) indicates the inhibitor inhibits glucose uptake, and a (—) indicates no inhibition. N.D. = not determined.

for (or inhibitors of) Na^+ or Cl^-, respectively (Pappas and Read, 1972a; Read *et al.*, 1974; Pappas and Hansen, 1977; Uglem and Prior, 1980). Large increases in the internal Na^+ and Cl^- concentrations occur in *H. diminuta* during glucose accumulation, and there is a net accumulation of Cl^- prior to that of Na^+ (Pappas and Hansen, 1977; Uglem and Prior, 1980). Concomitant with net Cl^- accumulation is a hyperpolarization of the external membrane from —45 mV to nearly —60 mV, and this increased potential difference might be responsible for the observed accumulation of Na^+ during long-term incubation of *H. diminuta* in media containing glucose (Uglem and Prior, 1980). The coupled fluxes of Na^+, Cl^- and glucose in *H. diminuta* may involve an asymmetric carrier which co-transports Na^+, Cl^- and glucose into the worm, while only Na^+ is transported out of the worm (Uglem and Prior, 1980).

The energy for "active transport" of glucose by *H. diminuta* and other species of tapeworms may be derived partially from the maintenance of the Na^+-gradient across the external membrane (reviewed by Pappas and Read, 1975). Maintenance of such a gradient may play a role in providing energy for glucose accumulation, an assumption supported by the observation that apparent Na^+-K^+-activated ATPase occurs in the proximal tegumentary membranes of *H. diminuta* (Lumsden and Murphy, 1980). However, it appears that transport of glucose across the external membranes of *H. diminuta* is more closely linked to "chemical energy" (ATP) than the Na^+-gradient (Podesta, 1979; Uglem and Prior, 1980).

Glycerol uptake by *H. diminuta* (Pittman and Fisher, 1972; Uglem *et al.*, 1974) and *T. taeniaeformis* larvae and adults (von Brand *et al.*, 1966) is mediated (saturable). It is not known whether *H. diminuta* accumulates glycerol, but *T. taeniaeformis* does not due to the rapid metabolism of this solute. Glycerol uptake in both species is inhibited partially in Na^+-free media, but unaffected by phlorizin (Pittman and Fisher, 1972; von Brand *et al.*, 1966); glycerol uptake by *H. diminuta* is inhibited by phloretin (Uglem *et al.*, 1974).

Two distinct glycerol transport sites occur in *H. diminuta*, one of which is Na^+-insensitive and another which is Na^+-sensitive. Both systems display saturation kinetics and are inhibited by glycerol and phloretin, while only the Na^+-sensitive transport system is inhibited by 1,2-propandiol. Despite the observation that glycerol uptake is Na^+-sensitive, it does not appear to be Na^+-coupled (Uglem *et al.*, 1974).

Some monosaccharides enter cestodes by diffusion. Fructose enters *T. crassiceps* larvae by diffusion (Pappas *et al.*, 1973b), while 3-0-methyl glucose enters *C. verticillatum* by diffusion (Fisher and Read, 1971).

C. Amino acids

The mechanism of amino acid uptake has been examined in a number of tapeworm species; with the exception of proline and glutamate which apparently enter *T. crassiceps* larvae by diffusion (Pappas *et al.*, 1973c), amino acids enter cestodes *via* specific, mediated systems (diffusion may contribute at high substrate concentrations). In *H. diminuta* and *H. citelli*, amino acid uptake is not inhibited by di- or tripeptides (Read *et al.*, 1963; Senturia, 1964), and glucose has no effect on short-term amino acid uptake by *H. diminuta* or *C. verticillatum* (Kilejian, 1966a; Read *et al.*, 1960a). In both *H. diminuta* and *T. crassiceps* larvae, an unsubstituted α-amino group is essential for maximal interaction of the solute and transport site (Read *et al.*, 1963; Laws and Read, 1969; Pappas *et al.*, 1973c); however, the amino acid transport sites of *H. diminuta* adults (Pappas and Gamble, 1980) and larvae (Arme and Coates, 1973), *T. crassiceps* larvae (Haynes and Taylor, 1968; Pappas *et al.*, 1973c) and *H. citelli* adults (Senturia, 1964) are unable to discriminate between D- and L-isomers of amino acids.

Cestodes accumulate amino acids against a concentration difference; this is indicative of "active transport". *H. diminuta* adults accumulate methionine (Read *et al.*, 1963; Pappas *et al.*, 1974), proline (Kilejian, 1966b), histidine (Woodward and Read, 1969), cycloleucine (Harris and Read, 1968), and phenylalanine, tyrosine and tryptophan (Pappas and Gamble, 1980), while both adults and cysticercoids accumulate α-aminoisobutyric acid (Harris and Read, 1968; Arme and Coates, 1971, 1973). Additionally, *H. citelli* accumulates methionine (Senturia, 1964), *C. verticillatum* accumulates valine and leucine (Read *et al.*, 1960b), *T. crassiceps* larvae accumulate methionine and phenylalanine (Pappas *et al.*, 1973c), and *Cotugnia digonopora* accumulates leucine (Roy and Srivastava, 1981).

The role of ions in amino acid uptake and accumulation by cestodes is not clear. Initial uptake rates of amino acids by *H. diminuta* (Read *et al.*, 1963; Pappas *et al.*, 1974; Pappas and Gamble, 1980) and *T. crassiceps* larvae (Pappas *et al.*, 1973c) appear independent of Na^+ in the external medium. However, Lussier *et al.* (1979) reported inhibition of amino acid uptake by *H. diminuta* in Na^+-free media, but even in this instance inhibition was only partial. *H. diminuta* (Pappas *et al.*, 1974) and *T. crassiceps* larvae (Pappas *et al.*, 1973c) accumulate amino acids against a concentration difference in Na^+-free media, and neither amino acid uptake nor accumulation by *H. diminuta* is affected in Cl^--free media (Pappas *et al.*, 1974). Apparently, ion gradients play little if any role in providing the necessary "energy" for amino acid uptake and accumulation by cestodes. It has been suggested that amino acid uptake by *H. diminuta* may be linked in some manner to the membrane-bound ATPase activity, since ATP in the external medium stimulates methionine and leucine uptake (Lussier *et al.*, 1978), but this phenomenon has not been investigated further.

As early as 1957 it was suggested that amino acids enter *H. diminuta* through more than one site (Daugherty, 1957), and subsequent studies have implicated at least six amino acid transport sites in this species (Pappas and Read, 1975; MacInnis *et al.*, 1976) (Table V). *H. diminuta* cysticercoids also have more than one amino acid transport site (Arme and Coates, 1971, 1973), and multiple sites apparently occur in other species of cestodes, as follows: three sites (one for basic and two for neutral amino acids) in *T. crassiceps* larvae (Haynes and Taylor, 1968; Pappas and Read, 1973; Pappas *et al.*, 1973c); two sites (one each for neutral and basic amino acids) in *C. verticillatum* (Read *et al.*, 1960a, 1960b); and two undefined sites in *C. diagnopora* (Roy and Srivastava, 1981). Further experiments are necessary to determine whether the above species possess additional amino acid transport sites.

D. Purines, pyrimidines and nucleosides

Absorption of purines, pyrimidines and nucleosides by cestodes occurs, most commonly, through distinct mediated systems. In almost every case, however, significant diffusion components are present. *T. crassiceps* larvae absorb purine bases (adenine and hypoxanthine) and nucleosides (adenosine and uridine) *via* separate mediated systems, while pyrimidine bases enter larvae primarily by diffusion. Hypoxanthine uptake by larvae displays sigmoid kinetics suggesting an allosteric (co-operative binding) mechanism. However, this departure from typical saturation kinetics may reflect an effect of absorbed hypoxanthine on metabolic processes and/or cytoplasmic binding rather than co-operative binding to a transport locus (Uglem and Levy, 1976).

Purines and pyrimidines enter *H. diminuta via* mediated systems, and at least three distinct systems have been implicated; two of these systems bind several different purine and pyrimidine bases, while the third binds only hypoxanthine and adenine. Two of these transport sites may possess multiple binding loci for substrates since uptake of some bases (thymine, adenine and hypoxanthine) displays sigmoid kinetics (MacInnis and Ridley, 1969; Pappas *et al.*, 1973d). However, as noted above for *T. crassiceps* larvae, sigmoid kinetics may be a reflection of metabolism and/or cytoplasmic binding rather than co-operative binding.

Nucleosides (uridine, thymidine, adenosine, guanosine and deoxyadenosine) are absorbed by *H. diminuta* through a mediated system which displays a diurnal periodicity and which is distinct from the purine and pyrimidine transport sites (Page and MacInnis, 1975; Page *et al.*, 1977). Uptake of nucleosides is Na^+-dependent, and in the absence of Na^+ nucleosides enter the worm by diffusion only (McCracken *et al.*, 1975). It is not known whether Na^+ and nucleoside fluxes are coupled in *H. diminuta.*

Table V. A summary of the major amino acid transport systems in *Hymenolepis diminuta* (from MacInnis *et al.*, unpublished (published in Pappas and Read, 1975, 1976)).

	Transport system					
	Dicarboxylic	Glycine	Serine	Leucine	Phenylalanine	Dibasic
Major amino acids interacting	Aspartic Glutamic Methionine	Glycine Methionine	Serine Alanine Threonine Methionine Valine Proline	Leucine Isoleucine Methionine	Phenylalanine Tyrosine Histidine Methionine	Arginine Lysine
Overlapping amino acids	Serine Alanine Glycine	Serine Threonine Alanine	Glycine	Glycine Serine Threonine Alanine Valine	Leucine Isoleucine	Histidine

E. Lipids

Fatty acids are absorbed by *H. diminuta* adults and larvae by a combination of mediated transport and diffusion. Short-chain fatty acids enter larvae and adults through a system which is specific for short-chain fatty acids (Arme and Read, 1968; Arme *et al.*, 1973), while long-chain fatty acids enter adults *via* a separate system (Chappell *et al.*, 1969). Uptake of palmitate by *H. diminuta* may be stimulated *or* inhibited in the presence of other fatty acids, depending on the inhibitor:substrate ratio (Chappell *et al.*, 1969), but this phenomenon has not been studied in detail.

Cholesterol, a major constituent of cestode membranes, enters hydatid cysts of *Echinococcus granulosus* by simple diffusion (Bahr *et al.*, 1979), while adults of *H. diminuta* absorb cholesterol, in part, through a mediated system which is specific for sterols (Johnson and Cain, 1980). The mediated system for cholesterol uptake in *H. diminuta* has not been characterized.

F. Vitamins

Vitamins are required for the normal growth and development of tapeworms (see Roberts, 1980, for a review), and tapeworms are readily permeable to water-soluble vitamins. The most interesting examples of vitamin absorption are those involving cobalamine (B_{12}). Adult *Diphyllobothrium latum* (Brante and Ernberg, 1957, 1958; Nyberg, 1958) and spargana of *Spirometra mansonoides* (Tkachuck *et al.*, 1976a) are readily permeable to cobalamine, and in the latter species uptake occurs through a saturable (mediated) system (Tkachuck *et al.*, 1976a). The presence of enzymes in cestodes requiring cobalamine as a cofactor (Tkachuck *et al.*, 1976b, 1977) is directly related to the ability of cestodes to absorb and accumulate this vitamin, and the presence or absence of such enzymes may be a key factor in determining some end-products of metabolism (Schaefer *et al.*, 1978).

H. diminuta absorbs thiamine (Pappas and Read, 1972b) and riboflavin (Pappas and Read, 1972c) *via* separate, mediated systems which are saturated at very low substrate concentrations. Pyridoxine (Pappas and Read, 1972c) and nicotinamide (Pappas, 1972) enter *H. diminuta* by diffusion, and *H. diminuta* is not permeable to cobalamine (Tkachuck *et al.*, 1976a).

VI. ADSORPTION

The glycocalyx of the cestode tegument consists primarily of polyionic proteins. Both cationic and anionic binding sites occur in the cestode glycocalyx, but anionic sites preponderate (see Fig. 2 for example), so the cestode surface bears a net negative charge (reviewed by Lumsden, 1975; Lumsden and Murphy, 1980). It is not surprising, therefore, that various ionic species in the external medium are adsorbed to the glycocalyx by electrostatic interactions. Adsorption may also occur by non-ionic interactions.

Adsorption of a variety of ionic molecular species to the surfaces of cestodes during *in vitro* incubations is demonstrable. *H. diminuta* (Lumsden *et al.*, 1970b; Lumsden, 1972) and *Lacistorhynchus tenuis* adsorb colloidal iron, and adsorption of $^{45}Ca^{++}$ to the surface of *H. diminuta* has been demonstrated by several methods (Lumsden, 1973; Lumsden and Berger, 1974). Adsorption of Ca^{++} is pH-dependent (more Ca^{++} is adsorbed as the pH of the incubation medium increases) (Lumsden, 1973), and adsorbed Ca^{++} is removed by treating tapeworms with chelating agents (EDTA or EGTA) (Lumsden, 1972, 1973). Ca^{++} adsorption by *H. diminuta* is inhibited by other cations (Zn^{++}, Mg^{++} and La^{+++}), and La^{+++} apparently has the highest affinity for the glycocalyx (Lumsden, 1973; Lumsden and Berger, 1974). Adsorption of Zn^{++}, Ca^{++} and Mg^{++}, as well as Co^{++}, is also suggested by the observations that treatment of *H. diminuta* with chelating agents (EDTA, EGTA or *o*-phenanthroline) decreases alkaline phosphohydrolase activity, and that activity is partially restored by treating worms with these same divalent cations (Lumsden and Berger, 1974).

Higher molecular weight polyionic species are also adsorbed by cestodes. *H. diminuta* adsorbs polylysine and polyglutamate; as is expected, more polylysine (a polycation) than polyglutamate is adsorbed to the anionic tapeworm surface (Lumsden, 1972). Some enzymes may also be adsorbed to the surface of *H. diminuta*, but the nature of the interactions between the enzymes and the cestode surface are not fully understood in most instances. Adsorption of radioactively labelled trypsin to the surface of *H. diminuta*, and inhibition of adsorption by other ionic species, has been demonstrated conclusively. Adsorption of trypsin (a cation at physiological pH) to the electronegative surface of *H. diminuta* apparently involves cationic binding sites on the worm's surface (rather than anionic sites which predominate), since adsorption is inhibited by polyglutamate (a polyanion) but not polylysine (Schroeder and Pappas, 1980). Adsorption of trypsin to the surface of *H. diminuta* is apparently not involved in the phenomenon of enzyme inactivation by this species (Schroeder and Pappas, 1980; Schroeder *et al.*, 1981) (see Section VII).

Adsorption of amylase (and a resulting increase in amylolytic activity) to the surface of *H. diminuta* by electrostatic interactions has been reported (Read, 1973), however, these findings have been questioned recently (Thomas and Turner, 1980) (see Section VII). Lipase also interacts with the surface of *H. diminuta*, but this interaction is very weak and probably involves forces other than electrostatic interactions (Ruff and Read, 1973).

Immunoglobulins are adsorbed to the surfaces of *H. diminuta* and *H. microstoma in vivo*, and it has been suggested that this binding represents specific antigen-antibody interactions at the worms' surfaces (Befus, 1977; Threadgold and Befus, 1977). Moreover, *H. diminuta* grown in rats or mice sometimes have readily visible "darkened areas" which are associated with abnormalities in the underlying tegument and cytoplasm; these "darkened areas", which may represent immune damage, disappear when the worms are incubated *in vitro* for a short period of time (30 min) (Befus and Threadgold, 1975). It is uncertain whether antibodies adversely affect intestinal dwelling cestodes, but it is known that antibodies may alter the permeability characteristics of larval *T. crassiceps* and *T. taeniaeformis* (Murrell, 1971; Hustead and Williams, 1977b) (see Section VII).

The bile salts sodium taurocholate and sodium glycocholate are adsorbed to the surfaces of *H. diminuta* and *H. microstoma* (Surgan and Roberts, 1976a). The adsorptive mechanism is not known, but the two bile salts are apparently adsorbed at different sites on the worms' surfaces (Surgan and Roberts, 1976a, 1976b), and adsorption may affect uptake of low molecular weight solutes by the worms *in vitro* (Surgan and Roberts, 1976b). Con A is also adsorbed to the surface of *H. diminuta*, probably through non-ionic binding of the lectin to D-glucose and related monosaccharide residues in the glycocalyx (Lumsden, 1973; McCracken and Lumsden, 1975b); although Con A adsorption may alter uptake of low molecular weight solutes *in vitro* (McCracken and Lumsden, 1975b), it is not known if this is of any significance *in vivo*.

VII. INTEGRATION

A variety of biochemical processes occur at the host–parasite interfaces of cestodes, including absorption, digestion, "protection" and "information transfer". Certainly, each of these processes plays a role in the maintenance of the parasitic life-style of cestodes, but more importantly, it is the integration of these processes that has led to the ability of cestodes to compete effectively with hosts for resources.

Cestodes demonstrate a range of permeabilities to various types of solutes. In many instances solutes not only move across the external membrane, but movement (influx) is enhanced *via* stereospecific, mediated systems. Such systems seem to be important in facilitating influx when the solute is present in low concentrations, and in allowing for regulation of internal metabolite pools when the external concentrations of metabolites vary.

Mediated transport systems of cestodes have an affinity for a variety of structurally related and metabolically important compounds, thus minimizing the amount of "information" which must be coded and synthesized during membrane biogenesis. It has also been suggested that transport systems with broad specificities allow parasites some "flexibility" in adjusting to new or altered physiological environments, as might be encountered in a new host species. Such "flexibility" is apparent when amino acid uptake by *H. diminuta* reared in rats or hamsters is compared (Read *et al.*, 1963). A more subtle regulatory role of the transport systems in modulating permeability and internal metabolite pools is also indicated. In *H. diminuta*, kinetic parameters describing amino acid (Read *et al.*, 1963; Kilejian, 1966a), glycerol (Pittman and Fisher, 1972) and glucose (Roberts, 1980) uptakes vary with worm age, and nucleoside uptake by *H. diminuta* displays a distinct diurnal periodicity (Page *et al.*, 1977). These variations in uptake rates may reflect (or affect) changes in internal metabolite pools and/or metabolic processes which accompany the development of the tapeworm in the definitive host or the diurnal migration of the tapeworm within the host's small intestine.

Solute absorption may be modulated by additional factors, although not necessarily in favour of the tapeworm. Adsorption of the bile salt sodium taurocholate inhibits glucose uptake in *H. diminuta* and *H. microstoma* (Surgan and Roberts, 1976b), and treatment of cysticerci of *T. taeniaeformis* or *T. crassiceps* with immune serum increases the permeability of larval membranes to low and high molecular weight substrates (Murrell, 1971; Hustead and Williams, 1977b). Solute absorption may even be affected by the density of the tapeworm infection, and therefore play a role in regulating intraspecific competition; this is suggested by the observation of Insler and Roberts (1980) that glucose uptake by *H. diminuta in vitro* is inhibited in worms grown under "crowded conditions".

The selectively permeable nature of the external surfaces of cestodes is important in excluding non-metabolizable substrates or allowing for the absorption of necessary cofactors. For example, *H. diminuta* is unable to metabolize fructose (Read and Simmons, 1963), but this is not surprising since *H. diminuta* is virtually impermeable to this monosaccharide (Arme and Read, 1970). The reason why fructose is excluded from the hexose transport system of *H. diminuta* and other species of cestodes (Table IV) is not clear. *Spirometra mansonoides* contains at least one enzyme which utilizes cobalamine as a cofactor, and this tapeworm absorbs large amounts of cobabamine *via* a specific, mediated system and converts it to the metabolically active adenosyl-derivative. *H. diminuta*, on the other hand, has no need for cobalamine and is impermeable to the vitamin (Tkachuck *et al.*, 1976a,b, 1977). In fact, Schaefer *et al.* (1978) report that "all representative species of the Orders Tetraphyllidea, Trypanorhyncha,

Caryophyllidea, Diphyllidea, and almost all Proteocephalidea contain high B_{12} concentrations", while no species tested from the Orders Cyclophyllidea and Pseudophyllidea (the latter defined by Wardle *et al.*, 1974) accumulate this vitamin. A most provocative question arises in this regard; were the enzymes utilizing cobalamine derivatives and the ability of cestodes to absorb cobalamine lost at the same time (possibly through deletion of linked genes), or was the loss of the enzymes a secondary effect resulting from a change in membrane permeability (or vice versa)? This question cannot be answered presently, but, as pointed out by Schaefer *et al.* (1978), such a distinct division in the ability of different groups of cestodes to absorb and accumulate cobalamine serves as an important biochemical parameter for interpreting systematic and evolutionary relationships of cestodes.

The significance of the ability of some cestodes to absorb high molecular weight substrates and/or colloidal particles (see Section V.A.) is unclear. Absorption of host's antibodies may represent a protective function, while other absorbed macromolecules may be of metabolic importance (Threadgold and Hopkins, 1981).

Enzymes of worm and/or host origin play an essential role in determining the availability of permeable solutes. *H. diminuta* is not permeable to phosphorylated monosaccharides or nucleotides. However, these esters are hydrolysed at the worm's surface and the hydrolysis products (monosaccharides or nucleosides) rapidly absorbed (Dike and Read, 1971b; Pappas and Read, 1974; Kuo, 1979); similar situations occur for glycerophosphate hydrolysis and glycerol uptake by *H. diminuta* (Uglem *et al.*, 1974) and adenosine monophosphate hydrolysis and nucleoside uptake by *T. crassiceps* larvae (Uglem and Levy, 1976). Thus, the membrane-bound enzymes (see Section IV) and mediated transport systems (see Section V) are located in the brush border membrane in such a manner that a "kinetic advantage" exists; hydrolysis products do not diffuse away from the worm's surface, but instead are rapidly absorbed. A representation of the interactions of membrane-bound enzymes and transport systems in *H. diminuta* is presented in Fig. 3. Clearly, the hydrolysis and subsequent transport of solutes at the worm's surface is an advantage to the tapeworm in competing for a variety of solutes.

The glycocalyx of the cestode tegument very likely plays an important role in the functioning of membrane-bound enzymes. Many of these enzymes require divalent cations for maximum activity (see Section IV) and divalent cations are strongly adsorbed to the glycocalyx (see Section VI). Thus, the glycocalyx may act as a "cation exchange resin" by concentrating those cations necessary to maintain maximum enzyme activity (Lumsden, 1972). Additionally, the glycocalyx may provide a physical barrier which prevents products of hydrolysis (and other solutes)

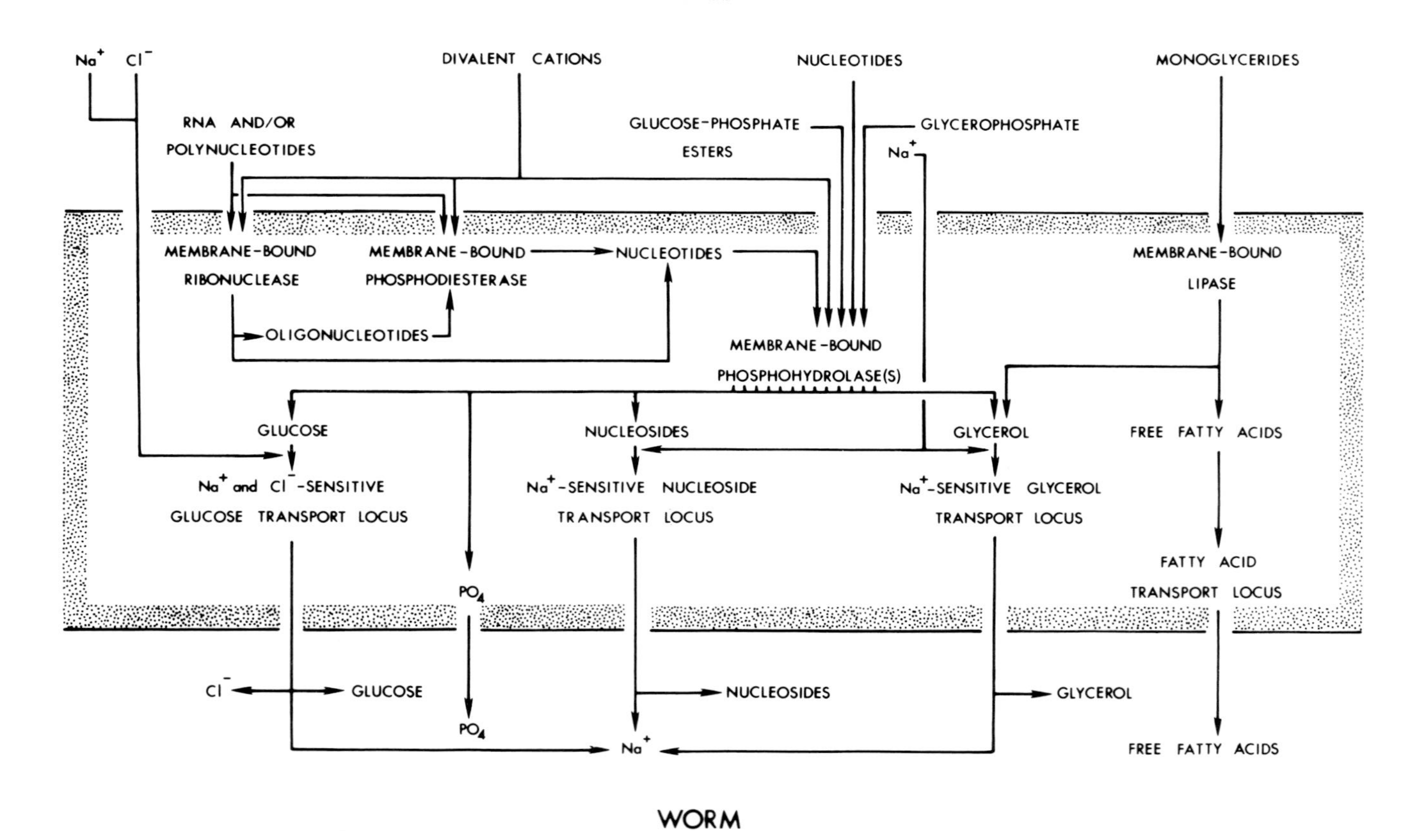

Fig. 3. A pictorial representation of the possible interactions of membrane-bound enzymes, cations and mediated transport systems in the tapeworm *Hymenolepis diminuta.* The brush border membrane is enclosed within the stippled area (modified after Pappas, 1980b).

from mixing rapidly with the external environment. This unstirred layer is, in fact, a barrier which must be crossed by solutes entering the worm from the external medium, and movement of solutes across the "unstirred layer" may (in some instances) be the limiting step in solute transport (Podesta, 1977; 1980). Thus, it is reasonable to assume that the "unstirred layer" is also a barrier against the diffusion of hydrolysis products formed by the action of membrane-bound enzymes, so these solutes are absorbed, rather than diffusing away from the worm's surface.

Adsorption of host's digestive enzymes to the glycocalyx also occurs. Trypsin (Schroeder and Pappas, 1980), amylase (Taylor and Thomas, 1968; Read, 1973) and lipase (Ruff and Read, 1973) are adsorbed by *H. diminuta in vitro*, but the significance of this phenomenon is uncertain. Trypsin and chymotrypsin are irreversibly inactivated by *H. diminuta* (Pappas and Read, 1972d, 1972e), but adsorption of trypsin apparently plays no role in inactivation of this enzyme (Schroeder and Pappas, 1980). Lipase is inhibited in the presence of *H. diminuta* (Ruff and Read, 1973), but the significance (if any) of this phenomenon in terms of benefiting the worm is unclear (Pappas, 1980a). Adsorption of amylase to tapeworms reportedly results in an enhancement of amylolytic activity (Taylor and Thomas, 1968; Read, 1973), but even this phenomenon of "contact digestion" is of dubious significance in increasing the amount of carbohydrate available to the tapeworm. The action of amylase on amylose produces a mixture of mono-, di- and trisaccharides, but only monosaccharides are utilized by most species of cestodes, including *H. diminuta* (Read and Simmons, 1963; Read, 1967). Also, the rates of carbohydrate digestion and absorption by the host's small intestine in *H. diminuta*-infected and uninfected rats do not differ significantly (Mead and Roberts, 1972), indicating no enhancement of amylolytic activity by *H. diminuta in vivo*. Moreover, Thomas and Turner (1980) have questioned the validity of previous experiments demonstrating "contact digestion" in cestodes, since their data indicate that the apparent stimulation of amylase activity is an artifact of the amylase assay procedure. Thus, the phenomenon of "contact digestion" by tapeworms needs to be re-evaluated.

Antibodies seem to have no deleterious effect on many cestodes, even though they may be adsorbed directly to the tapeworms' surfaces (see Section VI), so the external surface of the cestodes may protect the parasite against damage from the host's immune system. One protective mechanism which has been suggested for some cestodes is the absorption of intact immunoglobulins by the parasite (Threadgold and Hopkins, 1981). However, many cestodes are unable to absorb macromolecular solutes, and in these cases the rapid turnover of the tegumental membranes (Oaks and Lumsden, 1971) may be important in shedding adsorbed antibodies.

Potentially destructive proteolytic enzymes are inactivated by some cestodes *in vitro*. *H. diminuta* inactivates trypsin and chymotrypsin (Pappas and Read, 1972d, 1972e), but the mechanism of inactivation is not completely understood. Adsorption of the enzyme to the worm's surface is apparently not involved (Schroeder and Pappas, 1980), and no inhibitor of worm origin is demonstrable (Schroeder *et al.*, 1981). The mechanism of inactivation is, however, specific for trypsin and chymotrypsin since other proteolytic enzymes (papain, pepsin and subtilisin) are not affected (Schroeder *et al.*, 1981). It is not known whether inactivation of proteolytic enzymes by *H. diminuta* occurs *in vivo*, since attempts to demonstrate differences in the tryptic and total proteolytic activities of infected and uninfected hosts were unsuccessful (Pappas, 1978). However, the small amounts of enzyme inactivated by *H. diminuta* could alter the "microenvironment" of the worm in the host's small intestine while the relatively large quantities of these enzymes normally present in the host's small intestine would mask the inactivated enzyme (Pappas, 1980a).

Some cestodes display circadian rhythms (reviewed by Arai, 1980), rhythms which are established and maintained by changes in the parasite's environment (i.e. the host). Notable examples of such rhythms are the migration of *H. diminuta* in the rat's small intestine (Read and Kilejian, 1969; Chappell *et al.*, 1970), and diurnal variations in the amino acid pool (Defraites *et al.*, 1976) and uridine uptake (Page *et al.*, 1977) by *H. diminuta*. In fact, tapeworms provide strong support for the "optimal site-selection" hypothesis of Holmes (1973) in that they maximize the amount of worm tissue in the most favourable portion of the host's small intestine (Bailey, 1971; Turton, 1971). Thus, tapeworms must be able to sense changes in their environment and translate these changes into some usable form. Various types of "sensory" structures would be useful in this regard, but cestodes possess very few such structures (see Lumsden and Specian, 1980, for a review). However, adenyl cyclase activity has been demonstrated in the brush border membrane of one species of tapeworm (*H. microstoma*) (Conway-Jones and Rothman, 1978), and this activity may play a role in regulating metabolic and/or behavioural processes in cestodes.

The various functions of, and biochemical processes taking place at the host–parasite interface of cestodes are affected by a variety of parameters, and it is difficult to compare interfaces of various species of cestodes. Clearly, the nature of the external environment is important in determining the nature of the host–parasite interface, but the environments of different cestodes vary dramatically. For example, the interfaces between *H. diminuta*, *H. microstoma* and *C. verticillatum*, and the larvae of *T. crassiceps*, *T. taeniaeformis* and *E. granulosus* (all of which are mentioned in this chapter) and their respective hosts must be dissimilar in many respects, for these tapeworms live in different areas of their hosts' bodies and are subjected to quite different environmental conditions. The nature of the interface between a single tapeworm species and its host probably varies as well. *H. diminuta* undergoes a circadian migration in the host's small intestine, and the host–parasite interface (at least at the biochemical level) would change as the parasite migrates. Similarly, *H. microstoma* resides partially in the host's bile duct and partially in the host's small intestine, two environments with dissimilar chemical characteristics. Thus, care must be taken when comparing data from different species of cestodes, and when extrapolating data obtained from *in vitro* experiments to conditions *in vivo*.

VIII. REFERENCES

Arai, H. P. (1980). Migratory activity and related phenomena in *Hymenolepis diminuta. In* "Biology of the tapeworm *Hymenolepis diminuta*", (H. P. Arai, ed.), pp. 615–637. Academic Press, New York and London.

Arme, C. and Coates, A. (1971). Active transport of amino acids by cysticercoid larvae of *Hymenolepis diminuta. J. Parasit.* **57**, 1369–1370.

Arme, C. and Coates, A. (1973). *Hymenolepis diminuta:* active transport of α-aminoisobutyric acid by cysticercoid larvae. *Int. J. Parasit.* **3**, 553–560.

Arme, C. and Read, C. P. (1968). Studies on membrane transport. II. The absorption of acetate and butyrate by *Hymenolepis diminuta* (Cestoda). *Biol. Bull. mar. biol. Lab., Woods Hole* **135**, 80–91.

Arme, C. and Read, C. P. (1970). A surface enzyme in *Hymenolepis diminuta* (Cestoda). *J. Parasit.* **56**, 514–516.

Arme, C., Middleton, A. and Scott, J. P. (1973). Absorption of glucose and sodium acetate by cysticercoid larvae of *Hymenolepis diminuta. J. Parasit.* **59**, 214.

Bahr, J. M., Frayha, G. J. and Hajjar, J.-J. (1979). Mechanism of cholesterol absorption by the hydatid cysts of *Echinococcus granulosus* (Cestoda). *Comp. Biochem. Physiol.* **62A**, 485–490.

Bailey, G. N. A. (1971). *Hymenolepis diminuta:* circadian rhythm in movement and body length in the rat. *Expl. Parasit.* **29**, 285–291.

Bailey, H. H. and Fairbairn, D. (1968). Lipid metabolism in helminth parasites. V. Absorption of fatty acids and monoglycerides from micellary solution by *Hymenolepis diminuta* (Cestoda). *Comp. Biochem. Physiol.* **26**, 819–836.

Befus, A. D. (1977). *Hymenolepis diminuta* and *H. microstoma:* mouse immunoglobulins binding to the tegumental surface. *Expl. Parasit.* **41**, 242–251.

Befus, A. D. and Threadgold, L. T. (1975). Possible immunological damage to the tegument of *Hymenolepis diminuta* in mice and rats. *Parasitology* **71**, 525–534.

Belton, C. M. (1977). Freeze-fracture study of the tegument of larval *Taenia crassiceps. J. Parasit.* **63**, 306–313.

Brand, T., von and Gibbs, E. (1966). Aerobic and anaerobic metabolism of larval and adult *Taenia taeniaeformis.* III. Influence of some cations on glucose uptake, glucose leakage, and tissue glucose. *Proc. helminth. Soc. Wash.* **33**, 1–4.

Brand, T., von, MacMahon, P., Gibbs, E. and Higgins, H. (1964). Aerobic and anaerobic metabolism of larval and adult *Taenia taeniaeformis.* II. Hexose leakage and absorption: tissue glucose and polysaccharides. *Expl. Parasit.* **15**, 410–429.

Brand, T., von, Churchwell, F. and Higgins, H. (1966). Aerobic and anaerobic metabolism of larval and adult *Taenia taeniaeformis*. IV. Absorption of glycerol; relations between glycerol absorption and glucose absorption and leakage. *Expl. Parasit.* **19**, 110–123.

Brante, G. and Ernberg, T. (1957). The *in vitro* uptake of vitamin B_{12} by *Diphyllobothrium latum* and its blockage by intrinsic factor. *Scand. J. clin. Lab. Invest.* **9**, 313–314.

Brante, G. and Ernberg, T. (1958). The mechanism of pernicious tapeworm anemia studied with ^{60}Co-labelled vitamin B_{12}. *Acta med. scand.* **160**, 91–98.

Cain, G. D., Johnson, W. J. and Oaks, J. A. (1977). Lipids from subcellular fractions of the tegument of *Hymenolepis diminuta*. *J. Parasit.* **63**, 486–491.

Chappell, L. H. (1980). The biology of the external surfaces of helminth parasites. *Proc. R. Soc. Edinb.* **79B**, 145–171.

Chappell, L. H., Arme, C. and Read, C. P. (1969). Studies on membrane transport. V. Transport of long chain fatty acids in *Hymenolepis diminuta* (Cestoda). *Biol. Bull. mar. biol. Lab., Woods Hole*, **136**, 313–326.

Chappell, L. H., Arai, H. P., Dike, S. C. and Read, C. P. (1970). Circadian migration of *Hymenolepis* (Cestoda) in the host intestine. I. Observations on *H. diminuta* in the rat. *Comp. Biochem. Physiol.* **34**, 31–46.

Chordi, A. and Kagan, I. G. (1965). Identification and characterization of antigenic components of sheep hydatid fluid by immunoelectrophoresis. *J. Parasit.* **51**, 63–71.

Coltorti, E. A. and Varela-Díaz, V. M. (1972). IgG levels and host specificity in hydatid cyst fluid. *J. Parasit.* **58**, 753–756.

Coltorti, E. A. and Varela-Díaz, V. M. (1974). *Echinococcus granulosus:* penetration of macromolecules and their localization on the parasite membranes of cysts. *Expl. Parasit.* **35**, 225–231.

Coltorti, E. A. and Varela-Díaz, V. M. (1975). Penetration of host IgG molecules into hydatid cysts. *Z. ParasitKde*, **48**, 47–51.

Conder, G. A., Marchiondo, A. A., Williams, J. F. and Andersen, F. L. (1981). Freeze-etch characteristics of the tegument of three larval cestodes. *In* "Program and Abstracts, 56th Annual Meeting, The American Society of Parasitologists", pg. 66.

Conway-Jones, P. B. and Rothman, A. H. (1978). *Hymenolepis microstoma:* ultrastructural localization of adenyl cyclase in the tegument. *Expl. Parasit.* **46**, 152–156.

Daugherty, J. W. (1957). The active absorption of certain metabolites by certain helminths. *Am. J. trop. Med. Hyg.* **6**, 464–472.

Defraites, R. F., Newport, G. R. and Page, C. R., III (1976). Diurnal variation in levels of free pool amino acids in *Hymenolepis diminuta*. *In* "Program and Abstracts, 51st Annual Meeting, The American Society of Parasitologists", p. 49.

Dike, S. C. and Read, C. P. (1971a). Tegumentary phosphohydrolases of *Hymenolepis diminuta*. *J. Parasit.* **57**, 81–87.

Dike, S. C. and Read, C. P. (1971b). Relation of tegumentary phosphohydrolases and sugar transport in *Hymenolepis diminuta*. *J. Parasit.* **57**, 1251–1255.

Esch, G. W. and Kuhn, R. E. (1971). The uptake of ^{14}C-*Chlorella* protein by larval *Taenia crassiceps* (Cestoda). *Parasitology*, **62**, 27–29.

Fisher, F. M., Jr and Read, C. P. (1971). Transport of sugars in the tapeworm *Calliobothrium verticillatum*. *Biol. Bull. mar. biol. Lab., Woods Hole*, **140**, 46–62.

Friedman, P. A., Weinstein, P. P., Davidson, L. A. and Mueller, J. F. (1980). Polypeptides and glycoproteins from the tegument of *Spirometra mansonoides*. *In* "Program and Abstracts, 55th Annual Meeting, The American Society of Parasitologists", p. 56.

Friedman, P. A., Weinstein, P. P. and Mueller, J. F. (1981). Characterization of a high affinity receptor for vitamin B_{12} from tegumental membranes of the cestode *Spirometra mansonoides*. *In* "Program and Abstracts, 56th Annual Meeting, The American Society of Parasitologists", p. 34.

Gamble, H. R. and Pappas, P. W. (1980). Solubilization of membrane-bound ribonuclease (RNase) and alkaline phosphatase from the isolated brush border of *Hymenolepis diminuta* (Cestoda). *J. Parasit.* **66**, 434–438.

Gamble, H. R. and Pappas, P. W. (1981a). Type I phosphodiesterase in the isolated brush border membrane of *Hymenolepis diminuta*. *J. Parasit.* **67**, 617–622.

Gamble, H. R. and Pappas, P. W. (1981b). Partial characterization of ribonuclease (RNase) activity from the isolated and solubilized brush border of *Hymenolepis diminuta*. *J. Parasit.* **67**, 372–377.

Harris, B. G. and Read, C. P. (1968). Studies on membrane transport. III. Further characterization of amino acid systems in *Hymenolepis diminuta* (Cestoda). *Comp. Biochem. Physiol.* **26**, 545–552.

Haynes, W. D. G. and Taylor, A. E. R. (1968). Studies on the absorption of amino acids by larval tapeworms (Cyclophyllidea: *Taenia crassiceps*) *Parasitology*, **58**, 47–59.

Holmes, J. C. (1973). Site selection by parasitic helminths: interspecific interactions, site segregation, and their importance in the development of helminth communities. *Can. J. Zool.* **51**, 333–347.

Hopkins, C. A., Law, L. M. and Threadgold, L. T. (1978). *Schistocephalus solidus:* pinocytosis by the plerocercoid tegument. *Expl. Parasit.* **44**, 161–172.

Hustead, S. T. and Williams, J. F. (1977a). Permeability studies on taeniid metacestodes: I. Uptake of proteins by larval stages of *Taenia taeniaeformis*, *T. crassiceps*, and *Echinococcus granulosus*. *J. Parasit.* **63**, 314–321.

Hustead, S. T. and Williams, J. F. (1977b). Permeability studies on taeniid metacestodes: II. Antibody mediated effect on membrane permeability of *Taenia taeniaeformis* and *Taenia crassiceps*. *J. Parasit.* **63**, 322–326.

Insler, G. D. and Roberts, L. S. (1980). Developmental physiology of cestodes. XVI. Effects of certain excretory products on incorporation of ^{3}H-thymidine into DNA of *Hymenolepis diminuta*. *J. exp. Zool.* **211**, 55–61.

Johnson, W. J. and Cain, G. D. (1980). Partial saturation of the uptake of cholesterol in *Hymenolepis diminuta*. *In* "Program and Abstracts, 55th Annual Meeting, The American Society of Parasitologists", pg 55.

Kilejian, A. (1966a). Permeation of L-proline in the cestode, *Hymenolepis diminuta*. *J. Parasit.* **52**, 1108–1115.

Kilejian, A. (1966b). Formation of the L-proline pool in the cestode, *Hymenolepis diminuta*. *Expl. Parasit.* **19**, 358–365.

Knowles, W. J. and Oaks, J. A. (1979). Isolation and partial biochemical characterization of the brush border plasma membrane from the cestode, *Hymenolepis diminuta*. *J. Parasit.* **65**, 715–731.

Kuo, M. (1979). Hydrolysis and transport of nucleotides by *Hymenolepis diminuta. J. Chin. Soc. Vet. Sci.* **5**, 9–18.

Laws, G. F. and Read, C. P. (1969). Effect of the amino carboxy group on amino acid transport in *Hymenolepis diminuta* (Cestoda). *Comp. Biochem. Physiol.* **30**, 129–132.

Love, R. D. and Uglem, G. L. (1978). Estimation of the coupling coefficient for glucose and sodium transport in *Hymenolepis diminuta. J. Parasit.* **64**, 426–430.

Lumsden, R. D. (1966). Cytological studies on the absorptive surfaces of cestodes. II. The synthesis and intracellular transport of protein in the strobilar integument. *Z. ParasitKde*, **28**, 1–13.

Lumsden, R. D. (1972). Cytological studies on the absorptive surfaces of cestodes. VI. Cytochemical evaluation of electrostatic charge. *J. Parasit.* **58**, 229–234.

Lumsden, R. D. (1973). Cytological studies on the absorptive surfaces of cestodes. VII. Evidence for the function of the tegument glycocalyx in cation binding by *Hymenolepis diminuta. J. Parasit.* **59**, 1021–1030.

Lumsden, R. D. (1974). Relationship of extrinsic polysaccharides to the tegument glycocalyx of cestodes. *J. Parasit.* **60**, 374–375.

Lumsden, R. D. (1975). Surface ultrastructure and cytochemistry of parasitic helminths. *Expl. Parasit.* **37**, 267–339.

Lumsden, R. D. and Berger, B. (1974). Cytological studies on the absorptive surfaces of tapeworms, VIII. Phosphohydrolase activity and cation adsorption by the glycocalyx of *Hymenolepis diminuta. J. Parasit.* **60**, 774–751.

Lumsden, R. D. and Murphy, W. A. (1980). Morphological and functional aspects of the cestode surface. *In* "Cellular Interactions in Symbiosis and Parasitism", (C. B. Cook, P. W. Pappas and E. D. Rudolph, eds), pp. 95–130. The Ohio State University Press, Columbus.

Lumsden, R. D. and Specian, R. (1980). The morphology, histology, and fine structure of the adult stage of the cyclophyllidean tapeworm *Hymenolepis diminuta. In* "Biology of the tapeworm *Hymenolepis diminuta*", (H. P. Arai, ed.), pp. 157–280. Academic Press, New York.

Lumsden, R. D., Gonzalez, G., Mills, R. R. and Viles, J. M. (1968). Cytological studies on the absorptive surfaces of cestodes. III. Hydrolysis of phosphate esters. *J. Parasit.* **54**, 524–535.

Lumsden, R. D., Threadgold, L. T., Oaks, J. A. and Arme, C. (1970a). On the permeability of cestodes to colloids: an evaluation of the transmembranosis hypothesis. *Parasitology*, **60**, 185–195.

Lumsden, R. D., Oaks, J. A. and Alworth, W. L. (1970b). Cytological studies on the absorptive surfaces of cestodes. IV. Localization and cytochemical properties of membrane fixed cation binding sites. *J. Parasit.* **56**, 736–747.

Lussier, P. E., Podesta, R. B. and Mettrick, D. F. (1978). Effect of ATP on amino acid transport by *Hymenolepis diminuta. J. Parasit.* **64**, 1139–1140.

Lussier, P. E., Podesta, R. B. and Mettrick, D. F. (1979). *Hymenolepis diminuta:* Na^+-dependent and Na^+-independent components of neutral amino acid transport. *J. Parasit.* **65**, 842–848.

MacInnis, A. J. and Ridley, R. K. (1969). The molecular configurations of pyrimidines that causes allosteric activation of uracil transport in *Hymenolepis diminuta. J. Parasit.* **55**, 1134–1140.

MacInnis, A. J., Graff, D. J., Kilejian, A. and Read, C. P. (1976). Specificity of amino acid transport in the tapeworm *Hymenolepis diminuta* and its rat host. *In* "Studies in Parasitology", (J. E. Byram and G. L. Stewart, eds), pp. 183–204. William Marsh Rice University, Houston.

McCracken, R. O. and Lumsden, R. D. (1975a). Structure and function of parasite surface membranes. I. Mechanism of phlorizin inhibition of hexose transport by the cestode *Hymenolepis diminuta*. *Comp. Biochem. Physiol.* **50B**, 153–157.

McCracken, R. O. and Lumsden, R. D. (1975b). Structure and function of parasite surface membranes. II. Concanavalin A adsorption by the cestode *Hymenolepis diminuta*. *Comp. Biochem. Physiol.* **52B**, 331–337.

McCracken, R. O., Lumsden, R. D. and Page, C. R., III (1975). Sodium sensitive nucleoside transport by *Hymenolepis diminuta*. *J. Parasit.* **61**, 999–1005.

Mead, R. W. and Roberts, L. S. (1972). Intestinal digestion and absorption of starch in the intact rat: effects of cestode (*Hymenolepis diminuta*) infection. *Comp. Biochem. Physiol.* **41A**, 749–760.

Murrell, K. D. (1968). Respiration studies and glucose absorption kinetics of *Taenia crassiceps* larvae. *J. Parasit.* **54**, 1147–1150.

Murrell, K. D. (1971). The effect of antibody on the permeability control of larval *Taenia taeniaeformis*. *J. Parasit.* **57**, 875–880.

Norman, L., Kagan, I. G. and Chordi, A. (1964). Further studies on the analysis of sheep hydatid fluid by agar gel methods. *Am. J. trop. Med. Hyg.* **13**, 816–821.

Nyberg, W. (1958). The uptake and distribution of Co^{60}-labeled vitamin B_{12} by the fish tapeworm, *Diphyllobothrium latum*. *Expl. Parasit.* **7**, 178–190.

Oaks, J. A. and Lumsden, R. D. (1971). Cytological studies on the absorptive surfaces of cestodes. V. Incorporation of carbohydrate containing macromolecules into tegument membranes. *J. Parasit.* **57**, 1256–1268.

Oaks, J. A. and Mueller, J. F. (1981). Location of carbohydrate in the tegument of the procercoid of *Spirometra mansonoides*. *J. Parasit.* **67**, 325–331.

Oaks, J. A., Knowles, W. J. and Cain, G. D. (1977). A simple method of obtaining an enriched fraction of tegumental brush border from *Hymenolepis diminuta*. *J. Parasit.* **63**, 476–485.

Oaks, J. A., Cain, G. D. and Knowles, W. J. (1981). Dynamics of molecular components within tegumental fractions of *Hymenolepis diminuta*. *In* "Program and Abstracts, 56th Annual Meeting, The American Society of Parasitologists", pg. 63.

Page, C. R., III and MacInnis, A. J. (1975). Characterization of nucleoside transport in hymenolepidid cestodes. *J. Parasit.* **61**, 281–290.

Page, C. R., III, MacInnis, A. J. and Griffith, L. M. (1977). Diurnal periodicity of uridine uptake by *Hymenolepis diminuta*. *J. Parasit.* **63**, 91–95.

Pappas, P. W. (1972). *Hymenolepis diminuta:* absorption of nicotinamide. *Expl. Parasit.* **32**, 403–406.

Pappas, P. W. (1978). Tryptic and protease activities in the normal and *Hymenolepis diminuta*-infected rat small intestine. *J. Parasit.* **64**, 562–564.

Pappas, P. W. (1980a). Enzyme interactions at the host–parasite interface. *In* "Cellular Interactions in Symbiosis and Parasitism", (C. B. Cook, P. W. Pappas and E. D. Rudolph, eds), pp. 145–172. The Ohio State University Press, Columbus.

Pappas, P. W. (1980b). Structure, function and biochemistry of the cestode tegumentary membrane and associated glycocalyx. *In* "Endocytobiology", (W. Schwemmler and H. E. A. Schenk, eds), pp. 587–603. Walter de Gruyter and Co., New York and Berlin.

Pappas, P. W. (1980c). Phosphohydrolase activity of the isolated brush-border membrane of *Hymenolepis diminuta* (Cestoda) following sodium dodecyl sulfate (SDS)-polyacrylamide gel electrophoresis. *J. Parasit.* **66**, 914–919.

Pappas, P. W. (1981). *Hymenolepis diminuta:* partial characterization of membrane-bound nucleotidase activities (ATPase and 5′-nucleotidase) in the isolated brush border membrane. *Expl. Parasit.* **51**, 209–219.

Pappas, P. W. (1982a). Solubilization of the membrane-bound enzymes of the brush border plasma membrane of *Hymenolepis diminuta* using non-ionic detergents. *J. Parasit.* **68**, 588–592.

Pappas, P. W. (1982b). *Hymenolepis diminuta:* partial characterization of membrane-bound and solubilized alkaline phosphohydrolase activities of the isolated brush border plasma membrane. *Expl. Parasit.* **54**, 80–86.

Pappas, P. W. and Freeman, B. A. (1975). Sodium dependent glucose transport in the mouse bile duct tapeworm, *Hymenolepis microstoma. J. Parasit.* **61**, 434–439.

Pappas, P. W. and Gamble, H. R. (1978). *Hymenolepis diminuta:* action of worm RNase against RNA. *Expl. Parasit.* **46**, 256–261.

Pappas, P. W. and Gamble, H. R. (1980). Membrane transport of aromatic amino acids by *Hymenolepis diminuta* (Cestoda). *Parasitology*, **81**, 395–403.

Pappas, P. W. and Hansen, B. D. (1977). Chloride-sensitive glucose transport in *Hymenolepis diminuta. J. Parasit.* **63**, 800–804.

Pappas, P. W. and Narcisi, E. M. (1982). A comparison of membrane-bound enzymes in the isolated brush border plasma membranes of the cestodes *Hymenolepis diminuta* and *H. microstoma. Parasitology*, **84**, 391–396.

Pappas, P. W. and Read, C. P. (1972a). Sodium and glucose fluxes across the brush border of a flatworm (*Calliobothrium verticillatum*, Cestoda). *J. comp. Physiol.* **81**, 215–228.

Pappas, P. W. and Read, C. P. (1972b). Thiamine uptake by *Hymenolepis diminuta. J. Parasit.* **58**, 235–239.

Pappas, P. W. and Read, C. P. (1972c). The absorption of pyridoxine and riboflavin by *Hymenolepis diminuta. J. Parasit.* **58**, 417–421.

Pappas, P. W. and Read, C. P. (1972d). Trypsin inactivation by intact *Hymenolepis diminuta. J. Parasit.* **58**, 864–871.

Pappas, P. W. and Read, C. P. (1972e). Inactivation of α- and β-chymotrypsin by intact *Hymenolepis diminuta* (Cestoda). *Biol. Bull. mar. biol. Lab., Woods Hole*, **143**, 605–616.

Pappas, P. W. and Read, C. P. (1973). Permeability and membrane transport in the larva of *Taenia crassiceps. Parasitology*, **66**, 33–42.

Pappas, P. W. and Read, C. P. (1974). Relation of nucleoside transport and surface phosphohydrolase activity in *Hymenolepis diminuta. J. Parasit.* **60**, 447–452.

Pappas, P. W. and Read, C. P. (1975). Membrane transport in helminth parasites: a review. *Expl. Parasit.* **37**, 469–530.

Pappas, P. W., Uglem, G. L. and Read, C. P. (1973a). Ribonuclease activity associated with intact *Hymenolepis diminuta. J. Parasit.* **59**, 824–828.

Pappas, P. W., Uglem, G. L. and Read, C. P. (1973b). *Taenia crassiceps:* absorption of hexoses and partial characterization of Na^+-dependent glucose absorption by larvae. *Expl. Parasit.* **33**, 127–137.

Pappas, P. W., Uglem, G. L. and Read, C. P. (1973c). Mechanisms and specificity of amino acid transport in *Taenia crassiceps* larvae. *Int. J. Parasit.* **3**, 641–651.

Pappas, P. W., Uglem, G. L. and Read, C. P. (1973d). The influx of purines and pyrimidines across the brush border of *Hymenolepis diminuta. Parasitology*, **66**, 525–538.

Pappas, P. W., Uglem, G. L. and Read, C. P. (1974). Anion and cation requirements for glucose and methionine accumulation in *Hymenolepis diminuta* (Cestoda). *Biol. Bull. mar. biol. Lab., Woods Hole*, **146**, 56–66.

Phifer, K. O. (1960a). Permeation and membrane transport in animal parasites: further observations on the uptake of glucose by *Hymenolepis diminuta. J. Parasit.* **46**, 137–144.

Phifer, K. O. (1960b). Permeation and membrane transport in animal parasites: the absorption of glucose by *Hymenolepis diminuta. J. Parasit.* **46**, 51–62.

Pittman, R. G. and Fisher, F. M., Jr (1972). The membrane transport of glycerol by *Hymenolepis diminuta. J. Parasit.* **58**, 742–749.

Podesta, R. B. (1977). *Hymenolepis diminuta:* unstirred layer thickness and effects on active and passive transport kinetics. *Expl. Parasit.* **43**, 12–24.

Podesta, R. B. (1979). Cellular Na^+ and ATP effects on galactose influx by tissue slices of *Hymenolepis diminuta. J. Parasit.* **65**, 669–671.

Podesta, R. B. (1980). Concepts of membrane biology in *Hymenolepis diminuta. In* "Biology of the tapeworm *Hymenolepis diminuta*" (H. P. Arai, ed.), pp. 505–549. Academic Press, New York and London.

Rahman, M. S., Mettrick, D. F. and Podesta, R. B. (1981a). Use of saponin in the preparation of brush border from a parasitic flatworm. *Can. J. Zool.* **59**, 911–917.

Rahman, M. S., Mettrick, D. F. and Podesta, R. B. (1981b). Properties of a Mg^{2+}-dependent, and Ca^{2+}-inhibited ATPase localized in the brush border of the surface epithelial syncytium of a parasitic flatworm. *Can. J. Zool.* **59**, 918–923.

Read, C. P. (1961). Competitions between sugars in their absorption by tapeworms. *J. Parasit.* **47**, 1015–1016.

Read, C. P. (1967). Carbohydrate metabolism in *Hymenolepis* (Cestoda). *J. Parasit.* **53**, 1023–1029.

Read, C. P. (1973). Contact digestion in tapeworms. *J. Parasit.* **59**, 672–677.

Read, C. P. and Kilejian, A. (1969). Circadian migratory behavior of a cestode symbiote in the rat host. *J. Parasit.* **55**, 574–578.

Read, C. P. and Rothman, A. H (1958). The role of carbohydrates in the biology of cestodes. IV. The carbohydrates metabolized *in vitro* by some cyclophyllidean cestodes. *Expl. Parasit.* **7**, 217–223.

Read, C. P. and Simmons, J. E., Jr (1963). Biochemistry and physiology of tapeworms. *Physiol. Rev.* **43**, 263–305.

Read, C. P., Simmons, J. E., Jr and Rothman, A. H. (1960a). Permeation and membrane transport in animal parasites: amino acid permeation into cestodes from elasmobranchs. *J. Parasit.* **46**, 33–41.

Read, C. P., Simmons, J. E., Jr, Campbell, J. W. and Rothman, A. H. (1960b).

Permeation and membrane transport in animal parasites: studies on a tapeworm-elasmobranch symbiosis. *Biol. Bull. mar. biol. Lab., Woods Hole*, **119**, 120–133.

Read, C. P., Rothman, A. H. and Simmons, J. E., Jr (1963). Studies on membrane transport, with special reference to parasite–host integration. *Ann. N.Y. Acad. Sci.* **113**, 154–205.

Read, C. P., Stewart, G. L. and Pappas, P. W. (1974). Glucose and sodium fluxes across the brush border of *Hymenolepis diminuta* (Cestoda). *Biol. Bull. mar. biol. Lab., Woods Hole*, **147**, 146–162.

Roberts, L. S. (1980). Development of *Hymenolepis diminuta* in its definitive host. *In* "Biology of the Tapeworm *Hymenolepis diminuta*", (H. P. Arai, ed.), pp. 357–423. Academic Press, New York and London.

Rothman, A. H. (1966). Ultrastructural studies on enzyme activity in the cestode cuticle. *Expl. Parasit.* **19**, 332–338.

Rothman, A. H. (1967). Colloid transport in the cestode *Hymenolepis diminuta. Expl. Parasit.* **21**, 133–136.

Roy, T. K. and Srivastava, V. M. L. (1981). *Cotugnia digonopora:* transport of leucine. *Expl. Parasit.* **51**, 21–27.

Ruff, M. D. and Read, C. P. (1973). Inhibition of pancreatic lipase by *Hymenolepis diminuta. J. Parasit.* **59**, 105–111.

Schaefer, F. W., III, Weinstein, P. P. and Coggins, J. R. (1978). Vitamin B_{12} as a biochemical marker for cestode evolution. *In* "Program and Abstracts, 53rd Annual Meeting, The American Society of Parasitologists", pg. 52.

Schroeder, L. L. and Pappas, P. W. (1980). Trypsin adsorption by *Hymenolepis diminuta* (Cestoda). *J. Parasit.* **66**, 49–52.

Schroeder, L. L., Pappas, P. W. and Means, G. E. (1981). Trypsin inactivation by intact *Hymenolepis diminuta* (Cestoda); some characteristics of the inactivated enzyme. *J. Parasit.* **67**, 378–385.

Senturia, J. B. (1964). Studies on the absorption of methionine by the cestode, *Hymenolepis citelli. Comp. Biochem. Physiol.* **12**, 259–272.

Singer, S. J. and Nicolson, G. (1972). The fluid mosaic model of the structure of cell membranes. *Science, N.Y.* **175**, 720–731.

Sosa, A., Gonzales-Angula, A., Calzada, L. and Alva, S. (1978). Presence of ATPase on the vesicular membrane of *Cysticercus cellulosae. Experientia*, **34**, 175–177.

Starling, J. A. (1975). Tegumental carbohydrate transport in intestinal helminths: correlation between mechanisms of membrane transport and the biochemical environment of absorptive surfaces. *Trans. Am. microsc. Soc.* **94**, 508–523.

Surgan, M. H. and Roberts, L. S. (1976a). Adsorption of bile salts by the cestodes *Hymenolepis diminuta* and *H. microstoma. J. Parasit.* **62**, 78–86.

Surgan, M. H. and Roberts, L. S. (1976b). Effect of bile salts on the absorption of glucose and oleic acid by the cestodes *Hymenolepis diminuta* and *H. microstoma. J. Parasit.* **62**, 87–93.

Taylor, E. W. and Thomas, J. N. (1968). Membrane (contact) digestion in three species of tapeworm *Hymenolepis diminuta, Hymenolepis microstoma* and *Moniezia expansa. Parasitology*, **58**, 535–546.

Thomas, J. N. and Turner, S. G. (1980). A reinterpretation of the evidence for contact digestion in the tapeworm, *Hymenolepis diminuta. J. Physiol., Lond.* **301**, 79P–80P.

Threadgold, L. T. and Befus, A. D. (1977). *Hymenolepis diminuta:* ultrastructural localization of immunoglobulin-binding sites on the tegument. *Expl. Parasit.* **43**, 169–179.

Threadgold, L. T. and Hopkins, C. A. (1981). *Schistocephalus solidus* and *Ligula intestinalis:* pinocytosis by the tegument. *Expl. Parasit.* **51**, 444–456.

Tkachuck, R. D', Weinstein, P. P. and Mueller, J. F. (1976a). Comparison of the uptake of vitamin B_{12} by *Spirometra mansonoides* and *Hymenolepis diminuta* and the functional groups of B_{12} analogs affecting uptake. *J. Parasit.* **62**, 94–101.

Tkachuck, R. D., Weinstein, P. P. and Mueller, J. F. (1976b). Isolation and identification of a cobamide coenzyme from the tapeworm *Spirometra mansonoides. J. Parasit.* **62**, 948–950.

Tkachuck, R. D., Weinstein, P. P. and Mueller, J. F. (1977). Metabolic fate of cyanocobalamin taken up by *Spirometra mansonoides* spargana. *J. Parasit.* **63**, 694–700.

Trimble, J. J., III and Lumsden, R. D. (1975). Cytochemical characterization of tegument membrane-associated carbohydrates in *Taenia crassiceps* larvae. *J. Parasit.* **61**, 665–676.

Turton, J. A. (1971). Distribution and growth of *Hymenolepis diminuta* in the rat, hamster, and mouse. *Z. ParasitKde.* **37**, 315–329.

Uglem, G. L. (1976). Evidence for a sodium ion exchange carrier linked with glucose transport across the brush border of a flatworm (*Hymenolepis diminuta,* Cestoda). *Biochim. biophys. Acta,* **443**, 126–136.

Uglem, G. L. and Levy, M. G. (1976). Absorption kinetics of some purines, pyrimidines and nucleosides in *Taenia crassiceps* larvae. *In* "Studies in Parasitology", (J. E. Byram and G. L. Stewart, eds), pp. 225–236. William Marsh Rice University, Houston.

Uglem, G. L. and Love, R. D. (1977). *Hymenolepis diminuta:* properties of phlorizin inhibition of glucose transport. *Expl. Parasit.* **43**, 94–99.

Uglem, G. L. and Prior, D. J. (1980). *Hymenolepis diminuta:* chloride fluxes and membrane potentials associated with sodium-coupled glucose transport. *Expl. Parasit.* **50**, 287–294.

Uglem, G. L., Pappas, P. W. and Read, C. P. (1973). Surface aminopeptidase in *Moniliformis dubius* and its relation to amino acid uptake. *Parasitology,* **67**, 185–195.

Uglem, G. L., Pappas, P. W. and Read, C. P. (1974). Na^+-dependent and Na^+-independent glycerol fluxes in *Hymenolepis diminuta* (Cestoda). *J. comp. Physiol.* **93**, 157–171.

Uglem, G. L., Love, R. D. and Eubank, J. H. (1978). *Hymenolepis diminuta:* membrane transport of glucose and β-methylglucoside. *Expl. Parasit.* **45**, 88–92.

Varela-Díaz, V. M. and Coltorti, E. A. (1972). Further evidence of the passage of host immunoglobulins into hydatid cysts. *J. Parasit.* **58**, 1015–1016.

Wardle, R. A., McLeod, J. A. and Radinovsky, S. (1974). "Advances in the Zoology of Tapeworms 1950–1970", 274 pp. University of Minnesota Press, Minneapolis.

Woodward, C. K. and Read, C. P. (1969). Studies on membrane transport-VIII. Transport of histidine through two distinct systems in the tapeworm, *Hymenolepis diminuta*. *Comp. Biochem. Physiol.* **30**, 1161–1177.

Chapter 8

Protein Metabolism

B. G. Harris

I. INTRODUCTION

One cannot review the literature on protein metabolism in cestodes without quickly realizing that this is an almost totally neglected field. Sporadic attempts have been made to investigate the synthesis and degradation of tapeworm proteins, but these efforts have not resulted in a uniform picture of protein metabolism in the worms. In this brief review, three areas will be covered: protein synthesis, protein degradation and interconversion of amino acids. The earlier literature in this area has been reviewed by Read and Simmons (1963), von Brand (1973; 1979) and Barrett (1981). However, the amino acids and amines involved in neurotransmitter activity will not be covered since this area has been reviewed recently by Rew (1978) and by Van den Bossche (1980).

BIOLOGY OF THE EUCESTODA Vol. 2
ISBN 0-12-062102-9

II. PROTEIN SYNTHESIS

a. Cell-free systems

Agosin and Repetto (1967), working on larval *Echinococcus granulosus* scoleces, were the first to construct a cell-free protein synthesizing system with extracts of cestodes. Their work was followed several years later by the development of cell-free systems in *Hymenolepis diminuta* (Parker and MacInnis, 1977) and larval *Taenia crassiceps* (Naquira *et al.*, 1977; Agosin and Naquira, 1978). Generally, these workers have demonstrated that protein synthesis in cestodes is similar to that in mammals in that it requires polysomes, amino acid adenylates, aminoacyl-tRNAs, pH 5 fraction, ATP, GTP, magnesium and either sodium or potassium ions. However, the systems do show a certain degree of specificity. For example, *E. granulosus* ribosomes accepted the homologous pH 5 fraction and not the rat liver fraction, but the rat liver pH 5 fraction stimulated amino acid incorporation into *E. granulosus* ribosomes in the presence of *E. granulosus* sRNA (Agosin and Repetto, 1967). In the *H. diminuta* system (Parker and MacInnis, 1977), homologous ribosomes accepted heterologous tRNA, but this resulted in synthesis at a lower rate. With *T. crassiceps* (Naquira *et al.*, 1977), polysomes and the pH 5 fraction were exchangeable with these same fractions from mouse liver, but again the activity was somewhat lower when *T. crassiceps* polysomes were mixed together with the mouse liver pH 5 fraction. On the other hand, rat liver polysomes were not active with *T. crassiceps* pH 5 fraction, but the opposite mixture, i.e. *T. crassiceps* polysomes plus rat liver pH 5 fraction, had "significant activity" (Naquira *et al.*, 1977). Agosin and Naquira (1978) have used a cell-free system from reticulocytes to translate purified mRNA from *T. crassiceps*. In this study, the authors developed antisera against *T. crassiceps* antigens and used these antisera to demonstrate that a small portion of the protein synthesized in this heterologous system appeared to be actual *T. crassiceps* protein antigens. The molecular weights of these proteins were generally low, in the 13 000–22 000 range, but one protein of 260 000 was also produced (Agosin and Naquira, 1978).

Other than these few works, little has appeared in the literature regarding detailed mechanisms of protein synthesis in cestodes. Virtually nothing has been published on the aminoacyl-tRNA synthetases, initiation, elongation or termination factors or ribosomal proteins. Another very important area that has been neglected is that involving possible control mechanisms in protein synthesis in cestodes. This area is of particular importance as will be discussed in the next section.

B. *In vitro* incorporation studies

Protein synthesis has also been studied in cestodes with amino acid incorporation studies into whole worms *in vitro*. The results of these studies provide numerous incentives for projects on the mechanisms and regulation of protein synthesis in tapeworms. Harris and Read (1969) utilized ^{14}C-lysine uptake and incorporation studies in *H. diminuta* to demonstrate the very close relationship between the rates of energy metabolism (energy availability?) and protein synthesis. A direct correlation was drawn between the presence of glucose and CO_2 on the ability of starved worms to incorporate ^{14}C-lysine into protein. These workers further showed that 2,4-dinitrophenol, an uncoupler of electron-transport associated phosphorylation (Scheibel *et al.*, 1968), inhibited incorporation but not uptake of lysine.

In a carefully designed project, Bolla and Roberts (1971) studied carbohydrate levels, RNA, DNA and protein synthesis along the developing strobila of *H. diminuta*. These workers showed that protein synthesis (measured by incorporation of radioactive amino acids into trichloroacetic acid insoluble material) was highest in the gravid proglottides. They also demonstrated that crowding of worms produced a decrease in carbohydrate levels in the mature and gravid proglottides that could be correlated with a severe inhibition of protein synthesis ($>90\%$) in these two areas in "crowded" worms.

Culbreth *et al.* (1972) studied the relationship between size and the uptake and incorporation of ^{14}C-leucine in two different strains of larval *T. crassiceps*. An inverse relationship between larval biomass and uptake and incorporation of leucine was found, i.e. the greater the mass, the lower the level of protein synthesis was found. In addition, it was found that the ORF strain grew at a greater rate than did the KBS strain, and the same relationship was found in the rate of protein synthesis. The two strains appear to be different on the basis of a genetic mutation in the ORF strain (Dorais and Esch, 1969). One question that immediately arises from this study is, is this genetic mutation involved in the regulation of protein synthesis? If this were the case, then this system would offer an excellent model for such studies.

III. PROTEIN DEGRADATION

Although the presence of several enzymes has been demonstrated on the surface of cestodes (see Chapter 7), none of these appears to have proteolytic activity. Pappas and Read (1972) were unable to demonstrate proteolytic activity toward azoalbumin or casein with intact *H. diminuta*.

On the other hand, Sisova-Kasatockina and Dubovskaja (1975) found proteolytic activity in the tegument of *Schistocephalus solidus*, and stated that this proved that the "tegument participates in protein assimilation in cestodes". However, the "tegument" preparation was made by the use of a scalpel, and it is doubtful that this tissue containing proteolytic activity was only microtriches. The surface layer may contain proteolytic enzymes, but whether these enzymes contribute to proteolytic digestion of external protein was not demonstrated in this paper. Douch (1978) has demonstrated the presence of a leucyl-B-naphthylamidase ("leucine aminopeptidase") in *Moniezia expansa*. This enzyme had a molecular weight of about 200 000 as measured by Sephadex G-200 gel filtration, and was stimulated by Mn^{++}, Fe^{++} and Co^{++}. Other divalent cations such as Ni^{++}, Cd^{++}, Hg^{++} and Zn^{++} inhibited the enzyme. In addition, the enzyme was neither inhibited by nor effective on acetanilide, benzamide, benzanilide or salicylamide, structures that are common in anthelmintic drugs. Therefore, the physiological substrate of this enzyme is not known. Matskasi and Nemeth (1979) have identified proteolytic activity in extracts of *Ligula intestinalis* plerocercoids. The activity was effective against N-benzoyl-L-tyrosine ethyl ether and Azocoll. Two peaks of activity were separated on Sephadex G-100; a very high molecular weight component and a 60 000–65 000 peak. The lower molecular weight enzyme was activated by Ca^{++}, Mg^{++} and Mn^{++}. The only compound tested that inhibited the enzyme was phenylmethylsulfonyl fluoride, suggesting that it was probably a serine protease. Extracts of the plerocercoids also contained trypsin- and chymotrypsin-inhibitor activity in the range of 6700–7200 molecular weight. These inhibitors were not affected by heat or trichloroacetic acid, and neither inhibitor affected the proteolytic activity in the extract. Again, however, no physiological function was suggested for the protease or protease inhibitors.

IV. AMINO ACID INTERCONVERSIONS

As pointed out by Read and Simmons (1963), the cestodes studied thus far have a very limited ability to metabolize amino acids. This is presumably due to the fact that all the required amino acids are provided in a rather constant ratio by the host (Read, 1970), and the parasite has developed numerous transport systems to acquire these compounds (Pappas and Read, 1975). The few amino acids that are metabolized appear to participate in transamination reactions (alanine, aspartate, glutamate (Read and Simmons, 1963)) or in portions of the urea cycle generating ornithine and thus proline for collagen synthesis (Janssens and Bryant, 1969; Paltridge and Janssens, 1971; Fodge and Harris, unpublished observations). Fisher and Starling (1970) have demonstrated that valine was metabolized to α-ketoisovaleric acid in *Calliobothrium verticillatum*. However, the method of conversion, i.e. oxidative deamination, transamination, etc., was not studied. Recently, Mustafa *et al.* (1978) have partially purified and characterized a cytosolic glutamate dehydrogenase from *H. diminuta*. The enzyme was NAD^+-specific, and the apparent K_a for α-ketoglutarate was decreased by lowering the pH. The authors speculated that the physiological role of the enzyme in the parasite may be involved in maintenance of redox potential. They state that *H. diminuta* excretes alanine as a product of carbohydrate metabolism which could be derived from pyruvate by transamination with glutamate. The α-ketoglutarate thus formed could be reductively aminated to glutamate by glutamate dehydrogenase and thereby maintain the cytosolic $NADH/NAD^+$–redox balance (Mustafa *et al.*, 1978). This could also be a mechanism to "detoxify" ammonia produced by other reactions in the cell. Perhaps this work will stimulate research into the area of amino acids as possible end-products of carbohydrate metabolism in cestodes (Barrett, 1981).

V. CONCLUSIONS

As pointed out earlier and also by Barrett (1981), mechanisms of protein metabolism have not been extensively studied in parasitic helminths. This is probably due to the fact that in the majority of adult parasites protein synthesis does not appear to be critical, because most cells in the adult do not divide. However, protein synthesis is extremely important in the reproductive tissues of the adult (Harris *et al.*, 1972), and in cestodes differentiation, and therefore protein synthesis, is occurring continually along the length of the strobila. The strobila of the tapeworm offers an excellent model system of development. As exemplified by the work of Bolla and Roberts (1971), the entire sequence of development is present in one tapeworm. An adequate source of energy appears to be an important aspect in regulating the protein synthesizing machinery of the parasite. How is this regulated? What signals "adequately-fed" cells to utilize the energy for protein synthesis? What regulatory mechanisms are present in one set of maturing proglottides that are absent in other sets? It appears that cestodes should provide a very good model system to pursue basic studies in the control of protein synthesis.

VI. REFERENCES

Agosin, M. and Naquira, C. (1978). Translation of *Taenia crassiceps* mRNA in cell-free heterologous systems. *Comp. Biochem. Physiol.* **60B**, 183–187.

Agosin, M. and Repetto, Y. (1967). Studies on the metabolism of *Echinococcus granulosus* IX. Protein synthesis in scolices. *Expl. Parasit.* **21**, 195–208.

Barrett, J. (1981). "Biochemistry of Parasitic Helminths", University Park Press, Baltimore, MD.

Bolla, R. I. and Roberts, L. S. (1971). Developmental physiology of cestodes—X. The effect of crowding on carbohydrate levels and on RNA, DNA and protein synthesis in *Hymenolepis diminuta. Comp. Biochem. Physiol.* **40A**, 777–787.

Culbreth, K. L., Esch, G. W. and Kuhn, R. E. (1972). Growth and development of larval *Taenia crassiceps* (Cestoda) III. The relationship between larval biomass and the uptake and incorporation of ^{14}C-leucine. *Expl. Parasit.* **32**, 272–281.

Dorais, F. J. and Esch, G. W. (1969). Growth rate of two *Taenia crassiceps* strains. *Expl. Parasit.* **25**, 395–398.

Douch, P. G. C. (1978). L-leucyl-B-naphthylamidases of the cestode, *Moniezia expansa*, and the nematode, *Ascaris suum. Comp. Biochem. Physiol.* **60**, 63–66.

Fisher, F. J., Jr and Starling, J. A. (1970). The metabolism of L-valine by *Calliobothrium verticillatum* (Cestoda: Tetraphyllidea): Identification of α-ketoisovaleric acid. *J. Parasit.* **56**, 103–107.

Harris, B. G. and Read, C. P. (1969). Factors affecting protein synthesis in *Hymenolepis diminuta* (Cestoda). *Comp. Biochem. Physiol.* **28**, 645–654.

Harris, B. G., Talent, J. M. and Fodge, D. W. (1972). Amino acid transport in reproductive and muscle tissue of *Ascaris suum. J. Parasit.* **58**, 541–545.

Janssens, P. A. and Bryant, C. (1969). The ornithine-urea cycle in some parasitic helminths. *Comp. Biochem. Physiol.* **30**, 261–272.

Matskasi, I. and Nemeth, I. (1979). *Ligula intestinalis* (Cestoda: Pseudophyllidae): Studies on the properties of proteolytic and protease inhibitor activities of plerocercoid larvae. *Int. J. Parasit.* **9**, 221–228.

Mustafa, T., Komuniecki, R. and Mettrick, D. F. (1978). Cytosolic glutamate dehydrogenase in adult *Hymenolepis diminuta* (Cestoda). *Comp. Biochem. Physiol.* **61B**, 219–222.

Naquira, C., Paulin, J. and Agosin, M. (1977). *Taenia crassiceps:* Protein synthesis in larvae. *Expl. Parasit.* **41**, 359–369.

Paltridge, R. W. and Janssens, P. A. (1971). A reinvestigation of the status of the ornithine-urea cycle in adult *Ascaris lumbricoides. Comp. Biochem. Physiol.* **40B**, 503–513.

Pappas, P. W. and Read, C. P. (1972). Trypsin inactivation by intact *Hymenolepis diminuta. J. Parasit.* **58**, 864–871.

Pappas, P. W. and Read, C. P. (1975). Membrane transport in helminth parasites: a review. *Expl. Parasit.* **37**, 469–530.

Parker, R. D., Jr and MacInnis, A. J. (1977). *Hymenolepis diminuta:* Isolation, purification, and reconstruction *in vitro* of a cell-free system for protein synthesis. *Expl. Parasit.* **41**, 2–16.

Read, C. P. (1970). Some physiological and biochemical aspects of host–parasite relations. *J. Parasit.* **56**, 643–652.

Read, C. P. and Simmons, J. E., Jr (1963). Biochemistry and physiology of tapeworms. *Physiol. Rev.* **43**, 263–305.

Rew, Robert S. (1978). Mode of action of common anthelmintics. *J. Vet. Pharmac. Ther.* **1**, 183–198.

Scheibel, L. W., Saz, H. J. and Bueding, E. (1968). The anaerobic incorporation of ^{32}P into adenosine triphosphate by *Hymenolepis diminuta. J. Biol. Chem.* **243**, 2229–2235.

Sisova-Kasatockina, O. A. and Dubovskaja, A. Ja. (1975). Proteinase activity in certain cestode species parasitizing vertebrates of different classes. *Acta parasit. pol.* **23**, 389–392.

Van den Bossche, H. (1980). Peculiar targets in anthelmintic chemotherapy. *Biochem. Pharm.* **29**, 1981–1990.

von Brand, T. (1973). "Biochemistry of Parasites", Second Edition. Academic Press, New York and London.

von Brand, T. (1979). "Biochemistry and Physiology of Endoparasites", Elsevier/North Holland Biomedical Press, Amsterdam, The Netherlands.

Chapter 9

Carbohydrate Metabolism

L. S. Roberts

I. INTRODUCTION

In recent years it has become increasingly clear that carbohydrates—their acquisition, storage, and catabolism—play a crucial role in the biology of tapeworms, as well as numerous other endoparasitic animals. Though the full dimensions of the role are as yet incompletely known, the importance of carbohydrates is undoubtedly related to the worms' inability to oxidize the glucose molecule completely, and the excretion by the worms of relatively reduced compounds as end-products of their energy metabolism. As a consequence, tapeworms store large quantities of glycogen to mobilize for energy at times when a plentiful supply of exogenous glucose is not available. Detailed studies on the carbohydrate metabolism of cestodes have been carried out on only a very few species; therefore, we must be suitably cautious in our generalizations. However, there are many similarities between what is known of the metabolism of cestodes and that

BIOLOGY OF THE EUCESTODA Vol. 2
ISBN 0-12-062102-9

of trematodes and even nematodes. Thus, we may expect that many features of their carbohydrate metabolism will be common to a large number of cestodes.

Leaving specific aspects of carbohydrate absorption and of electron transport to other chapters, we will here consider (1) the requirements for exogenous carbohydrate, (2) glycogen and its metabolism, and (3) the catabolism of carbohydrate. Furthermore, carbohydrate metabolism of tapeworms has been covered in greater or lesser detail in a number of reviews (Barrett, 1976; von Brand, 1952, 1960, 1966, 1973, 1979; Bryant, 1975, 1978; Chappell, 1979; Fioravanti and Saz, 1980; Read and Simmons, 1963; Saz, 1972, 1981a,b; Smyth, 1969). Exhaustive treatment of all the material in these previous reviews is neither necessary nor possible in the space available for the present work.

II. REQUIREMENT FOR EXOGENOUS CARBOHYDRATE

Read (1959) pointed out that most free-living organisms can meet their caloric requirements for growth and maintenance by ingestion of sufficient protein or fat and protein, but that the "meager data available strongly suggest that many parasites have a *requirement for carbohydrate*" [his emphasis]. This requirement is now firmly established in the case of *Hymenolepis diminuta*, but experimental evidence for other cestodes remains scant and indirect.

Absence of, or restriction in, host dietary carbohydrate results in reduced establishment and stunting in size of *H. diminuta* (Chandler, 1943; Mettrick and Munro, 1965; Read and Rothman, 1957a,c; Read *et al.*, 1958; Roberts, 1966). If the diet of the host is changed from one adequate with respect to carbohydrate to one suboptimal in carbohydrate, the worms dramatically reduce their size (Chandler *et al.*, 1950; Roberts and Platzer, 1967). Because *H. diminuta* can absorb and ferment such a limited variety of carbohydrates (only glucose and galactose (Laurie, 1957)), it is of interest that of all host diets tested, any diet with less than a sufficient content of *starch* is suboptimal with respect to carbohydrate. Replacement of the starch in the diet by glucose, sucrose, fructose, or dextrins-maltose is deleterious to worm growth (Roberts, 1966; Roberts and Platzer, 1967; Read and Rothman, 1957a,c). Such diets not only produce stunted worms with inhibition of proglottis production, but the quantity of carbohydrate stored by the worms is lower, and their lipid content is higher (Roberts, 1966; Roberts and Platzer, 1967; Komuniecki and Roberts, 1975). It should be emphasized that all of these diets are nutritionally adequate for the rat hosts, even those formulations that are carbohydrate-free. If the carbohydrate component of the diet is a mono- or disaccharide, the host intestine apparently can absorb it so quickly that the worms are essentially starved of carbohydrate. If the diets are nutritionally inadequate for the rats, e.g. protein-free, this may not be the case, however, and a glucose diet may produce larger worms than a starch-containing one (Dunkley and Mettrick, 1969).

Though not as thoroughly investigated, effects similar to those in *H. diminuta* have been reported in *H. citelli* and *Vampirolepis (Hymenolepis) nana* when the hosts were fed diets inadequate in carbohydrate. The effects were observed only when *V. nana* was in its period of active growth, however, and not when the worms were becoming senescent (Read *et al.*, 1958).

Evidence that other tapeworms have a carbohydrate requirement is more indirect. Restriction in host diet causes *Davainea proglottina* and *Raillietina cesticillus*, cestodes of poultry, to destrobilate (Levine, 1938; Reid, 1942). When Read (1957) starved dogfish sharks (*Mustelus canis*) for eight days, the *Lacistorhynchus tenuis* in their spiral valves were drastically reduced in numbers and in size. The worm loss and weight reduction were much less if the sharks were given glucose every other day by stomach tube, and the administration of starch in like manner prevented the effects.

Davainea, *Raillietina*, and the hymenolepidids are all parasites of herbivores, and the host diets in nature would be expected to contain considerable quantities of starch. Furthermore, it appears that all adult cestodes store a large amount of glycogen under normal circumstances (Table I; figures representing total carbohydrate are not distinguished from those reported as "glycogen" in Table I because almost all the anthrone-sensitive, total carbohydrate appears to be glycogen (Fairbairn *et al.*, 1961)). It is not obvious how the tapeworms found in carnivores (*Lacistorhynchus*, above, and *Hydatigera*, *Proteocephalus*, and *Triaenophorus* in Table I) could obtain sufficient carbohydrate if their requirement is comparable to that of *H. diminuta*. Perhaps amino acids are much more glycogenic in some cestodes than is apparently the case with *H. diminuta*. In the case of metacestodes, it is probable that the glucose in the hosts' body fluids is adequate to nourish the worms located in their hosts' tissues.

Tapeworms living in the bile duct of their host are yet to be investigated. In one experiment on *Hymenolepis microstoma* (Roberts, unpublished), worms were much less severely affected by a host diet containing sucrose (RF diet of Roberts and Platzer, 1967, with sucrose substituted for starch) than *H. diminuta* would have been on a similar regimen. On the other hand, a carbohydrate-free diet (CF diet of Roberts and Platzer, 1967) almost completely suppressed growth.

Table I. Glycogen or total carbohydrate concentrations in cestodes.[a]

	Host(s)	Glycogen (% dry weight)	References
CYCLOPHYLLIDEA			
METACESTODES:			
Echinococcus granulosus protoscoleces	sheep	18.9	Agosin *et al.*, 1957
E. granulosus protoscoleces	sheep strain	16.9	McManus and Smyth, 1978
E. granulosus protoscoleces	horse strain	17.7	McManus and Smyth, 1978
E. multilocularis protoscoleces	cotton rats	12.0	McManus and Smyth, 1978
Hydatigera taeniaeformis strobilocerci	mice	19.7	von Brand and Bowman, 1961
H. taeniaeformis strobilocerci	mice	43.3	Hopkins, 1960
H. taeniaeformis strobilocerci	rats	24.9[b]	von Brand and Bowman, 1961
H. taeniaeformis strobilocerci	mice	25.4–31.4[b]	von Brand *et al.*, 1968
Taenia crassiceps cysticerci	mice	27.5	Taylor *et al.*, 1966
ADULTS:			
Hydatigera taeniaeformis	cats	23.2[b]	von Brand and Bowman, 1961
H. taeniaeformis	cats	9.4–21.0[b]	von Brand *et al.*, 1968
Hymenolepis citelli	*Citellus* at room temperature	40–56	Ford, 1972
H. citelli	*Citellus* maint. 6–10 days at 6°C	34–43	Ford, 1972
H. diminuta	rats	8.7–43	Roberts, 1961; Fairbairn *et al.*, 1961; Goodchild, 1961; Read, 1967
Moniezia expansa	sheep	24–32	Weinland, 1901[c]; von Brand, 1933; Wardle, 1937; López-Gorge and Monteoliva, 1965[c]
Raillietina cesticillus	chickens	22.4	Reid, 1942

			1935, Smorodintsev and Bebeshin, 1935[c]
Thysaniezia giardii	ruminants		
mature proglottides		11.2–49.7	
gravid proglottides		9.3–38.6	Singh *et al.*, 1977
PROTEOCEPHALATA			
METACESTODES:			
Proteocephalus ambloplitis plerocercoids	fish (*Micropterus*)	68.8	Marra and Esch, 1970
ADULTS:			
Proteocephalus ambloplitis	fish (*Micropterus*)	16.29–22.9	Marra and Esch, 1970
PSEUDOPHYLLIDEA			
METACESTODES: (all plerocercoids)			
Diphyllobothrium dendriticum	fish	33–36	Reuter, 1967, 1971
D. latum	fish	17.9	Markov, 1939[c]
Diphyllobothrium sp.	fish	31.5	Archer, and Hopkins, 1958
Ligula intestinalis	fish	38–52	Markov, 1939[c]
Schistocephalus solidus	fish	50.9	Hopkins, 1950
ADULTS:			
Diphyllobothrium latum	fish-eating mammals	20	Smorodintsev and Bebeshin, 1935[c]
Diphyllobothrium sp.	rats (experimental)	36.2	Archer and Hopkins, 1958
Eubothrium rugosum	fish (*Lota*)	22.8	Markov, 1939[c]
Schistocephalus solidus	water birds	28	Hopkins, personal communication to Smyth, 1969
Triaenophorus nodulosus	fish (*Esox*)	13.8	Markov, 1939 (cited by Markov, 1961 in Dogiel *et al.*)

[a] No distinction is made here between glycogen and total carbohydrates (see text).
[b] Calculated on basis of DW/FW = 0.22, von Brand and Bowman, 1961; for mice; DW/FW = 0.267 for adults from cats.
[c] Cited in von Brand, 1973 and/or Smyth, 1969.

III. GLYCOGEN STORAGE AND MOBILIZATION

A. Physicochemical characteristics of tapeworm glycogen

Electron microscopy has shown that glycogen in a variety of cells is not morphologically homogeneous, but occurs both as discrete granules (β-particles) and as aggregates or rosettes (α-particles) of granules. Lumsden (1965) described β-particles measuring 20–40 nm and α-particles measuring 60–200 nm from both *Hymenolepis diminuta* and *Lacistorhynchus tenuis.* Similar configurations have now been reported from a variety of other adults and metacestodes, and they are undoubtedly of general occurrence (Collin, 1970; Nieland and Weinbach, 1968; Baron, 1968; Sakamoto and Sugimura, 1969; von Bonsdorff *et al.*, 1971; Lumsden and Byram, 1967; Bråten, 1968). The particles can be extracted from the tissues and recovered intact if a mild procedure is used (Bueding and Orrell, 1964; Orrell *et al.*, 1964). Whether the product isolated by the mild procedure is "native" glycogen or is artefactual has been argued in the past, but there is evidence that the aggregations have metabolic importance (Colucci *et al.*, 1966; Roberts *et al.*, 1972), at least in the case of tapeworm glycogen. It is, however, somewhat misleading to speak of only two types of particles or two molecular weight components because the glycogen is very polydisperse, and apparently exists as varying combinations of particles to make up a wide range of molecular weights (Orrell *et al.*, 1966; Roberts *et al.*, 1972). The molecular weights range from several hundred thousand to over 1000 million (Orrell *et al.*, 1966). The nature of the bonds combining the aggregations is still unknown, but the bonds are clearly sensitive to strong acids and alkali. They are not ruptured by prolonged incubation with agents that break protein or hydrogen bonds: 8 M urea, 8 M guanidine, 2 M thiocyanate, 8 M LiBr, or 1% non-ionic, anionic or cationic detergents (Orrell *et al.*, 1966). Though the method of Bueding and Orrell (1964) frees the particles from protein, they have proteins associated with them in the cell, including some of the enzymes of glycogen metabolism. Barber *et al.* (1967) found that more α-glycan phosphorylase was bound to lower molecular weight particles of rat liver glycogen, while glycogen synthase was mostly associated with the higher molecular weight particles.

The physicochemical characteristics of *Moniezia expansa* glycogen have been examined in more detail than those of other cestode glycogens (Orpin *et al.*, 1976). *M. expansa* glycogen is a highly branched structure consisting of α-1, 4- and α-1, 6-linked glucopyranose units. Orpin *et al.* (1976) reported an average chain length of 12.9 glucose units, with an average inner chain length of 2.9 units and an average outer chain length of 9.0 units. Worms extracted with a mild method (Drochmans, 1962) yielded a particulate product with very high sedimentation coefficients (major component, $s_{20,w}=910$; minor component, $s_{20,w}=600$), compared with alkali-extracted glycogen ($s_{20,w}=61$). Electron microscopy showed α- and β-particles and possible γ-particles (small spheres seen at the periphery of β-particles, occasionally forming filaments up to 10 nm long) (Orpin *et al.*, 1976). In most respects the *M. expansa* glycogen was similar to other glycogens, but the α-particles were more stable to acidic conditions than rat liver glycogen; dissociation to their constituent β-particles occurred between pH 2.5 and 1.7. This may have been due to the greater protein (enzymes?) associated with the *M. expansa* glycogen (0.9% N, compared to 0.08–0.3% N for rat liver glycogen isolated under the same conditions).

Although glycogens isolated by mild methods from a variety of sources are polydisperse (rabbit liver and muscle, *Fasciola hepatica*, *Ascaris lumbricoides* muscle, *Phormia regina* flight muscle), the molecular weight (MW) spectrum of glycogen from *Hymenolepis diminuta* differs from the others in that about 30% of the total is very high in MW (average 900 million) (Orrell *et al.*, 1964). About 60% of the glycogen averages 25–60 million, while intermediate MW components are present in much smaller amounts (Roberts *et al.*, 1972). Colucci *et al.* (1966) incubated *H. diminuta* in ^{14}C-glucose, isolated the glycogen, then divided it into two components according to molecular weight by differential centrifugation. They found incorporation of label into the higher MW fraction at a relatively much higher rate than into the low MW fraction. Roberts *et al.* (1972) performed similar experiments, but they subdivided the glycogen into a much larger number of MW fractions by rate-zonal centrifugation. Those fractions showing the highest specific activity were those present in smallest amount i.e. those intermediate in MW (Fig. 1). This is a clear indication that glycogen particles are disassembled and reassembled within the cestode's cells. Otherwise, the initial incorporation would parallel the abundance of the respective fractions in the worm. Furthermore, the finding that glucose is initially incorporated into glycogen of intermediate MW, then assembled subsequently into higher and lower MW fractions, implies that the α- and β-particles themselves have a metabolic significance.

B. Distribution and localization in the worms

Distribution of glycogen in tapeworms has been studied both as a function of the tissues and cells in which it is deposited and as a function of position down the length of the strobila. From the reports on the glycogen concentration in different parts of the strobila, the general conclusion is that differences in metabolic state within the same worm may cause differences in glycogen level. Reid (1942) found higher glycogen in the anterior halves of *Raillietina cesticillus* than the posterior halves, or the same concentrations,

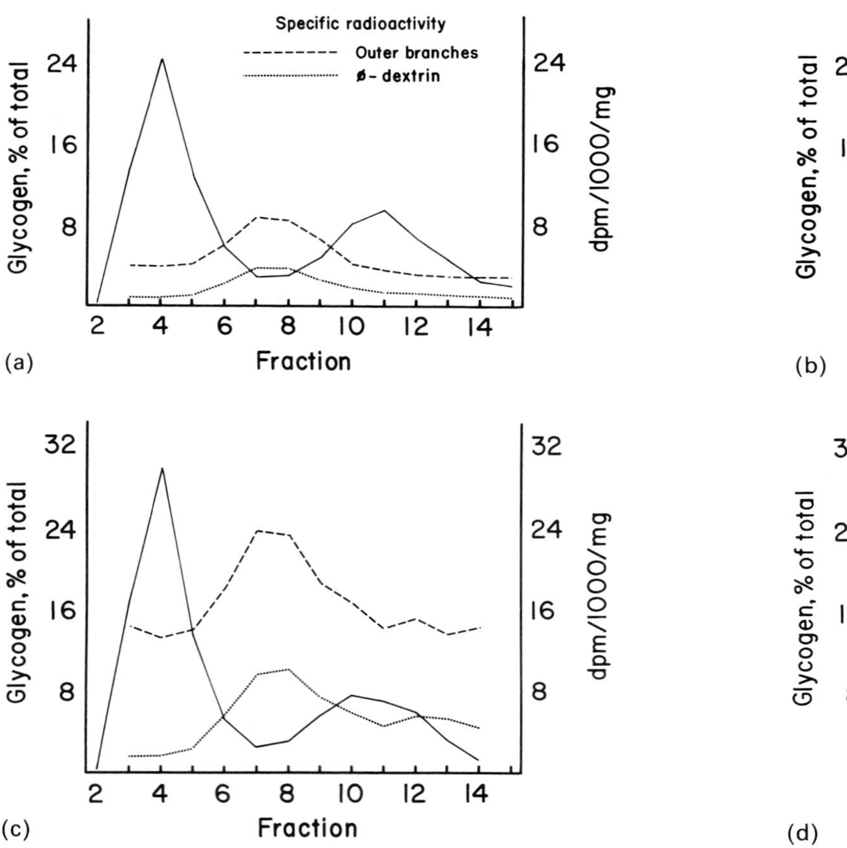

Specific radioactivity
Outer branches
ø- dextrin
Glycogen, % of total
dpm/1000/mg
Fraction
(a)
(c)

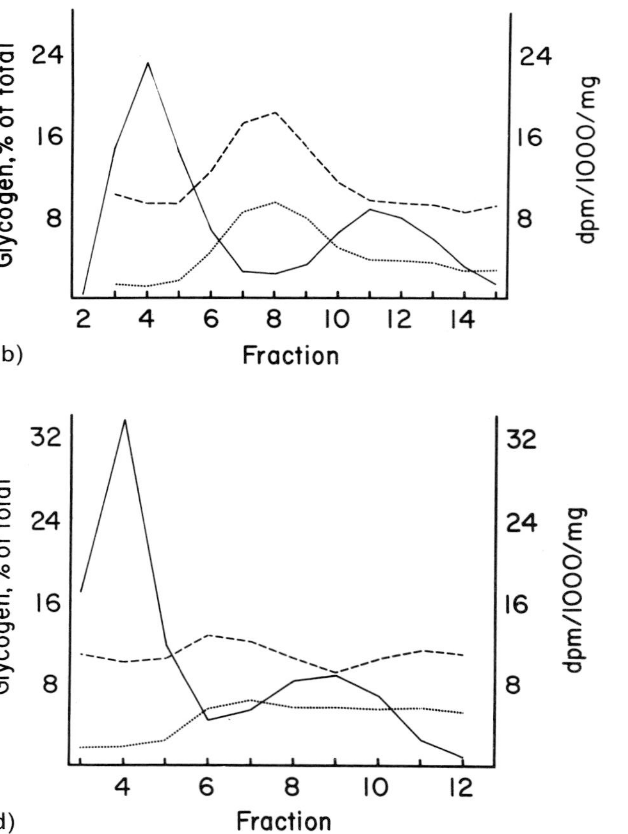

Glycogen, % of total
dpm/1000/mg
Fraction
(b)
(d)

depending on the time of day the worms were removed from the hosts. Read (1956) found that the glycogen level in the anterior quarters of *Hymenolepis diminuta* was about half of that found in the second and third quarters. On the other hand, Fairbairn *et al.* (1961) and Bolla and Roberts (1971) found that the carbohydrate concentration in the immature proglottides of normal *H. diminuta* was not lower than that in their mature regions. Fairbairn *et al.* (1961) reported that the terminal 50 mm (containing infective oncospheres) had considerably less glycogen than the rest of the worms. Bolla and Roberts (1971) found that carbohydrate in the mature and gravid regions of crowded *H. diminuta* was much lower than that in the immature proglottides.

A large number of papers are available that report the distribution of glycogen in cells and tissues of adult cestodes as observed by light microscopy (PAS positive, diastase labile material) (see for example, Read and Simmons, 1963; Hedrick and Daugherty, 1957; Mayberry and Tibbitts, 1971; Kwa, 1972; Moczoń, 1975) and by electron microscopy (references above, p. 348). These histochemical results all indicate that the highest glycogen concentrations are in the "parenchymal" cells and in and around the muscles. The ovaries and oocytes do not accumulate glycogen. Testes have little or none, but the sperm of some species store glycogen, while others apparently do not (see Davis and Roberts, 1983a; Rybicka, 1967). Vitellaria in cestodes accumulate glycogen during their development, and a considerable amount is borne by the vitelline cell as it joins the oocyte (see Davis and Roberts, 1983b). Histochemical distribution of glycogen in the developing embryo of *H. diminuta* was described by Rybicka (1967) and in the cysticercoids by Heyneman and Voge (1957) and Moczoń (1977).

After study of *H. diminuta* by electron microscopy, Reissig and Colucci (1968) concluded that only two types of cells comprised the parenchyma, namely, the cytons of the tegument and of the muscles. The existence of a separate, specifically parenchymal cell type could not be confirmed, and the tegumentary cytons did not bear glycogen. Therefore, the main storage cell for glycogen in this tapeworm is the myocyton (Fig. 2). Reissig and Colucci (1968) observed that the myocytons were connected to the contractile portions of the cells by thin stalks, and that both the cytons and fibres bore evaginations or sacs filled with glycogen particles. Lumsden and

---

Fig. 1. Effect of worm age on incorporation of ^{14}C-glucose into glycogen as a function of molecular weight of the glycogen in *Hymenolepis diminuta.* (a) 9 days post-infection of the rat host. (b) 12 days post-infection. (c) 15 days post-infection. (d) 84 days post-infection. Solid lines are the percentage of the total glycogen in the respective molecular weight fractions; dotted lines are the specific activities in the phosphorylase limit-dextrins of the fractions; and dashed lines are the specific activities in the outer branches of the fractions (from Roberts *et al.*, 1972).

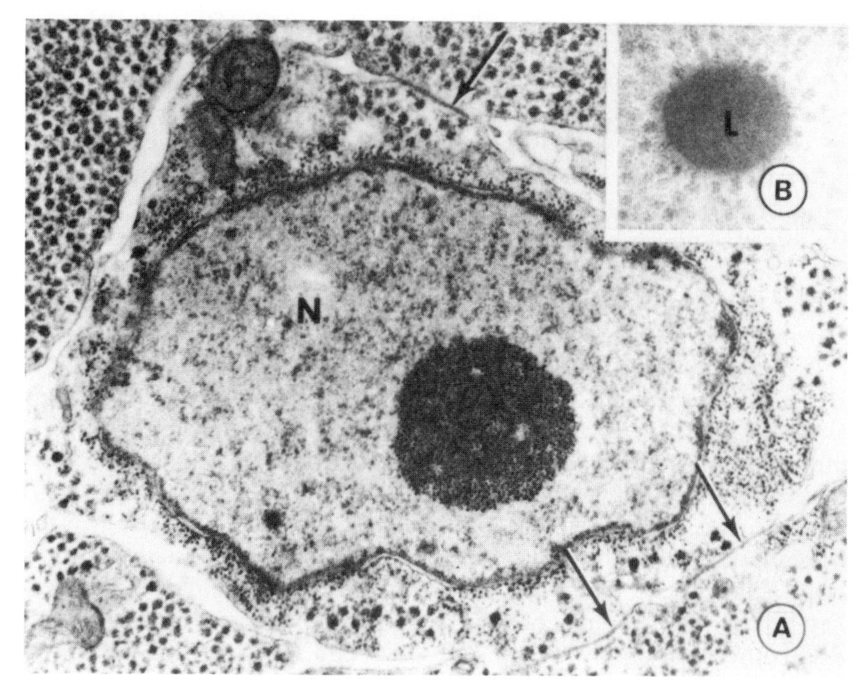

Fig. 2. Medullary myocyton characterized by nucleus (N) with prominent nucleolus. Perinuclear cytoplasm filled with free ribosomes, peripheral mitochondria and abundant α-glycogen. Cell junctions (arrows) join adjacent myocytons (×28 000). Inset: Peripheral cytoplasm of medullary myocyton containing lipid (L) surrounded by α-glycogen rosettes (×30 000) (from Lumsden and Specian, 1980).

Specian (1980) concurred in these views. It is not yet known whether the same situation obtains in other cestodes, or perhaps even trematodes, but it seems likely that what have been traditionally called parenchymal cells are in reality myocytons in all these worms. Von Bonsdorff *et al.* (1971), for example, found that glycogen was abundant in parenchymal cells and muscle cells of plerocercoids of *Diphyllobothrium dendriticum*, but that the cells identified as parenchymal were "remarkably few", and that they were "not easy to distinguish from muscle cells". Lumsden and Specian (1980) also observed glycogen in the perikarya of nerve cells (β-particles), the cytoplasm of sensory endings (β-particles), spermatozoa, and vitelline cells, but not in any other cells in adult *H. diminuta*. The various ducts (excretory, reproductive) were frequently invested with processes from the myocytons that bore abundant glycogen (Fig. 3).

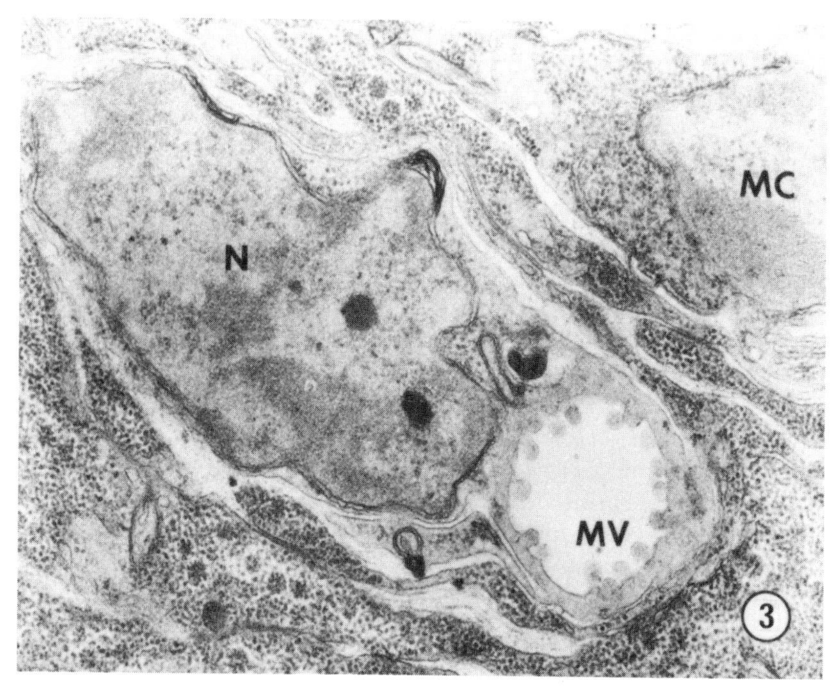

Fig. 3. Collecting duct of excretory system. Nucleus (N) directly adjacent to syncytial cytoplasm of duct. Ductal cytoplasm extended as short clavate microvilli (MV) at free surface. Carbohydrate-storing myocytons (MC) surround duct (× 43 900) (from Lumsden and Specian, 1980).

C. Concentration of glycogen in cestodes

The content of glycogen in cestodes can range up to 50% or more of dry weight (Table I). The data in Table I should not be interpreted as representing characteristic glycogen levels for each of the species listed, but only that tapeworms characteristically store high levels of glycogen. The glycogen concentration can vary with the feeding cycle of the host, the host diet, the numbers of cestodes present, and the stage of development of the worms. Reid (1942) found that the glycogen concentration of *Raillietina cesticillus* was reduced by 45% overnight when the host was not feeding. Glycogen concentration in *H. diminuta* is reduced by 60% in 24 h and 80% in 48 h during host fasting (Fairbairn *et al.*, 1961; Colucci *et al.*, 1966). Glycogen depleted during host starvation is rapidly resynthesized upon refeeding (Read and Rothman, 1957b). Host diets that are carbohydrate-free or containing sucrose or glucose as the sole carbohydrate result in very much lower storage of worms glycogen compared to worms from hosts fed high-starch diets (Roberts, 1966; Roberts and Platzer, 1967; Komuniecki and Roberts, 1975). The glycogen level of *Hymenolepis citelli* declines in hypothermic ground squirrels, but apparently not nearly so rapidly as it would if the hosts were fasted at room temperature (Ford, 1972).

Glycogen concentrations of hexacanth larvae and metacestodes of few species have been determined. Fairbairn *et al.* (1961) reported a carbohydrate concentration of 12% of dry weight in "infective eggs" of *H. diminuta*. Of those whose metacestodes attain considerable mass in the tissues of the intermediate hosts, the glycogen levels are usually quite high (Table I). These apparently fall to some degree after infection of the definitive host (Hopkins, 1950; Archer and Hopkins, 1958). Archer and Hopkins (1958) reported that *Diphyllobothrium* sp. decreased in glycogen content between 18 and 54 h postinfection of the definitive host, and that the concentration then rose to a level greater than that found in the plerocercoid. The lowest carbohydrate level recorded by Roberts (1961) in *H. diminuta* at four days after infection of the rat was 8.7% of dry weight, which then rose to over 40% in worms from low population densities (one to 10 worms per host).

The carbohydrate content of *H. diminuta* diminishes with increasing numbers of worms present in the host intestine (Roberts, 1961). This is one of the manifestations of the "crowding effect", an inverse correlation of size attained by the worms with increasing population density. The crowding effect has been reported in *Raillietina cesticillus*, *Hymenolepis microstoma* and *Vampirolepis nana* (see Roberts, 1980), but it has been studied most extensively in *H. diminuta*. Read (1959) concluded that the crowding effect "may be interpreted in terms of competition for utilizable carbohydrate by the individual worms in the populations". Present evidence indicates that, while competition for host dietary carbohydrate plays some role in the crowding effect, the phenomenon is a complicated one and probably cannot be explained by a single, simple mechanism (Roberts, 1980; Insler and Roberts, 1980a).

D. Glycogen synthesis

Glycogenesis in a variety of tapeworms has been observed after incubation *in vitro* in simple saline solutions supplemented by glucose (Read and Simmons, 1963). In *H. diminuta* the process is markedly enhanced by the presence of CO_2, and some carbon from HCO_3^- in the medium is incorporated into glycogen (Fairbairn *et al.*, 1961; Read, 1967; Scheibel and Saz, 1966). Other substrates, including fumarate, malate, succinate, glutamate, aspartate and galactose do not lead to *net* glycogen synthesis

(Read, 1967; Komuniecki and Roberts, 1977b), although radioactivity from ^{14}C-galactose is incorporated into glycogen (Komuniecki and Roberts, 1977b). The enzymes necessary for galactose incorporation, namely, galactokinase, galactose 1-phosphate uridyl transferase and UDPgalactose 4-epimerase, have been reported in *H. diminuta*, but their specific activities are low (Komuniecki and Roberts, 1977c). The enhancement of glycogenesis by the presence of CO_2 is undoubtedly related to the fixation of CO_2 into phosphoenolpyruvate (PEP) to form oxaloacetate, catalysed by phosphoenolpyruvate carboxykinase (PEPCK) in the worms, and the reactions subsequent to this step (see below), i.e. CO_2 leads to enhanced energy generation, which favours biosynthesis. The exact pathway of incorporation of the bicarbonate carbon into glycogen has not been determined, but it may involve a reversal of glycolysis beginning with PEP (Prescott and Campbell, 1965). In *H. diminuta*, fasting of the host stimulates both glycogenesis and incorporation of CO_2 into glycogen when the worms are subsequently incubated *in vitro* with glucose (Fairbairn *et al.*, 1961; Prescott and Campbell, 1965). Prescott and Campbell (1965) found a sevenfold increase in the specific activity of PEPCK *in the direction of PEP formation* during the first 6 h of host starvation, but the activity of this enzyme in the direction of oxaloacetate formation decreased in parallel with the decline in the glycogen concentration of the worm.

Because the glycogen concentrations in tapeworms are potently affected by physiological conditions, it would be expected that the crucial enzymes of glycogen synthesis, especially glycogen synthase, would be highly regulated. In fact this is the case, and control of the enzyme differs in some respects from the mammalian isofunctional enzymes that have been studied. As in mammalian systems, glycogen synthase exists in at least one form that is active at physiological concentrations of glucose 6-phosphate (G6P) and one or more forms that are inactive at physiological G6P concentrations, and these forms can be interconverted (Dendinger and Roberts, 1977a; Mied and Bueding, 1974, 1979a). The active form is inactivated by phosphorylation, but whether the enzyme has multiple phosphorylation sites, as in mammalian systems (Soderling, 1979; Roach, 1981), is unknown.

In any case, the proportion of synthase present in each form in the worm at a given time may be stabilized during isolation by homogenizing the worms in a buffer containing EDTA to chelate Mg^{++} and block the kinase reaction, and NaF to inhibit the phosphatase reaction. The percentage of total enzyme activity in the active form would reflect the synthetic activity for glycogen at a given time. Thus, Dendinger and Roberts (1977b) found that the percentage of active synthase declined when the rat hosts of *Hymenolepis diminuta* were fasted 24 h, then increased rapidly when the rats were re-fed. The effect was probably due to the glucose made available as the starch in the meal was digested since feeding the host a meal containing only cellulose had no effect. Furthermore, if worms were removed from fasted hosts, then incubated in a medium containing glucose, there was a rapid conversion of inactive to active synthase. This observation was confirmed by Mied and Bueding (1979b). In accordance with the observation that glycogen concentration increases during the prepatent period of *H. diminuta* (Roberts, 1961), the total synthase activity and the percentage of active synthase were found to increase during development of the worms (Dendinger and Roberts, 1977a). Interestingly, though the glycogen content of crowded worms is much lower than that of non-crowded *H. diminuta*, the percentage of active synthase in 100-worm populations is generally about twice that in worms from 10-worm populations.

Mied and Bueding (1979a) studied metabolic regulation of glycogen synthase from *H. diminuta*. They distinguished three allosteric sites of activation in the active form. In contrast, inactive synthase showed only one allosteric activation site (G6P site), which responded to a much smaller number of activators than the G6P site of active synthase, and was characterized by a much lower affinity for the activators. The inactive tapeworm synthase was not activated by P_i; indeed, P_i inhibited the activation by G6P. On the other hand, the active synthase was activated by P_i, PP_i, and a wide variety of sugar phosphates, including intermediates of the glycolytic pathway; however, intermediates of the tricarboxylic acid cycle were without effect. The latter finding is in contrast to the case in active synthase from vertebrate tissues. Another difference between the active synthase of the tapeworm and the isofunctional enzymes of some mammalian sources, and of yeast, is that the cestode enzyme exhibits characteristics of both a "K system" and a "V system", i.e. the positive effector, G6P, both decreases the K_m for the substrate (UDPG) and raises the V_{max}, whereas the V_{max} is usually unaffected in the other synthases. Activation of the active synthase of *Hymenolepis* is consistent with the physiological needs for glycogen synthesis of the worm. Thus, a plentiful supply of exogenous glucose corresponds to higher levels of glycolytic intermediates, especially G6P, and glycogen synthesis would be thus stimulated. A high level of UTP would be correlated with high ATP and energy charge, and UTP was found to be a positive activator. UMP was an inhibitor, and a high UMP level in the worm would be correlated with a lower energy charge when the glucose moities would need to be degraded through glycolysis, rather than being stored.

Finally, the efficacy of the various molecular weight species of glycogen found in *Hymenolepis* (see above), as primers for glycogen synthase, was studied by Mied and Bueding (1979b). As would be expected from the results of Roberts *et al.* (1972), Mied and Bueding (1979b) found that the intermediate molecular weight fractions were most effective in stimulating activity of the tapeworm synthase (Fig. 4). The specificity of this effect was shown by extracting and fractionating rabbit liver glycogen by molecular weight, then testing these molecular fractions as primers. Surprisingly, the most effective primers of the *rabbit* liver glycogen were the fractions of very high molecular weight, and those fractions were much higher in olecular weight than the intermediate fractions from *Hymenolepis*. This specificity could not be accounted for by differences in branch length and degree of branching of the glycogen polymers (Berman and Bueding, quoted by Mied and Bueding, 1979b). While the physiological significance of these

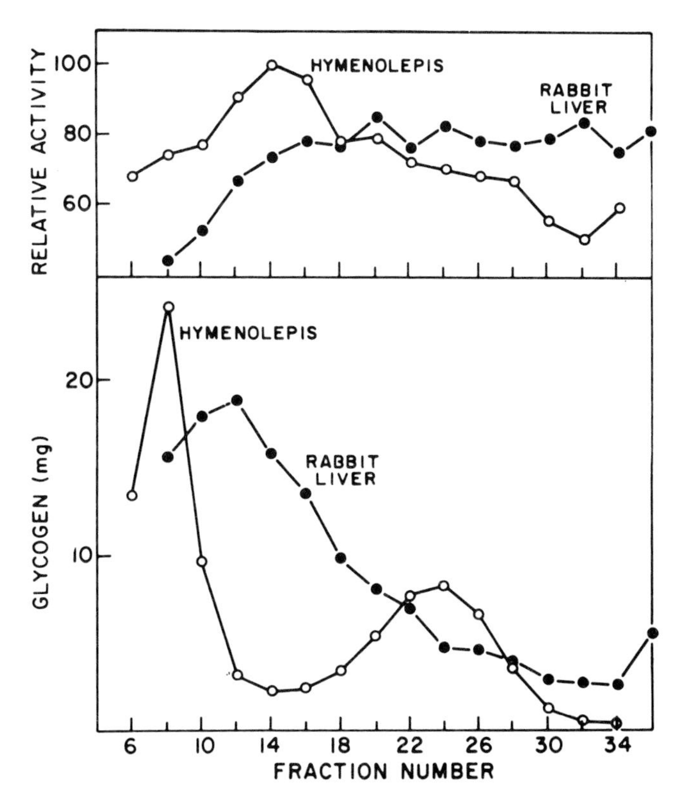

Fig. 4. Glycogen synthase activity as a function of molecular weight of primer glycogen from *Hymenolepis diminuta* and from rabbit liver. Synthase was in its less active form from *H. diminuta* taken from starved hosts (10 mM UDPG, 10 mM G6P, pH 7.5). The lower portion shows the molecular weight distributions of the two glycogens obtained by rate-zonal centrifugation; molecular weight increases with fraction number. Upper portion shows synthase activity in presence of a saturating level (0.04%) of the corresponding glycogen fraction. The synthase activity in the presence of *H. diminuta* fraction 14 is used as a reference and assigned a relative activity of 100 (from Mied and Bueding, 1979b).

observations must remain speculative, it could be related to the way in which the glycogen is packaged, stored and mobilized in the tapeworm myocytons.

E. Glycogen phosphorolysis

The glucose residues in glycogen apparently are mobilized by the enzyme glycogen phosphorylase, which exists in both an active phosphorylated form (phosphorylase a) and a less active dephosphorylated form (phosphorylase b). The reactions leading to the activation and deactivation of phosphorylase are presumably similar in tapeworms to those described in vertebrate tissues, although they have not been investigated adequately.

The presence of phosphorylase in *Hymenolepis diminuta* was first demonstrated by Read (1951). Platzer and Roberts (1970) studied the effect of host pyridoxine deficiency on phosphorylase activity in *H. diminuta.* They homogenized the worms in the presence of NaF and EDTA, thus presumably inhibiting the phosphorylase phosphatase and kinase reactions and stabilizing the proportions of phosphorylase a and b. The phosphorylase a requires a coenzyme, pyridoxal phosphate. Platzer and Roberts (1970) found that the specific activity of phosphorylase was 31% lower in worms from pyridoxine limited or deficient hosts. The activity in worms fed complete diets declined with worm maturation between 9 and 15 days post-infection. Adenosine 5′-phosphate (AMP), which is a positive allosteric effector for phosphorylase b in other systems, had no effect on *H. diminuta* phosphorylase activity under the conditions used by Platzer and Roberts (1970).

Determinations of other enzymes of glycogen metabolism in *H. diminuta* have been by histochemical means, and the results must be interpreted with suitable caution. Moczoń (1975, 1977) assessed phosphorylase activity in the direction of glycogen synthesis by evaluating the intensity of iodine staining of the product. On the basis of these studies (Table II), Moczoń found evidence for the presence of phosphorylase a and b, phosphorylase a phosphatase, phosphorylase b kinase, a cAMP-dependent protein kinase, and 1,4-α-glucan branching enzyme. He concluded that the mechanisms regulating phosphorylase activity in *H. diminuta* were similar to those operating in the tissues of vertebrates. Enzyme activity was not observed in the tegument, ovaries, or vitellaria. Sperm were positive, and activity was high in the parenchymal tissues of the adult. Activity was observed in the inner envelope, scolex and cercomere of the cysticercoid and in the muscle cells of the oncospheres. Although vitelline cells had considerable quantities of glycogen, they did not show phosphorylase activity until after they joined the zygotes and the embryos began to cleave.

Table II. Histochemical studies of reactions associated with glycogen phosphorlyase in *Hymenolepis diminuta*, modified from Moczoń (1975).[a]

Medium to detect:	Additions/subtractions	Reaction intensity *H. diminuta*	rat muscle	rat liver
Phosphorylase a activity	None	+ + +	+ + +	+
	−NaF	+ +	+ +	illegible[b]
	−EDTA	+ + +	+ + +	±
	+pyridoxal 5-phosphate (1 mM)	+ + + +	+ + + +	+ + +
	+cAMP (1 mM), $MgCl_2$ (1 mM), and −EDTA	+ + + +	+ + + +	+ + + +
	+cAMP, $MgCl_2$, −EDTA, and +caffeine (8 mM)	+ +	+ + +	+ +
	+insulin (1 IU/ml), −ethanol	+ + +	+ + +	+
	+phlorizin (1 mM)	"very strong apparent activation of polysaccharide synthesis"		
	+phlorizin (3 mM)			
	+D-glucose (0.1 M)	—	—	—
Phosphorylase a and phosphorylase b kinase activity	None	+ + + +	+ + + +	+ + +
	+cAMP (1 mM), $MgCl_2$ 1 mM)	+ + + + +	+ + + + +	+ + + +
	+pyridoxal 5-phosphate (1 mM)	+ + + +	+ + + +	+ + +
	+caffeine (8 mM)	+ +	+ +	+
	+$CaCl_2$ (1 mM), −$MgCl_2$	+ + + +	+ + + +	+ + +
	−polyvinyl pyrrolidone	illegible[b]	illegible[b]	illegible[b]

[a] Frozen sections, air-dried, incubation media overlaid; incubated 1 h at 23–24°C, pH 7; see Moczoń (1975) for further media details.
[b] Results could not be interpreted because of diffusion of products from sections.

IV. GLYCOLYSIS

The term "glycolysis" is usually applied to homolactic fermentation, the series of reactions that catabolize glucose to lactate. Indeed, some flatworms are homolactic fermenters, e.g. *Schistosoma mansoni* (Bueding, 1950; Schiller *et al.*, 1975). However, recent years have brought increasing realization that a variety of other compounds accumulate as end-products of carbohydrate dissimilation in various parasitic worms (including tapeworms), free-living molluscs, and even in some vertebrate tissues (Saz, 1981a; Gäde *et al.*, 1975; Hochachka *et al.*, 1975; Johnston, 1975). The pathways giving rise to these diverse end-products are of considerable interest as adaptations to energy production under hypoxic conditions.

A. Pathway to phosphoenolpyruvate (PEP)

Evidence for the operation of the glycolytic sequence (Embden-Meyerhof pathway) can be demonstration of the enzymes of the pathway, of metabolic intermediates, and of the incorporation of radiolabelled carbon from glucose into the end-products. These demonstrations have not been complete for any cestode. Yet, there can be little doubt that the pathway exists, is important in these worms, and is very likely a feature common to all species. Moreover, because the end-products of glucose catabolism are relatively reduced compounds, and essentially all energy derived from glucose carbon is via glycolysis and its terminal reactions, and because these pathways are energetically inferior to aerobic respiration, we may infer that the flux rate through the pathway must be comparatively quite high.

At least some enzymes of glycolysis have been demonstrated in *Hymenolepis diminuta*, *Echinococcus granulosus*, *Moniezia expansa*, *Taenia crassiceps*, *T. taeniaeformis*, *Schistocephalus solidus*, *Bothriocephalus gowkongensis*, *Triaenophorus crassus* and *Khawia sinensis* (see Agosin and Aravena, 1959b; Behm and Bryant, 1975a; von Brand, 1973, 1979; Körting, 1976). Körting and Barrett (1977) measured the activity of all the enzymes in homogenates of *Schistocephalus solidus* plerocercoids, phosphoglyceromutase and phosphopyruvate hydratase being measured jointly, and Beis and Barrett (1979) reported the levels of all phosphorylated intermediates in *S. solidus*. Read (1951) assayed several of the enzymes and phosphorylated intermediates in *Hymenolepis diminuta*, and Scheibel and Saz (1966) concluded, on the basis of radiolabel experiments, that the Embden-Meyerhof sequence occurred in this cestode.

Enzymatic regulation of glycolysis has been studied in *Moniezia expansa* and *Schistocephalus solidus* (Behm and Bryant, 1975a; Beis and Barrett, 1979). Indication of which particular enzymatic steps may be regulatory in a given metabolic pathway *in vivo* may be obtained by determining the mass action ratios of substrates and products for each reaction and then comparing these to the apparent equilibrium constants for the reactions. A reaction is a non-equilibrium reaction if the mass action ratio is 5% or less of the apparent equilibrium constant (Rolleston, 1972). Non-equilibrium reactions may or may not be regulatory; equilibrium reactions usually are not regulatory. Strong evidence that a non-equilibrium reaction is regulatory is provided if the rate of flow through a pathway changes in a direction opposite to that of the concentration of substrate for the enzyme, e.g. the substrate concentration increases when the flux rate through the pathway decreases.

Such considerations were applied to glycolysis in the anterior portions (anterior gram) of *M. expansa* by Behm and Bryant (1975a). They determined the levels of intermediates, both aerobically and anaerobically, and in the presence and absence of exogenous glucose, and the activities of selected glycolytic enzymes. Aerobic incubation (5% CO_2, 95% air) decreased the flux through the pathway compared with anaerobic incubation (5% CO_2, 95% N_2) in the absence of glucose, i.e. a Pasteur effect. Behm and Bryant (1975a) concluded that the reactions catalysed by phosphofructokinase and pyruvate kinase were probably regulatory sites under those conditions. In the presence of glucose, phosphofructokinase did not appear to be regulatory; however, an active fructose 1,6-diphosphatase was present, which complicated the interpretation of the results.

Similar studies were conducted using plerocercoids of *Schistocephalus solidus* (Beis and Barrett, 1979). Interestingly, an increase in ambient temperature from 25°C to 40°C stimulates the plerocercoid to mature sexually and begin egg development (Sinha and Hopkins, 1967). Such an increase in temperature occurs in nature when the fish intermediate host is eaten by the bird definitive host. As might be expected, dramatic changes in the energy metabolism of the worms accompany this activation. The glycolytic flux is greatly increased (Körting and Barrett, 1977). Beis and Barrett (1979) found that the steady state content of fructose 6-phosphate, fructose 1,6-disphosphate, 3-phosphoglycerate, PEP, and lactate decreased upon activation, while the remaining glycolytic intermediates were unchanged. Thus, Beis and Barrett (1979) concluded that phosphofructokinase, pyruvate kinase, and probably PEP carboxykinase were functioning as regulatory enzymes.

It is to be noted that the foregoing studies were conducted on whole worms that contain many different tissue types. The comparisons of mass action ratios and apparent equilibrium constants ignore possible internal compartmentalization, which may have considerable importance in the living worms.

B. Hexokinase(s)

The kinetics and other characteristics of individual enzymes in the glycolytic pathway of tapeworms have been little investigated. As hexokinase is the critical enzyme initiating the sequence, it was partially purified from *H. diminuta* by Komuniecki and Roberts (1977a). The enzyme exhibited broad hexose specificity and phosphorylated glucose (K_m=0.09mM), fructose (K_m=55.0 mM), and mannose (K_m=5.6 mM), but not galactose. A separate galactokinase activity is also reported (Komuniecki and Roberts, 1977b). However, *H. diminuta* apparently has no mediated systems to absorb fructose and mannose (Read and Simmons, 1963), and the high K_m's for these substrates suggest that they cannot be important energy sources *in vivo*. The hexokinase was non-competitively inhibited by glucose 6-phosphate ($K_i = 4.3 \times 10^{-4}$ M), and this inhibition could conceivably become a bottleneck in glycogen synthesis or in glycolytic degradation of glucose. However, the K_i for glucose 6-phosphate inhibition is much higher than that for three mammalian "low K_m" hexokinases described by others (Grossbard and Schimke, 1966), and the concentration of glucose 6-phosphate found in starved worms after refeeding (0.336 μmoles/g tissue) is considerably below the K_i. Thus, assuming no internal compartmentalization, it appears that the accumulation of glucose 6-phosphate is unlikely to inhibit the tapeworm enzyme *in vivo*. No "high K_m" glucokinase, which is less potently inhibited by G6P in rat liver, was found in the cestode.

In contrast, Agosin and Aravena (1959b) found four distinct hexokinases in protoscoleces of *Echinococcus granulosus*, catalysing specifically the phosphorylation of glucose, fructose, mannose and glucosamine. The K_m of the glucokinase (0.1 mM) was quite similar to that reported by Komuniecki and Roberts (1977a) for the hexokinase of *H. diminuta*. The gluco-, fructo- and mannokinase from *E. granulosus* were all inhibited by G6P (Agosin and Aravena, 1959b), but effects on glucose utilization were not investigated further.

C. Aldolase

Aldolase from *Taenia crassiceps* metacestodes was studied by Phifer (1958). The enzyme was not purified; a homogenate was used in the assays. In some characteristics, the tapeworm enzyme appeared to differ considerably from the isofunctional enzymes from other systems (yeast, bacteria, rabbit muscle, protozoa). The aldolase from *T. crassiceps* was not affected by ferric, cobaltous, magnesium, stannous or nickel salts, and it was neither activated nor inhibited in the presence of metal-binding agents. The lack of effect by iodoacetate indicated an absence of functional sulfhydryl groups on the enzyme.

D. Phosphofructokinase (PFK)

Behm and Bryant (1975b) concluded that possible controls of PFK from *M. expansa* were similar to those known for mammalian PFK's. Both Mg^{++} and Mn^{++} activated *M. expansa* PFK. The enzyme was activated by ATP up to about 0.25 mM and somewhat inhibited by the ATP above about 0.5 mM in the presence of $(NH_4)_2SO_4$. Without added NH_4^+, they observed a striking sigmoid inhibition above 0.05–0.01 mM ATP. Little or no evidence for co-operativity between fructose 6-phosphate and ATP was found in the presence of NH_4^+, but there was a strong positive co-operativity in its absence. GTP and ITP could also serve as substrates for the reaction, but neither was inhibitory at high concentrations. Thus, neither ITP nor GTP could serve the regulatory role of ATP. AMP partially relieved the inhibition caused by high ATP in both the presence and absence of NH_4^+, and the activation by AMP was apparently not synergistic with NH_4^+.

V. TERMINAL ENERGY-GENERATING REACTIONS

Cestodes and many other parasitic helminths exhibit a predominantly anaerobic energy metabolism, even in the presence of substantial amounts of molecular oxygen. Many exhibit no Pasteur effect, and oxygen uptake results in substantial peroxide formation. As discussed above, their initial glycolytic reactions are surprisingly similar to their aerobic counterparts, but the terminal energy-generating reactions are modified and result in the accumulation of a wide variety of reduced organic end-products. With few exceptions, the TCA cycle is absent or of doubtful physiological significance, and succinate, lactate, acetate, propionate and alanine are the major end-products of carbohydrate dissimilation. The presence or relative amount of each varies with the species, stage of development, the presence or absence of O_2 or CO_2, and even, apparently, the strain of the parasite (Saz, 1981a,b; Fairbairn *et al.*, 1961; Watts and Fairbairn, 1974; Ovington and Bryant, 1981). A variety of other reduced end-products are excreted by other parasitic helminths, especially nematodes (see Barrett, 1976), but space limitations preclude review here.

A. Cytosolic reactions

The reaction pathway in *H. diminuta* is summarized in Fig. 5. The cytosolic reactions have elicited considerable interest because of their importance in maintaining the cytosolic redox balance (reoxidizing the NADH generated in glycolysis) and because the activity of each branch determines the relative amounts of lactate and succinate (and/or acetate) produced.

In the "lactate branch", PEP is dephosphorylated to pyruvate by pyruvate kinase, generating ATP, and the pyruvate is reduced to lactate by lactate dehydrogenase, regenerating NAD^+. In the "malate branch" CO_2 is fixed by PEPCK to form oxaloacetate, generating ITP or GTP, and the oxalacetate is reduced to malate by malate dehydrogenase, regenerating NAD^+. Thus, each branch of the pathway is in apparent redox balance and is energetically equivalent. Therefore, the proportion of PEP going to lactate or malate would seem to be determined by other factors. Of relevance to this point, however, are the studies of Fields and Quinn (1981). They compared the lactate dehydrogenase and malate dehydrogenase systems by computer simulation. On the assumption that the dehydrogenases maintain their substrates and products at instantaneous equilibrium, the simulations permitted study of the redox ratio ($NADH/NAD^+$) as a function of lactate and malate concentrations. Accumulation of lactate led to higher redox ratios than accumulation of malate. Moreover, *in vivo*, little malate would accumulate because it is rapidly metabolized further to succinate or other products in many animals, e.g. tapeworms (Fig. 5). Therefore, in addition to the additional energy to be derived from the molecule metabolized through the "malate branch" (see Section B: Mitochondrial reactions), the two branches may not actually have equal impact on the cytosolic redox balance.

(1) Pyruvate kinase

The activity of pyruvate kinase (PK) was measured in a crude extract of *H. diminuta* by Bueding and Saz (1968), who found that it was only about 20% that of PEPCK from the same organism. These workers attributed the high ratio of succinate/lactate excreted by this parasite to the low PK/PEPCK ratio. It has become apparent in light of later results that this explanation may not be sufficient.

Bryant (1972b) isolated at least two different PK isozymes in *M. expansa* by ammonium sulphate precipitation. With the same cestode, he found that the ratio of lactate:succinate excreted changed dramatically with changes in incubation conditions, i.e. the presence or absence of oxygen, glucose and/or fumarate (Bryant, 1972a). He assumed that glucose oxidation would be more efficient and ATP production greater under

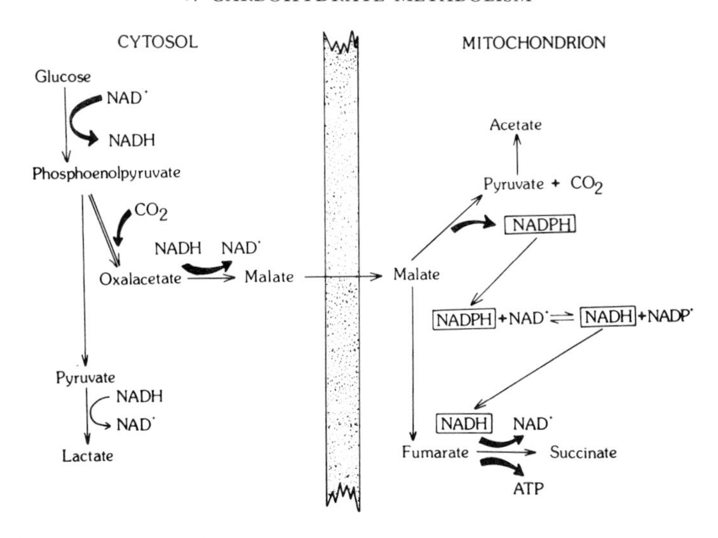

Fig. 5. Carbohydrate dissimilation by adult *Hymenolepis diminuta* (from Fioravanti and Saz, 1980).

aerobic conditions, and suggested that this would lead to increased levels of FDP, activation of PK, and greater lactate production. Under anaerobic conditions, the final electron acceptor would be fumarate, lowering the synthesis of ATP, and therefore the concentration of FDP, and decreasing the activity of PK. This would allow the PEPCK to compete more favourably for the PEP and increase succinate production (Bryant, 1972a,b). The increase in the malate pool via the succinate pathway would also serve to depress lactate production.

A large proportion of the glucose consumed by tetrathyridea of *Mesocestoides corti* was accounted for as lactate (Köhler and Hanselmann, 1974). None the less, PK activity was very low in these worms, even in the presence of the allosteric activator, FDP. Only about half as much succinate as lactate was excreted anaerobically, though the specific activity of the PEPCK measured was 28-fold greater than the PK. Neither glucose consumption nor lactate production was affected by the presence of oxygen, but succinate production decreased dramatically when oxygen was present. Scheibel and Saz (1966) reported a similar finding in *H. diminuta.* The decrease in succinate production could not be accounted for by increased CO_2 production, and the explanation for these observations remains unclear (Köhler and Hanselmann, 1974).

Körting (1976) stated that PK in *Bothriocephalus gowkongensis*, *Khawia sinensis*, *Triaenophorus crassus* and *Schistocephalus solidus* had a "strict requirement" for FDP. He also demonstrated lactate dehydrogenase,

PEPCK, and several other enzymes in crude homogenates of these cestodes.

PK from plerocercoids of *Ligula intestinalis* was studied by McManus (1975), who also found activation of PK by FDP and inhibition by ATP and malate. PEPCK was also present. McManus (1975) suggested that PK activity, and therefore proportions of acids excreted, was controlled by intracellular concentrations of ATP, FDP, malate and PEP.

Watts and Fairbairn (1974) reported that there was a developmental shift in relative proportion of acids excreted by *H. diminuta* from a molar ratio of 0.9:0.6:1 (succinate:acetate:lactate) at six days post-infection of the rat host to a ratio of 5.0:2.5:1 at 14 days post-infection. In a study of PK as a function of development in *H. diminuta*, Carter and Fairbairn (1975) found that the highest PK activity occurred in 4–6-day-old worms and in the most anterior regions of adult worms. They reported five isozymes of PK in adult worms, based on elution from DEAE cellulose columns with KCl. Only two isozymes occurred in the hexacanth larvae and cysticercoids, but by four days post-infection, another isozyme was present, and all five were found by ten days. Four of the isozymes were subject to strong allosteric activation by FDP. The fifth was unaffected by the presence of FDP but had a relatively low K_m for PEP. In contrast to the pyruvate kinases of other organisms, the pH optimum for none of the *H. diminuta* isozymes was affected by the presence of FDP, nor were the parasite enzymes inhibited by alanine, phenylalanine, proline or valine. (These observations correlate well with the lack of an anapleurotic TCA cycle and minimal amino acid metabolism in this worm, see sections VI and VII.) On the other hand, all five isozymes were inhibited by ATP and Ca^{++}, in common with mammalian and invertebrate PK's. Carter and Fairbairn (1975) pointed out that one of the isozymes (FDP insensitive, low K_m [PEP]) most resembled PK from rat muscle, while the remaining four isozymes were like those from liver (type L). However, type L PK's are typical of aerobic, gluconeogenic tissues in which PEP is synthesized from both 2-phosphoglycerate and oxaloacetate. In such tissues, alanine contributes considerable carbohydrate carbon and is a powerful allosteric inhibitor of type L PK. The essentially fermentative function of the tapeworm isozymes is indicated by the lack of an effect of alanine and the activation by FDP. Carter and Fairbairn (1975) suggested that the appearance of the additional isozymes during development of the worm was probably of some, as yet undetermined, adaptive significance. The greater production of lactate by the anterior regions of the worm and by very young worms might be related to the fact that lactic acid is a stronger acid than the others excreted by the cestodes, and the escape of the acid from the tissues would be facilitated by the greater surface to volume ratio of the small worms (Fairbairn, 1970). Also, accumulation of lactate in the tissues of the larger worms might lead to an increase in the cytosolic NADH/NAD ratio (Fields and Quinn, 1981).

Using extraction and assay methods somewhat different from those of Carter and Fairbairn (1975), Moon *et al.* (1977b) also studied PK from *H. diminuta.* The PK preparation used by Moon *et al.* (1977b) apparently corresponded to the "adult" isozyme of Carter and Fairbairn (1975) in that it was FDP insensitive, and it had a low K_m(PEP) (0.08 mM). However, Moon *et al.* (1977b) stated that their PK preparation was not sensitive to inhibition by ATP and Ca^{++}↑ was inhibited 50% by 6.3 mM lactate and 30 mM HCO_3^-, and was not cold labile. The K_m for PEP was increased 6-fold by 30 mM HCO_3^-, but the V_{max} was unaltered. The effect could not be fully explained by increased osmolarity, and Moon *et al.* (1977b) believed that the inhibition represented a specific modulator-enzyme interaction. Because HCO_3^- is a co-substrate for PEPCK, this enzyme was maximally activated under the conditions that inhibited PK, and they suggested that the shifts in ratio of lactate and succinate excreted were due to alterations in the poise of the HCO_3^-:CO_2 system in the post-prandial intestine. This question will be explored below.

(2) Lactate dehydrogenase

L(+)-lactate dehydrogenase (LDH) from a variety of animal tissues has been well studied, and the LDH from *Hymenolepis* spp. has attracted the attention of several investigators (Burke *et al.*, 1972; Walkey and Fairbairn, 1973; Logan *et al.*, 1977; Moon *et al.*, 1977a; Pappas and Schroeder, 1979). Burke *et al.* (1972) purified the enzyme from *H. diminuta* 128-fold, and based on its Michaelis' constants and the fact that it showed substrate inhibition by pyruvate, they concluded that the enzyme most resembled the H form of mammalian LDH. The enzyme was NAD-NADH and L(+)-lactate specific; no activity was shown with NADP, NADPH, or D(—)-lactate as substrates. Burke *et al.* (1972) were able to demonstrate only one form of the enzyme, either in crude extracts or purified preparations, using electrophoresis on cellulose acetate strips and polyacrylamide gels and by isoelectric focussing.

Walkey and Fairbairn (1973), however, were able to distinguish two isozymes of LDH from *H. diminuta* by isoelectric focussing. One isozyme predominated in hexacanths and cysticercoids, while the other predominated in the anterior portions (containing the scolex and some immature proglottides) of adult worms and in 7-day-old worms. The hexacanth-cysticercoid form was relatively more sensitive to heat inactivation, only slightly inhibited by excess pyruvate, and had an apparent K_m(pyr) of 0.28 mM. The young worm form was resistant to heat inactivation, strongly inhibited by pyruvate, and had an apparent K_m(pyr) of 0.028 mM. The K_m(lact) for both was 50–100 times their respective K_m's for pyruvate. Logan *et al.* (1977) confirmed the presence of two LDH isozymes in *H. diminuta* by starch-gel electrophoresis. They designated the isozyme characteristic of the anteriors and young worms as LDH-A, while that characteristic of the eggs was called LDH-B. They found LDH-A only in anterior ends of worms and in non-ovarian tissues of mature proglottides. Both isozymes were present in whole worms, hexacanths, whole posterior proglottides, and whole mature proglottides. If the LDH's of *H. diminuta* are tetramers, as indicated by the results of Burke *et al.* (1972), and if the isozymic subunits are encoded on separate gene loci, then it is curious that there are only two isozymes, rather than five as in L(+) LDH's of other organisms (Logan *et al.*, 1977). For whatever reason, it appears that polymerization of the subunits in *H. diminuta* is not random (Logan *et al.*, 1977).

Moon *et al.* (1977a) reported four electrophoretic bands in a crude preparation and two bands in a partially purified preparation of LDH from *H. diminuta.* They only observed one peak by isoelectric focussing. In agreement with Walkey and Fairbairn (1973), the kinetic parameters measured by Moon *et al.* (1977a) would greatly favour lactate over pyruvate formation. For example, they reported a K_m(lact) of 7.8 mM and a K_m(pyr) of 0.25 mM at pH 7.4 with a partially purified enzyme preparation.

Similar results were reported by Pappas and Schroeder (1979) with LDH from *Hymenolepis microstoma.* They found two isozymes by polyacrylamide gel electrophoresis and kinetics that would favour lactate formation (K_m[lact] = 3.8 mM; K_m[pyr] = 0.51 mM; K_m[NAD] = 0.17 mM; K_m[NADH] = 0.011 mM). In common with the enzyme from *H. diminuta* (see Walkey and Fairbairn, 1973; Burke *et al.*, 1972), that from *H. microstoma* was L(+)-lactate and NAD specific and had an active thiol group. In contrast to the *H. diminuta* enzyme, the LDH of *H. microstoma* did not show significant substrate (pyruvate and lactate) inhibition.

(3) Phosphoenolpyruvate carboxykinase

Phosphoenolpyruvate carboxykinase (PEPCK) has been reported from a number of cestodes, including *Hymenolepis diminuta*, *Schistocephalus solidus*, *Moniezia expansa*, *Echinococcus granulosus*, *Mesocestoides corti*, *Bothriocephalus gowkongensis*, *Khawia sinensis*, *Triaenophorus crassus*, and *Ligula intestinalis* (Prescott and Campbell, 1965; Moon *et al.*, 1977b; Körting and Barrett, 1977; Behm and Bryant, 1975c; Agosin and Repetto, 1963, 1965; Köhler and Hanselmann, 1974; Körting, 1976; McManus, 1975). Although the reaction catalysed by PEPCK is reversible, the enzyme's main role in cestodes seems to be opposite to that in vertebrate tissues, where it serves to form phosphoenolpyruvate for gluconeogenesis. Prescott and Campbell (1965) and Moon *et al.* (1977b) studied the enzyme from *H. diminuta* with somewhat differing results. Prescott and Campbell (1965) found a pH optimum of 6.3, that the enzyme required Mn^{++} (Mg^{++} was almost as effective), and that ITP/IDP was the most effective cofactor. Moon *et al.* (1977b) reported a pH optimum of 5.5, found that activity in the presence of Mg^{++} was much lower than with Mn^{++}, and that GTP/GDP was the preferred cofactor. They observed a K_m(PEP) of 0.42 mM at pH 5.5. The enzyme was inhibited by L-lactate and activated by HCO_3^-. Activity was measured in the direction of oxaloacetate (OAA) formation. Interestingly, Prescott and Campbell (1965) found that PEPCK activity in the direction of OAA formation decreased when food was withheld from the rat hosts, then increased rapidly upon refeeding. However, activity in the direction of PEP formation increased sevenfold within 6 h from the beginning of host starvation. They found some PEPCK activity associated with the particulate fraction, but most activity was in the cytosol. Moon *et al.* (1977) worked only with the cytosolic enzyme.

Behm and Bryant (1975c) investigated both the cytosolic and the mitochondrial enzymes in *M. expansa.* Though the cytosolic enzyme showed considerably more activity in the presence of Mn^{++} than Mg^{++}, there was no activation of the mitochondrial PEPCK by Mg^{++}; therefore, Behm and Bryant (1975c) concluded that they were indeed different enzymes. The functional significance of the difference, if any, remains unclear. Though Behm and Bryant (1975a,b) had reported evidence that PEPCK might be a regulatory enzyme in *M. expansa*, its ready reversibility led them to the conclusion that PEPCK in this cestode was controlled mainly by concentrations of its substrates (Behm and Bryant, 1975c). They found that ATP was a competitive inhibitor of IDP in the carboxylation reaction, but since ATP also inhibited pyruvate kinase in *M. expansa* (Behm and Bryant, 1975b), this nucleotide is unlikely to regulate the direction of flow at the PEP branchpoint.

Because of the differing primary functions of PEPCK in the cestodes and their hosts, Reynolds (1980) compared the two enzymes in the hope that the PEPCK from *H. diminuta* might be selectively inhibited and thus provide an avenue for anthelmintic attack. The enzymes from the two sources were similar, but the cestode enzyme showed greater affinities for PEP, metal ions and inhibitors. Quinolinate and 3-mercaptopicolinate were competitive inhibitors with PEP for both enzymes. The tapeworm preparation contained no ferroactivator, but ferroactivator from rat liver protected the tapeworm enzyme from inactivation by Fe^{++}. It was suggested, therefore, that the cestode enzyme may be structurally analogous to the host enzyme.

(4) Malate dehydrogenase

Malate dehydrogenase (MDH) from *H. diminuta* was studied by Moon *et al.* (1977a), and the enzyme from *H. microstoma* was examined by Pappas and Schroeder (1979). Moon *et al.* (1977a) found four bands by gel electrophoresis but only one peak by isoelectric focussing. Pappas and Schroeder (1979) reported three isozymes, shown by gel electrophoresis. Kinetic parameters would appear to favour OAA reduction by the *H. microstoma* enzyme: $K_m(\text{OAA}) = 0.0059$ mM; $K_m(\text{mal}) = 1.09$ mM; $K_m(\text{NADH}) = 0.017$ mM; $K_m(\text{NAD}) = 0.180$ mM. The K_m for malate was not reported by Moon *et al.* (1977), but it appears likely that OAA reduction would be favoured in *H. diminuta*, as well. The activity of MDH considerably exceeded that of LDH in both species. Both Moon *et al.* (1977a) and Pappas and Schroeder (1979) concluded that the proportion of lactate to succinate excreted by these worms probably depended more on regulation of PK and PEPCK than on LDH and MDH.

(5) The succinate:lactate question

As illustrated in Fig. 5, cytoplasmically generated malate is the source of excreted succinate in most cestodes. The proportion of succinate to lactate formed has been viewed as depending on control of the foregoing cytoplasmic reactions, i.e. the proportion of PEP carboxylated to OAA to that dephosphorylated to pyruvate. A theoretical contribution to help our understanding of these processes was made by Podesta *et al.* (1976). However, a recent report by Ovington and Bryant (1981) has posed some serious questions concerning the hypothesis of Podesta *et al.* (1976).

Podesta *et al.* (1976) reasoned that since there was, in fact, oxygen available in the host gut (Podesta and Mettrick, 1974), the adaptive value of an anaerobic energy metabolism in *H. diminuta* (and other parasitic helminths) must be other than an ability of the organisms to withstand hypoxia. Inasmuch as there is considerable release of CO_2 in the post-prandial intestine as the acidic chyme is passed into the duodenum, there is an acidification of worm tissues as the CO_2 diffuses in. The acidification is countered by worm secretion of H^+ and by mobilization of Ca^{++} from the calcareous corpuscles, thus releasing carbonate. The increased osmolality and greater HCO_3^- concentrations activate PEPCK and favour succinate over lactate secretion. Because succinate is a dicarboxylic acid, it is twice as effective as lactate in metabolic disposal of H^+. With a decrease in passage of stomach chyme into the intestine, there would be a decrease in acid stress on the worms, along with concomitant decrease in tissue HCO_3^- concentrations and osmolality. Under these conditions, the lactate branch would be favoured.

Ovington and Bryant (1981) noted that not all available evidence supported the hypothesis of Podesta *et al.* (1976), and they subjected the hypothesis to re-examination. Surprisingly, they found that the succinate excretion of *H. diminuta*, over a wide range of CO_2 concentrations and other incubation conditions, was only a small fraction of total acid excretion. Acetate excretion was unaffected by the various experimental variables to which the worms were subjected. Lactate excretion decreased with increasing HCO_3^- concentration, while succinate excretion increased only slightly. In these experiments the amount of succinate excreted would have been insufficient to account for H^+ removal, but the combined lactate and acetate excretion would have been sufficient to prevent tissue acidification. Not only do these results not support the hypothesis of Podesta *et al.* (1976), they are at variance with numerous reports of high succinate excretion by *H. diminuta* (Fairbairn *et al.*, 1961; Watts and Fairbairn, 1974; Coles and Simpkin, 1977; Insler and Roberts, 1980a,b). Some earlier workers, however, found high lactate with little succinate excretion (Read, 1956; Laurie, 1957). Ovington and Bryant (1981) could only explain these discrepancies on the basis of a possible strain difference in the cestode. If such is indeed the explanation for the results, potential for investigation of genetic variation in metabolic control is considerable. In any case, Ovington and Bryant (1981) argued that, rather than having an abundant oxygen supply, the tissues of *H. diminuta*, at least at times, must be subjected to very hypoxic conditions. They suggested, therefore, that the energy metabolism of cestodes is adaptive to fluctuating CO_2 and O_2 tensions. While conceding that the excretion of organic acids may serve to maintain tissue pH in the presence of high levels of ambient CO_2, they disagreed that high CO_2 tension causes a major shift from lactic to succinic acid excretion.

B. Mitochondrial reactions

While the adaptive significance of the excretion of organic acids by *H. diminuta* may be as suggested by Ovington and Bryant (1981), there is greater value in excretion of succinate over lactate, i.e. generation of additional ATP and maintenance of redox balance in a predominantly anaerobic environment (Fields and Quinn, 1981) (Fig. 5). The malate produced by the MDH reaction enters the mitochondria, there to undergo a dismutation, one half being oxidized to pyruvate by "malic" enzyme, and one half being dehydrated to fumarate by fumarase. Reduction of the fumarate to succinate is accompanied by an electron transport associated, net generation of ATP via a Site I phosphorylation. Thus, fumarate is the final electron acceptor, and reducing equivalents are provided by the "malic" enzyme reaction. Though the fumarate reductase is NAD^+-linked, and the "malic" enzyme is $NADP^+$-linked, a mitochondrial transhydrogenase couples the reactions (Fioravanti and Saz, 1980). It is possible that a further ATP may be produced by the catabolism of pyruvate to acetate in *H. diminuta*, although these reactions have been insufficiently studied. In *Spirometra mansonoides* the succinate is decarboxylated to propionate, with a substrate level generation of ATP (Pietrzak and Saz, 1981).

(1) "Malic" enzyme (Malate dehydrogenase, decarboxylating)

The presence of a mitochondrial, $NADP^+$-linked, "malic" enzyme (E.C. 1.1.1.40) in *H. diminuta* was demonstrated by Prescott and Campbell (1965) and confirmed by Saz *et al.* (1972). Li *et al.* (1972) purified the enzyme to apparent homogeneity. In common with $NADP^+$-dependent "malic" enzymes from other systems, the cestode enzyme was heat stable in the crude homogenate and also decarboxylated oxalacetate. It required Mn^{++} and also showed high activity with Mg^{++}, but it had a very high affinity for the divalent cation, since the presence of a chelating agent was necessary to demonstrate the ion requirement. Its molecular weight was about 120 000, which is one half or less the molecular weight of the enzyme reported from other systems. This enzyme differs from that found in the nematode, *Ascaris suum*, which is NAD^+-linked (E.C. 1.1.1.39) (Saz and Lescure, 1969; Saz, 1971). Interestingly, Fioravanti and Saz (1978) found that the "malic" enzyme of *S. mansonoides* was NAD^+-dependent, although about 60% of activity was shown in the presence of NADPH. *S. mansonoides*, in common with *A. suum* and in contrast with *H. diminuta*, excretes substantial quantities of propionate. Körting and Barrett (1977) found a very low "malic" enzyme activity in plerocercoids of *Schistocephalus solidus*, and the activity was detected in the cytosol, not the mitochondria. *S. solidus* also excretes propionate, but very little succinate (see below) (Körting and Barrett, 1977).

(2) Pyruvate dehydrogenase complex

The presence of a pyruvate dehydrogenase complex was demonstrated by Watts and Fairbairn (1974) in *H. diminuta*, accounting for the production of excreted acetate. It is possible that the oxidative decarboxylation of pyruvate to acetyl-CoA, followed by hydrolysis of acetyl-CoA to acetate, is coupled to ATP generation (Hochachka, 1980).

(3) Fumarase (fumarate hydratase)

The enzyme catalysing the dehydration of malate to fumarate (fumarase) has not been studied in detail, but its activity has been demonstrated in *H. diminuta* and *S. solidus* (Read, 1953; Fioravanti and Saz, 1980; Körting and Barrett, 1977).

(4) Fumarate reductase

This reaction normally functions in the more familiar opposite direction in the Krebs cycle, i.e. as a succinate dehydrogenase reaction. Evidence for its presence has been reported in *Vampirolepis nana* cysticercoids and adults, *H. diminuta* cysticercoids and adults, *Echinococcus granulosus* protoscoleces and adults, *Diphyllobothrium latum* adults and plerocercoids, *Schistocephalus solidus* plerocercoids, and *Spirometra mansonoides* adults (Goldberg and Nolf, 1954; Heyneman and Voge, 1960; Read, 1952; Scheibel *et al.*, 1968; Agosin and Repetto, 1965; Bryant and Morseth, 1968; Salminen, 1974; Körting and Barrett, 1977; Fioravanti and Saz, 1978).

The enzyme was measured in adults and plerocercoids of *D. latum* by Salminen (1974). He observed the reaction in the direction of succinate oxidation using phenazine methosulphate (PMS), cytochrome *c*, and coenzyme Q as electron acceptors. The enzyme could transfer electrons to PMS and coenzyme Q in both adults and plerocercoids, but the plerocercoid enzyme was unable to transfer electrons to cytochrome *c*.

The enzyme was studied by Fioravanti and Saz (1978, 1980) in the direction of fumarate reduction in *H. diminuta* and *S. mansonoides*. In agreement with Scheibel *et al.* (1968), they found that the *H. diminuta* enzyme was NADH-coupled; NADPH could not serve as a pyridine nucleotide hydride donor (Fioravanti and Saz, 1980). In contrast, the enzyme from *S. mansonoides* showed a similar activity with both electron donors (Fioravanti and Saz, 1978). In both tapeworm species, however, there was a substantial inhibition of activity in the presence of rotenone, an inhibitor of Site I phosphorylation. The dismutation of malate in *H. diminuta* can be completed because of the presence of a transhydrogenase, transferring reducing equivalents from the NADPH produced in the "malic" enzyme reaction to NADH for use in the fumarate reductase reaction (Fioravanti and Saz, 1980).

(5) Succinate decarboxylation to propionate

In mammals propionate is glycogenic by fixation of CO_2 into succinate via the following reversible reactions (see Tkachuck *et al.*, 1977; Pietrzak and Saz, 1981):

(1) $CO_2 + ATP + E(\text{propionyl-CoA carboxylase}) \leftrightarrows \text{“}CO_2\text{”} \sim E + + ADP + P_i$

(2) $\text{“}CO_2\text{”} \sim E + \text{propionyl-CoA} \leftrightarrows \text{methylmalonyl-CoA} + E$

(3) $\text{methylmalonyl-CoA} \xrightleftharpoons{\text{mutase + coenzyme } B_{12}} \text{succinyl-CoA}$

(4) $\text{succinyl-CoA} + \text{propionate} \xrightleftharpoons{\text{transferase}} \text{propionyl-CoA} + + \text{succinate}$

Sum (1–4): $CO_2 + ATP + \text{propionate} \leftrightarrows \text{succinate} + ADP + P_i$

At least some cestodes, e.g. *S. mansonoides*, *S. solidus*, and *E. granulosus*, as well as nematodes and trematodes (*Ascaris suum*, *Fasciola hepatica*) excrete propionate. Reversal of the foregoing reactions would be a possible avenue for propionate production from succinate and the generation of another ATP. The necessary enzymes (propionyl-CoA carboxylase, methylmalonyl-CoA mutase, and acyl-CoA transferase) have been demonstrated in *S. mansonoides* (Meyer *et al.*, 1978; Tkachuck *et al.*, 1977; Pietrzak and Saz, 1981). Pietrzak and Saz (1981) presented evidence that these reactions were indeed coupled to a substrate level phosphorylation of ADP to ATP. In *F. hepatica* they found that acetyl-CoA stimulated the reactions considerably more than propionyl-CoA, suggesting that acetyl-CoA may be the physiological CoA-donor.

Körting and Barrett (1977) found that the plerocercoids of *Schistocephalus solidus* secreted acetate and propionate in a ratio of 0.3 under anaerobic and 4.0 under aerobic conditions. They suggested that the propionate arose from decarboxylation of succinate but did not investigate the reactions. The worms excreted little or no succinate or lactate, despite the presence of an active LDH. Aerobically, only 22% of the glycogen catabolized was accounted for by organic acids, while anaerobically, 70% appeared as excreted acids. Thus, Körting and Barrett (1977) suggested that the mitochondrial redox balance was maintained mainly by conversion of carbohydrate to acetate under anaerobic conditions, and aerobically, molecular oxygen would be the final electron acceptor.

VI. ROLE OF AMINO ACIDS

In many organisms most amino acids can be glycogenic after conversion to pyruvate or to intermediates of the TCA cycle, and most amino acids can be oxidized through the TCA cycle after catabolism to acetyl-CoA. Though our knowledge of these reactions in cestodes is lamentably meagre, the use of amino acids either for glycogen carbon or as energy sources appears extremely limited in these organisms. In fact, alanine, at least, seems to be a true excretory product in *H. diminuta* (Wack *et al.*,

1983). Upon incubation in balanced salt solution, *H. diminuta* releases a wide variety of amino acids into the medium, though it is by no means clear that any or all of these are metabolic end-products (Zavras and Roberts, unpublished). In any case, alanine accounts for 40–50% or more on a molar basis of the amino acids released. The main products formed by adult *Echinococcus granulosus* from fumarate are succinate, malate and alanine (Bryant and Morseth, 1968).

Alanine aminotransferase and aspartate aminotransferase were reported from *H. diminuta* by Aldrich *et al.* (1954), who found that, of a number of amino acids tested, only alanine, aspartate and glutamate could act as amino donors. These two aminotransferases were also found in *H. diminuta* by Prescott and Campbell (1965), and in *H. diminuta* and *Raillietina cesticillus* by Foster and Daugherty (1959). Abu Senna (1966) reported that Alcopar inhibited alanine and aspartate amino-transferases from *Vampirolepis nana* by 45–50%, and cycloserine inhibited the alanine aminotransferase by 87%, but had no effect on the aspartate aminotransferase. The apparent absence of other aminotransferases in *H. diminuta*, *H. citelli*, and *V. nana* was confirmed by Wertheim *et al.* (1960). Wertheim *et al.* (1960) also observed a rather peculiar amino donor specificity among the three hymenolepidids. Alanine could act as amino donor to α-ketoglutarate in *H. diminuta* and *H. citelli*, but not in *V. nana*; aspartate and asparagine could act as donors to pyruvate in *H. diminuta*, but not in *H. citelli* or *V. nana*. Aspartate and asparagine could act as amino donors to α-ketoglutarate in all three species, and glutamate could act as donor to pyruvate in all three. These observations imply a unidirectionality in the alanine aminotransferase reaction of *V. nana*, i.e. in the direction of alanine production from pyruvate.

Platzer and Roberts (1970) found that the activity of alanine aminotransferase decreased with development of *H. diminuta* from 9 to 15 days post-infection.

Nations *et al.* (1973) reported an aspartate decarboxylase-like activity in extracts of *H. diminuta*, but they found that the CO_2 and alanine were actually formed as a result of other reactions: oxalacetate and glutamate were produced by aminotransferases from aspartate and α-ketoglutarate, then the oxaloacetate was decarboxylated to pyruvate and CO_2, and the alanine was a product of transamination between the pyruvate and glutamate.

The extremely limited transmination capabilities of *H. diminuta* suggest that this worm neither relies heavily on synthesis to satisfy amino acid requirements nor catabolizes amino acids to any significant extent for derivation of energy. A physiological advantage of alanine excretion was proposed by Mustafa *et al.* (1978) in their study of glutamate dehydrogenase (GDH) from *H. diminuta*. The cestode enzyme differed markedly from GDH's from most mammalian sources and resembled those reported from a fungus and some parasitic protozoa. It was found in the cytosol, was NAD-specific, was unaffected by the presence of added mono-, di-, or triphosphonucleotides, and was not activated by several amino acids that stimulate GDH from rat liver. The apparent K_m for α-ketoglutarate was lowered almost 10-fold by decreasing the assay pH from 7.4 to 6.5. Mustafa *et al.* (1978) suggested that the physiological value of the tapeworm GDH might lie in the diversion of pyruvate to alanine (by alanine aminotransferase) from lactate under conditions of tissue acidification. The GDH would permit formation of the less acidic alanine as an end-product by regenerating the α-ketoglutarate formed in the transamination reaction and simultaneously replacing the LDH in maintaining cytoplasmic redox balance.

Rasero *et al.* (1968) reported evidence that the γ-aminobutyrate pathway was present in *Moniezia expansa*. This pathway would yield succinate from glutamate, which is presumably readily available to the worms, and would also be an alternative source of reducing power for the fumarate reductase reaction. For the pathway to function, glutamate decarboxylase, γ-aminobutyrate-α-ketoglutarate aminotransferase, succinic semialdehyde dehydrogenase, and glutamate dehydrogenase are required. The existence of the pathway in *M. expansa* was reinvestigated by Cornish and Bryant (1975). They were unable to demonstrate the production of γ-aminobutyrate or the activity of any of the foregoing enzymes, and they concluded that the γ-aminobutyrate pathway was absent in *M. expansa*. They did show alanine and aspartate aminotransferase activities. In their experiments glucose was catabolized to succinate, lactate, alanine and malate in approximately the following proportions: 2:1:1 at 60 min and 2:1:0.2 at 120 min (succinate:lactate:alanine + malate). Most of the label from catabolized glutamate was found in succinate, with some in α-ketoglutarate, and small amounts in glutamine, malate, lactate, alanine and aspartate.

VII. TRICARBOXYLIC ACID CYCLE

Because of the apparent importance of short-chain acids as end-products of cestode energy metabolism, the possible presence of a tricarboxylic acid cycle (TCA; Krebs cycle) in tapeworms has received little attention.

Scheibel and Saz (1966) concluded that a TCA cycle was not active in *H. diminuta*, and Ward and Fairbairn (1970) could demonstrate only some of the necessary enzymes in this cestode. Agosin and Repetto (1963) reported that TCA cycle intermediates were oxidized by whole protoscoleces and/or homogenates of *Echinococcus granulosus*, and they suggested that failure to demonstrate catabolism of some intermediates in the past had been due to permeability barriers. They also demonstrated activity of the TCA cycle enzymes. They concluded that their evidence was strong that the cycle was "potentially operative in this organism". However, they showed a very active CO_2 fixation in *E. granulosus* and that about 63% of the CO_2 fixed ended up in succinate, which was excreted. They concluded that:

> in intact scolices either the enzymes leading to succinic acid are more active than those allowing its utilization or that an intracellular regulatory mechanism prevents the latter from working at full capacity. (Agosin and Repetto, 1963)

Davey and Bryant (1969) provided evidence that the TCA cycle was present in *M. expansa*. Incubation in ^{14}C-succinate or ^{14}C-malate led to labelling of malate, citrate, α-ketoglutarate, succinate and fumarate (as well as aspartate, glutamate, alanine, and lactate), and the authors considered this circumstantial evidence for the presence of succinic dehydrogenase, fumarase, malic dehydrogenase, condensing enzyme, aconitase, and isocitric dehydrogenase. Incubation with $NaH^{14}CO_3$ resulted in labelling only of malate, fumarate and succinate (of the TCA intermediates) as well as aspartate, alanine, lactate and pyruvate. Succinate and alanine were particularly heavily labelled. If any significant amount of succinate in this experiment were entering in the TCA cycle, the lack of label in citrate and α-ketoglutarate would be difficult to explain. Either the TCA cycle is not operating or the succinate, resulting from CO_2 fixation, is specifically sequestered (in the mitochondria) away from the loci of the TCA cycle enzymes.

McManus (1975) investigated the TCA cycle enzymes in the plerocercoids of *Ligula intestinalis*. Although he found all enzymes present, aconitase and malate dehydrogenase (in the direction of malate oxidation) were very low in activity. McManus (1975) concluded that the complete cycle was of questionable importance to the parasite.

In contrast, Körting and Barrett (1977) found that only 22% of the glycogen catabolized by plerocercoids of *S. solidus* was accounted for by excreted acids, and the activities of the TCA cycle enzymes were sufficient to account for the observed rates of carbohydrate catabolism. The levels of TCA cycle intermediates were determined by Beis and Barrett (1979) in *S. solidus*, and the intermediates increased two- to threefold upon activation of the plerocercoids.

VIII. OTHER METABOLIC PATHWAYS

A. Glyoxylate pathway

There have been few attempts to demonstrate the glyoxylate pathway in cestodes. Because they were unable to demonstrate isocitrate lyase and malate synthase in protoscoleces of *E. granulosus*, Agosin and Repetto (1963) concluded that the pathway did not function in this tapeworm. The glyoxylate pathway is best known in micro-organisms and plants where it is responsible for net biosynthesis of carbohydrate from fatty acids. It has also been demonstrated in nematodes (Rothstein and Mayoh, 1964). Inasmuch as the ability of cestodes to catabolize fatty acids appears extremely limited (Fairbairn *et al.*, 1961; Barrett and Körting, 1977; Körting and Barrett, 1977), the absence of the glyoxylate cycle is unsurprising.

B. Phosphogluconate pathway

Among the functions of the phosphogluconate pathway (hexose monophosphate shunt, pentose phosphate pathway) in animal tissues are the conversions of hexoses into pentoses, especially D-ribose-5-phosphate, for synthesis of nucleic acids, and the conversion of pentoses into hexoses to be degraded by glycolysis. Though these processes presumably are important to cestodes, this pathway has been little studied.

De Ley and Vercruysse (1955) reported that glucose 6-phosphate dehydrogenase (G6PDH) and 6-phosphogluconate dehydrogenase (6PGDH) were active in the following cestode species: *Anoplocephala perfoliata*, *Moniezia benedeni*, *Taeniarhynchus saginatus*, *Taenia pisiformis*, and *Dipylidium caninum* adults. Agosin and Aravena (1959a,b, 1960a) showed that G6PDH and 6PGDH both were active in *E. granulosus* protoscoleces. In addition, they found evidence for transketolase, transaldolase, phosphopentose isomerase, and, possibly, phosphoketopentose epimerase. The following substances were released into the incubation medium: glucose, fructose, ribose, sedoheptulose and ribulose. Some lactate was produced from ribose-5-phosphate by an acetone-powder extract. Because succinate produced after incubation with glucose-6-^{14}C was somewhat higher in activity than that produced after incubation with glucose labelled in the 1-^{14}C position, Agosin and Repetto (1965) suggested that a minor portion of glucose catabolized was via the phosphogluconate pathway. Agosin and Aravena (1960b) partially purified and studied the phosphopentose isomerase from this tapeworm. Thus, there is good evidence for the presence of the phosphogluconate pathway in *E. granulosus* protoscoleces, and the pathway is likely to be functional in other cestodes as well.

Because little $^{14}CO_2$ was produced from glucose-1-^{14}C or glucose-6-^{14}C by *H. diminuta*, Scheibel and Saz (1966) concluded that a "quantitatively significant" phosphogluconate pathway was not present in this cestode. However, since *H. diminuta* apparently can synthesize nucleotides from exogenous purine bases (Davis and Roberts, unpublished), the worms may be able to synthesize ribose. Thus the phosphogluconate pathway might be present and biologically important but quantitatively insignificant when compared to glycolysis.

IX. SUMMARY

The acquisition, storage and catabolism of carbohydrates are of crucial importance in the biology of cestodes. Tapeworms seem to be one of the few groups that requires an exogenous source of carbohydrate, probably because of their limited ability to degrade fatty and amino acids. Quantity and quality of carbohydrate in the host diet have a marked effect on tapeworm growth and development. The amount of glycogen stored at a given time is affected by a number of environmental variables, but large quantities are characteristically present. Glycogen is stored in the myocytons both as discrete particles (β-particles) and as an array of particulate aggregations (α-particles). Glucose moieties are initially incorporated into particles intermediate in molecular weight, which are then apparently disassembled and reassembled into glycogen of low molecular weight and very high molecular weight. This appears to be the result of the effectiveness of intermediate molecular weight glycogen as primer for glycogen synthase. The presence of CO_2 is necessary for optimal glycogen synthesis, and some of the ambient CO_2 is incorporated into the glycogen synthesized. Glycogen synthase is a highly regulated enzyme, existing in active (dephosphorylated) and inactive (phosphorylated) forms. Synthase is activated by the phosphorylated intermediates of glycolysis but not TCA cycle intermediates. Though not well studied, glycogen phosphorylase appears to be regulated in a way generally similar to isofunctional enzymes of other systems.

Glucose moieties from glycogen are phosphorylated by hexokinase and are degraded by classical glycolysis as far as phosphoenolpyruvate (PEP). Phosphofructokinase and pyruvate kinase seem to be regulatory enzymes in at least some species.

In some species, PEP (usually a minor amount) may be channelled into the tricarboxylic acid cycle to be degraded to CO_2 under aerobic conditions. In all cases known, however, a significant proportion of the carbon from glycolysis is excreted as a relatively reduced product, even when molecular oxygen is present. The most abundant end-products of carbohydrate metabolism are lactate, acetate, succinate, propionate and alanine. The amounts and proportions of each depend on a number of conditions, including age, ambient O_2 and CO_2 tension, species and apparently even the strain of cestode. Lactate arises in the cytosol upon dephosphorylation of PEP to pyruvate by pyruvate kinase, followed by reduction of pyruvate to lactate by lactate dehydrogenase. The succinate is produced in the mitochondria after CO_2 is fixed into PEP by PEP carboxykinase in the cytosol to form oxalacetate, which is reduced to malate by malate dehydrogenase. Each of these branches generates a high energy phosphate bond and reoxidizes the NADH formed in glycolysis, thus preserving the cytoplasmic redox balance. Malate enters the mitochondria and there undergoes a dismutation reaction. One half is oxidized and decarboxylated by "malic" enzyme to pyruvate, which is then either decarboxylated by the pyruvate dehydrogenase complex and excreted as acetate or transaminated and excreted as alanine. One half the malate is dehydrated to fumarate by fumarase, which is then reduced to succinate by fumarate reductase. Reducing equivalents for the fumarate reductase reaction are provided by the "malic" enzyme reaction. Reduction of the fumarate to succinate is accompanied by an electron transport associated net generation of ATP via a Site I phosphorylation. The catabolism of pyruvate to acetate probably also generates an ATP. Therefore, two more ATP are apparently generated anaerobically by the mitochondrial reactions than if glucose carbon were excreted solely as lactate. Other pathways may be available aerobically which would negate part of the energetic advantage of succinate excretion. It has been argued that succinate excretion is not an adaptation to hypoxia, but rather is related to avoidance of tissue acidification and metabolic disposal of H^+ under the high pCO_2 conditions of the post-prandial intestine. Discovery of a strain of *Hymenolepis diminuta* that produces little succinate under a wide variety of CO_2 concentrations raises serious questions for the foregoing hypothesis.

Some tapeworms excrete propionate formed by decarboxylation of succinate, and this generates additional ATP. Some species excrete much more propionate under anaerobic than aerobic conditions.

The role of amino acids in carbohydrate metabolism of cestodes appears small; these worms have little transamination capability. Alanine, formed by transamination from glutamate to pyruvate, is excreted as an end-product, possibly because it is a less acidic compound than lactate. In this case, it is likely that glutamate dehydrogenase serves to regenerate α-ketoglutarate and replaces lactic dehydrogenase in maintaining the redox balance of the cytosol.

The phosphogluconate pathway is probably important in cestodes, but the glyoxylate cycle probably does not occur.

Overall, the carbohydrate metabolism of cestodes is characterized by a high rate of glycogen storage and metabolism, a high rate of glycolysis, secretion of reduced end-products even under aerobic conditions, and a lesser importance of the TCA cycle than is usual in aerobic organisms. These characteristics can be viewed as adaptations to the host's intestinal environment, where food supply fluctuates and pO_2 varies from relatively high partial pressures to periods of severe hypoxia.

Acknowledgements

I deeply appreciate the comments and suggestions on the manuscript made by the following persons: Dr R. W. Komuniecki, Dr R. E. Davis and Dr P. W. Pappas.

X. REFERENCES

Abu Senna, H. O. (1966). [Transaminases in *Hymenolepis nana* and the effect on them of some compounds.] *Medskaya Parazit.* **35**, 490–491. (in Russian; English summary).

Agosin, M. and Aravena, L. (1959a). Enzimas del ciclo de las pentosas en *Echinococcus granulosus.* Comunicación preliminar. *Boln. chil. Parasit.* **14**, 30–32.

Agosin, M. and Aravena, L. (1959b). Studies on the metabolism of *Echinococcus granulosus.* III. Glycolysis, with special reference to hexokinases and related glycolytic enzymes. *Biochim. biophys. Acta*, **34**, 90–102.

Agosin, M. and Aravena, L. (1960a). Studies on the metabolism of *Echinococcus granulosus.* IV. Enzymes of the pentose phosphate pathway. *Expl Parasit.* **10**, 23–38.

Agosin, M. and Aravena, L. (1960b). Studies on the metabolism of *Echinococcus granulosus.* V. The phosphopentose isomerase of hydatid cyst scolices. *Enzymologia*, **22**, 281–294.

Agosin, M. and Repetto, Y. (1963). Studies on the metabolism of *Echinococcus granulosus.* VII. Reactions of the tricarboxylic acid cycle in *E. granulosus* scolices. *Comp. Biochem. Physiol.* **8**, 245–261.

Agosin, M. and Repetto, Y. (1965). Studies on the metabolism of *Echinococcus granulosus.* VIII. The pathway to succinate in *E. granulosus* scolices. *Comp. Biochem. Physiol.* **14**, 299–309.

Agosin, M., von Brand, T., Rivera, G. G. and McMahon, P. (1957). Studies on the metabolism of *Echinococcus granulosus.* I. General chemical composition and respiratory reactions. *Expl Parasit.* **6**, 37–51.

Aldrich, D. V., Chandler, A. C. and Daugherty, J. W. (1954). Intermediary protein metabolism in helminths. II. Effect of host castration on amino acid metabolism in *Hymenolepis diminuta. Expl Parasit.* **3**, 173–184.

Archer, D. M. and Hopkins, C. A. (1958). Studies on cestode metabolism. V. The chemical composition of *Diphyllobothrium* sp. in the plerocercoid and adult stages. *Expl Parasit.* **7**, 542–554.

Barber, A. A., Orrell, S. A., Jr and Bueding, E. (1967). Association of enzymes with rat liver glycogen isolated by rate-zonal centrifugation. *J. Biol. Chem.* **242**, 4040–4044.

Baron, P. J. (1968). On the histology and ultrastructure of *Cysticercus longicollis*, the cysticercus of *Taenia crassiceps* Zeder, 1800 (Cestoda, Cyclophyllidea). *Parasitology*, **58**, 497–513.

Barrett, J. (1976). Bioenergetics in helminths. *In* "Biochemistry of Parasites and Host–Parasite Relationships", (H. Van den Bossche, ed.), pp. 67–80. Elsevier/North-Holland Biomedical Press, Amsterdam.

Barrett, J., and Körting, W. (1977). Lipid catabolism in the plerocercoids of *Schistocephalus solidus* (Cestoda: Pseudophyllidea). *Int. J. Parasit.* **7**, 419–422.

Behm, C. A. and Bryant, C. (1975a). Studies of regulatory metabolism in *Moniezia expansa*: general considerations. *Int. J. Parasit.* **5**, 209–217.

Behm, C. A. and Bryant, C. (1975b). Studies of regulatory metabolism in *Moniezia expansa*: the role of phosphofructokinase (with a note on pyruvate kinase). *Int. J. Parasit.* **5**, 339–346.

Behm, C. A. and Bryant, C. (1975c). Studies of regulatory metabolism in *Moniezia expansa*: the role of phosphoenolpyruvate carboxykinase. *Int. J. Parasit.* **5**, 347–354.

Beis, I. and Barrett, J. (1979). The contents of adenine nucleotides and glycolytic and tricarboxylic acid cycle intermediates in activated and non-activated plerocercoids of *Schistocephalus solidus* (Cestoda: Pseudophyllidea). *Int. J. Parasit.* **9**, 465–468.

Bolla, R. I. and Roberts, L. S. (1971). Developmental physiology of cestodes—X. The effect of crowding on carbohydrate levels and on RNA, DNA and protein synthesis in *Hymenolepis diminuta. Comp. Biochem. Physiol.* **40A**, 777–787.

von Bonsdorff, C. H., Forssten, T., Gustafsson, M. K. S. and Wikgren, B. J. (1971). Cellular composition of plerocercoids of *Diphyllobothrium dendriticum* (Cestoda). *Acta zool. fenn.* **132**, 1–25.

von Brand, T. (1933). Untersuchungen über den Stoffbestand einiger Cestoden und den Stoffwechsel von *Moniezia expansa. Z. vergl. Physiol.* **18**, 592–596.

von Brand, T. (1952). "Chemical Physiology of Endoparasitic Animals", Academic Press, New York and London.

von Brand, T. (1960). Recent advances in carbohydrate biochemistry of helminths. *Helminth. Abstr.* **29**, 97–111.

von Brand, T. (1966). "Biochemistry of Parasites", Academic Press, New York and London.

von Brand, T. (1973). "Biochemistry of Parasites", ed. 2, Academic Press, New York and London.

von Brand, T. (1979). "Biochemistry and Physiology of Endoparasites", Elsevier/North-Holland Biomedical Press, Amsterdam.

von Brand, T. and Bowman, I. B. R. (1961). Studies on the aerobic and anaerobic metabolism of larval and adult *Taenia taeniaeformis. Expl Parasit.* **11**, 276–297.

von Brand, T., Churchwell, F. and Eckert, J. (1968). Aerobic and anaerobic metabolism of larval and adult *Taenia taeniaeformis*: V. Glycogen synthesis, metabolic endproducts, and carbon balances of glucose and glycerol utilization. *Expl Parasit.* **23**, 309–318.

Bråten, T. (1968). The fine structure of the tegument of *Diphyllobothrium latum* (L.). A comparison of the plerocercoid and the adult stages. *Z. ParasitKde.* **30**, 104–112.

Bryant, C. (1972a). The utilization of carbon dioxide by *Moniezia expansa*: aspects of metabolic regulation. *In* "Comparative Biochemistry of Parasites", (H. Van den Bossche, ed.), pp. 49–79. Academic Press, New York and London.

Bryant, C. (1972b). Metabolic regulation in *Moniezia expansa* (Cestoda): the role of pyruvate kinase. *Int. J. Parasit.* **2**, 333–340.

Bryant, C. (1975). Carbon dioxide utilisation, and the regulation of respiratory metabolic pathways in parasitic helminths. *In* "Advances in Parasitology", (B. Dawes, ed.), Vol. 13, pp. 35–69. Academic Press, New York and London.

Bryant, C. (1978). The regulation of respiratory metabolism in parasitic helminths. *In* "Advances in Parasitology", (W. H. R. Lumsden, R. Muller, and J. R. Baker, eds), Vol. 16, pp. 311–331. Academic Press, New York and London.

Bryant, C., and Morseth, D. J. (1968). The metabolism of radioactive fumaric acid and some other substrates by whole adult *Echinococcus granulosus* (Cestoda). *Comp. Biochem. Physiol.* **25**, 541–546.

Bueding, E. (1950). Carbohydrate metabolism of *Schistosoma mansoni*. *J. gen. Physiol.* **33**, 475–495.

Bueding, E. and Orrell, S. A., Jr (1964). A mild procedure for the isolation of polydisperse glycogen from animal tissues. *J. Biol. Chem.* **239**, 4018–4020.

Bueding, E. and Saz, H. J. (1968). Pyruvate kinase and phosphoenolpyruvate carboxykinase activities of *Ascaris* muscle, *Hymenolepis diminuta*, and *Schistosoma mansoni*. *Comp. Biochem. Physiol.* **24**, 511–518.

Burke, W. F., Gracy, R. W. and Harris, B. G. (1972). Studies on enzymes from parasitic helminths. III. Purification and properties of lactate dehydrogenase from the tapeworm, *Hymenolepis diminuta*. *Comp. Biochem. Physiol.* **43B**, 345–359.

Carter, C. E. and Fairbairn, D. (1975). Multienzymic nature of pyruvate kinase during development of *Hymenolepis diminuta* (Cestoda). *J. expl Zool.* **194**, 439–448.

Chandler, A. C. (1943). Studies on the nutrition of tapeworms. *Amer. J. Hyg.* **37**, 121–130.

Chandler, A. C., Read, C. P. and Nicholas, H. O. (1950). Observations on certain phases of nutrition and host parasite relations of *Hymenolepis diminuta* in white rats. *J. Parasit.* **36**, 523–535.

Chappell, L. H. (1979). "Physiology of Parasites", John Wiley and Sons, New York.

Coles, G. C. and Simpkin, K. G. (1977). Metabolic gradient in *Hymenolepis diminuta* under aerobic conditions. *Int. J. Parasit.* **7**, 127–128.

Collin, W. K. (1970). Electron microscopy of postembryonic stanges of the tapeworm, *Hymenolepis citelli*. *J. Parasit.* **56**, 1159–1170.

Colucci, A. V., Orrell, S. A., Saz, H. J. and Bueding, E. (1966). Differential glucose incorporation into glycogen by *Hymenolepis diminuta*. *J. Biol. Chem.* **241**, 464–468.

Cornish, R. A. and Bryant, C. (1975). Studies of regulatory metabolism in *Moniezia expansa*: glutamate, and the absence of the γ-amino-butyrate pathway. *Int. J. Parasit.* **5**, 355–362.

Davey, R. A. and Bryant, C. (1969). The tricarboxylic acid cycle and associated reactions in *Moniezia expansa* (Cestoda). *Comp. Biochem. Physiol.* **31**, 503–511.

Davis, R. E. and Roberts, L. S. (1983a). Platyhelminthes—Eucestoda. *In* "Reproductive Biology of Invertebrates. Vol. II. Spermatogenesis and Sperm Function", (K. G. Adiyodi and R. G. Adiyodi, eds), John Wiley and Sons Ltd, Chichester, (in press).

Davis, R. E. and Roberts, L. S. (1983b). Platyhelminthes—Eucestoda. *In* "Reproductive Biology of Invertebrates. Vol. I. Oogenesis, Oviposition, and Oosorption", (K. G. Adiyodi and R. G. Adiyodi, eds), pp. 109–133. John Wiley and Sons Ltd, Chichester.

Dendinger, J. E. and Roberts, L. S. (1977a). Glycogen synthase in the rat tapeworm, *Hymenolepis diminuta*—I. Enzyme activity during development and with crowding. *Comp. Biochem. Physiol.* **58B**, 215–219.

Dendinger, J. E. and Roberts, L. S. (1977b). Glycogen synthase in the rat tapeworm, *Hymenolepis diminuta*—II. Control of enzyme activity by glucose and glycogen. *Comp. Biochem. Physiol.* **58B**, 231–236.

Drochmans, P. (1962). Morphologie du glycogen. Etude au microscope éléctronique de colorations négatives du glycogen particulaire. *J. Ultrastruct. Res.* **6**, 141–163.

Dunkley, L. C. and Mettrick, D. F. (1969). *Hymenolepis diminuta*: effect of quality of host dietary carbohydrate on growth. *Expl Parasit.* **25**, 146–161.

Fairbairn, D. (1970). Biochemical adaptation and loss of genetic capacity in helminth parasites. *Biol. Rev.* **45**, 29–72.

Fairbairn, D., Wertheim, G., Harpur, R. P. and Schiller, E. L. (1961). Biochemistry of normal and irradiated strains of *Hymenolepis diminuta. Expl Parasit.* **11**, 248–263.

Fields, J. H. A. and Quinn, J. F. (1981). Some theoretical considerations on cytosolic redox balance during anaerobiosis in marine invertebrates. *J. Theor. Biol.* **88**, 35–45.

Fioravanti, C. F. and Saz, H. J. (1978). "Malic" enzyme, fumarate reductase, and transhydrogenase systems in the mitochondria of adult *Spirometra mansonoides. J. expl Zool.* **206**, 167–177.

Fioravanti, C. F. and Saz, H. J. (1980). Energy metabolism of adult *Hymenolepis diminuta. In* "Biology of the Tapeworm *Hymenolepis diminuta*", (H. P. Arai, ed.), pp. 463–504. Academic Press, New York and London.

Ford, B. R. (1972). *Hymenolepis citelli*: development and chemical composition in hypothermic ground squirrels. *Expl. Parasit.* **32**, 62–70.

Foster, W. B. and Daugherty, J. W. (1959). Establishment and distribution of *Raillietina cesticillus* in the fowl and comparative studies on amino acid metabolism of *R. cesticillus* and *Hymenolepis diminuta. Expl. Parasit.* **8**, 413–426.

Gäde, G., Wilps, H., Kluytmans, J. H. F. M. and de Zwaan, A. (1975). Glycogen degradation and end-products of anaerobic metabolism in the fresh water bivalve *Anodonta cygnea. J. comp. Physiol.* **104**, 79–85.

Goldberg, E. and Nolf, L. O. (1954). Succinic dehydrogenase activity in the cestode *Hymenolepis nana. Expl. Parasit.* **3**, 275–284.

Goodchild, C. G. (1961). Carbohydrate content of the tapeworm *Hymenolepis diminuta* from normal, bileless, and starved rats. *J. Parasit.* **47**, 401–405.

Grossbard, L. and Schimke, R. T. (1966). Multiple hexokinases from rat tissues: Purification and comparison of soluble forms. *J. Biol. Chem.* **241**, 3546–3560.

Hedrick, R. M. and Daugherty, J. W. (1957). Comparative histochemical studies on cestodes. I. The distribution of glycogen in *Hymenolepis diminuta* and *Raillietina cesticillus*. *J. Parasit.* **43**, 497–504.

Heyneman, D. and Voge, M. (1957). Glycogen distribution in cysticercoids of three hymenolepidid cestodes. *J. Parasit.* **43**, 527–531.

Heyneman, D. and Voge, M. (1960). Succinic dehydrogenase activity in cysticercoids of *Hymenolepis* (Cestoda: Hymenolepididae) measured by the tetrazolium technique. *Expl. Parasit.* **9**, 14–17.

Hochachka, P. W. (1980). "Living Without Oxygen. Closed and Open Systems in Hypoxia Tolerance", Harvard University Press, Cambridge, Mass., and London, England.

Hochachka, P. W., Owen, T. G., Allen, J. F. and Whittow, G. C. (1975). Multiple end-products of anaerobiosis in diving vertebrates. *Comp. Biochem. Physiol.* **50B**, 17–22.

Hopkins, C. A. (1950). Studies on cestode metabolism. I. Glycogen metabolism in *Schistocephalus solidus in vivo*. *J. Parasit.* **36**, 384–390.

Hopkins, C. A. (1960). Studies on cestode metabolism. VI. Analytical procedures and their application to *Hydatigera taeniaeformis*. *Expl. Parasit.* **9**, 159–166.

Insler, G. D. and Roberts, L. S. (1980a). Developmental physiology of cestodes. XV. A system for testing possible crowding factors *in vitro*. *J. expl. Zool.* **211**, 45–54.

Insler, G. D. and Roberts, L. S. (1980b). Developmental physiology of cestodes. XVI. Effects of certain excretory products on incorporation of ^{3}H-thymidine into DNA of *Hymenolepis diminuta*. *J. expl. Zool.* **211**, 55–61.

Johnston, I. A. (1975). Anaerobic metabolism in the carp (*Carassius carassius* L.). *Comp. Biochem. Physiol.* **51B**, 235–241.

Köhler, P. and Hanselmann. (1974). Anaerobic and aerobic metabolism in the larvae (tetrahyridia) of *Mesocestoides corti*. *Expl. Parasit.* **36**, 178–188.

Komuniecki, R. and Roberts, L. S. (1975). Developmental physiology of cestodes. XIV. Roughage and carbohydrate content of host diet for optimal growth and development of *Hymenolepis diminuta*. *J. Parasit.* **61**, 427–433.

Komuniecki, R. W. and Roberts, L. S. (1977a). Hexokinase from the rat tapeworm, *Hymenolepis diminuta*. *Comp. Biochem. Physiol.* **57B**, 45–49.

Komuniecki, R. and Roberts, L. S. (1977b). Galactose utilization by the rat tapeworm, *Hymenolepis diminuta*. *Comp. Biochem. Physiol.* **57B**, 329–333.

Komuniecki, R. W. and Roberts, L. S. (1977c). Enzymes of galactose utilization in the rat tapeworm, *Hymenolepis diminuta*. *Comp. Biochem. Physiol.* **58B**, 35–38.

Körting, W. (1976). Metabolism in parasitic helminths of freshwater fish. *In* "Biochemistry of Parasites and Host–Parasite Relationships", (H. Van den Bossche, ed.), pp. 95–100. Elsevier/North Holland Biomedical Press, Amsterdam.

Körting, W. and Barrett, J. (1977). Carbohydrate catabolism in the plerocercoids of *Schistocephalus solidus* (Cestoda: Pseudophyllidea). *Int. J. Parasit.* **7**, 411–417.

Kwa, B. H. (1972). Studies on the sparganum of *Spirometra erinacei*. I. The histology and histochemistry of the scolex. *Int. J. Parasit.* **2**, 23–28.

Laurie, J. S. (1957). The *in vitro* fermentation of carbohydrates by two species of cestodes and one species of Acanthocephala. *Expl. Parasit.* **6**, 245–260.

Levine, P. P. (1938). Observations on the biology of the poultry cestode *Davainea proglottina* in the intestine of the host. *J. Parasit.* **24**, 423–431.

de Ley, J. and Vercruysse, R. (1955). Glucose-6-phosphate and gluconate-6-phosphate dehydrogenase in worms. *Biochim. biophys. Acta*, **16**, 615–616.

Li, T., Gracy, R. W. and Harris, B. G. (1972). Studies on enzymes from parasitic helminths. II. Purification and properties of malic enzyme from the tapeworm, *Hymenolepis diminuta. Arch. Biochem. Biophys.* **150**, 397–406.

Logan, J., Ubelaker, J. E. and Vrijenhoek, R. C. (1977). Isozymes of L(+) LDH in *Hymenolepis diminuta. Comp. Biochem. Physiol.* **57B**, 51–53.

Lumsden, R. D. (1965). Macromolecular structure of glycogen in some cyclophyllidean and trypanorhynch cestodes. *J. Parasit.* **51**, 501–515.

Lumsden, R. D. and Byram, J. (1967). The ultrastructure of cestode muscle. *J. Parasit.* **53**, 326–342.

Lumsden, R. D. and Specian, R. (1980). The morphology, histology, and fine structure of the adult stage of the cyclophyllidean tapeworm *Hymenolepis diminuta. In* "Biology of the Tapeworm *Hymenolepis diminuta*", (H. Arai, ed.), pp. 157–280. Academic Press, New York and London.

Markov, G. S. (1961). Physiology of fish parasites. *In* "Parasitology of Fishes", (V. A. Dogiel, G. K. Petrushevski and Yu. I. Polyanski, eds), pp. 117–139. Oliver and Boyd Ltd (Z. Kabata, translator), Edinburgh and London.

Marra, M. and Esch, G. W. (1970). Distribution of carbohydrates in adults and larvae of *Proteocephalus ambloplitis* (Leidy, 1887). *J. Parasit.* **56**, 398–400.

Mayberry, L. F. and Tibbitts, F. D. (1971). *Hymenolepis diminuta* (order Cyclophyllidea): histochemical localization of glycogen, neutral lipid, and alkaline phosphatase in developing worms. *Z. ParasitKde.* **38**, 66–76.

McManus, D. P. (1975). Tricarboxylic acid cycle enzymes in the plerocercoid of *Ligula intestinalis* (Cestoda: Pseudophyllidea). *Z. ParasitKde.* **45**, 319–322.

McManus, D. P. and Smyth, J. D. (1978). Differences in the chemical composition and carbohydrate metabolism of *Echinococcus granulosus* (sheep and horse strain) and *E. multilocularis. Parasitology*, **77**, 103–109.

Mettrick, D. F. and Munro, H. N. (1965). Studies on the protein metabolism of cestodes. I. The effect of host dietary constituents on the growth of *Hymenolepis diminuta. Parasitology*, **55**, 453–466.

Meyer, H., Mueller, J. and Meyer, F. (1978). Isolation of an acyl-CoA carboxylase from the tapeworm *Spirometra mansonoides. Biochem. biophys. Res. Commun.* **82**, 834–839.

Mied, P. A. and Bueding, E. (1974). Factors controlling the glycogen synthase activity of a tapeworm. *Fedn. Proc. Fedn. Am. Socs. expl. Biol.* **33**, 1414.

Mied, P. A. and Bueding, E. (1979a). Glycogen synthase of *Hymenolepis diminuta.* I. Allosteric activation and inhibition. *J. Parasit.* **65**, 14–24.

Mied, P. A. and Bueding, E. (1979b). Glycogen synthase of *Hymenolepis diminuta.* II. Nutritional state, interconversion of forms, and primer glycogen molecular weight as control factors. *J. Parasit.* **65**, 25–39.

Moczón, T. (1975). Histochemical studies on the enzymes of *Hymenolepis diminuta* (Rud., 1819) (Cestoda). V. Some enzymes of the synthesis and phosphorolytic degradation of glycogen in mature cestodes. *Acta parasit. pol.* **23**, 569–592.

Moczón, T. (1977). Histochemical studies on the enzymes of *Hymenolepis diminuta* (Rud., 1819) (Cestoda). VI. Some enzymes of the synthesis and phosphorolytic degradation of glycogen in oncospheres and cysticercoids. *Acta parasit. pol.* **24**, 275–282.

Moon, T. W., Hulbert, W. C., Mustafa, T. and Mettrick, D. F. (1977a). A study of lactate dehydrogenase and malate dehydrogenase in adult *Hymenolepis diminuta* (Cestoda). *Comp. Biochem. Physiol.* **56B**, 249–254.

Moon, T. W., Mustafa, T., Hulbert, W. C., Podesta, R. G. and Mettrick, D. F. (1977b). The phosphoenol-pyruvate branchpoint in adult *Hymenolepis diminuta* (Cestoda): a study of pyruvate kinase and phosphoenol-pyruvate carboxykinase. *J. expl Zool.* **200**, 325–336.

Mustafa, T., Komuniecki, R. and Mettrick, D. F. (1978). Cytosolic glutamate dehydrogenase in adult *Hymenolepis diminuta* (Cestoda). *Comp. Biochem. Physiol.* **63B**, 219–222.

Nations, C., Hicks, T. C. and Ubelaker, J. E. (1973). CO_2 production by extracts of *Hymenolepis diminuta* (Cestoda: Hymenolepididae) with aspartate and α-ketoglutarate as substrates. *J. Parasit.* **59**, 112–116.

Nieland, M. L. and Weinbach, E. C. (1968). The bladder of *Cysticercus fasciolaris*: electron microscopy and carbohydrate content. *Parasitology*, **58**, 489–496.

Orpin, C. G., Huskisson, N. S. and Ward, P. F. V. (1976). Molecular structure and morphology of glycogen isolated from the cestode, *Moniezia expansa*. *Parasitology*, **73**, 83–95.

Orrell, S. A., Bueding, E. and Colucci, A. V. (1966). Relationship between sedimentation coefficient distribution and glycogen level in the cestode *Hymenolepis diminuta*. *Comp. Biochem. Physiol.* **18**, 657–662.

Orrell, S. A., Jr, Bueding, E. and Reissig, M. (1964). Physical characteristics of undegraded glycogen. *In* "Control of Glycogen Metabolism", (W. J. Whelan and M. P. Cameron, eds), pp. 29–44. J. and A. Churchill Ltd, London.

Ovington, K. S. and Bryant, C. (1981). The role of carbon dioxide in the formation of end-products by *Hymenolepis diminuta*. *Int. J. Parasit.* **11**, 221–228.

Pappas, P. W. and Schroeder, L. L. (1979). *Hymenolepis microstoma:* lactate and malate dehydrogenases of the adult worm. *Expl. Parasit.* **47**, 134–139.

Phifer, K. (1958). Aldolase in the larval form of *Taenia crassiceps*. *Expl. Parasit.* **7**, 269–275.

Pietrzak, S. M. and Saz, H. J. (1981). Succinate decarboxylation to propionate and the associated phosphorylation in *Fasciola hepatica* and *Spirometra mansonoides*. *Molec. Bioch. Parasit.* **3**, 61–70.

Platzer, E. G. and Roberts, L. S. (1970). Developmental physiology of cestodes—part VII. Vitamin B_6 and *Hymenolepis diminuta*: vitamin levels in the cestode and effects of deficiency on phosphorylase and transaminase activities. *Comp. Biochem. Physiol.* **35**, 535–552.

Podesta, R. B. and Mettrick, D. F. (1974). Pathophysiology of cestode infections: effect of *Hymenolepis diminuta* on oxygen tensions, pH and gastrointestinal function. *Int. J. Parasit.* **4**, 277–292.

Podesta, R. B., Mustafa, T., Moon, T. W., Hulbert, W. C. and Mettrick, D. F. (1976). Anaerobes in an aerobic environment: role of CO_2 in energy metabolism of *Hymenolepis diminuta*. *In* "Biochemistry of Parasites and

Host–Parasite Relationships", (H. Van den Bossche, ed.), pp. 81–88. Elsevier/North-Holland Biomedical Press, Amsterdam.

Prescott, L. M. and Campbell, J. W. (1965). Phosphoenolpyruvate carboxylase activity and glycogenesis in the flatworm *Hymenolepis diminuta. Comp. Biochem. Physiol.* **14**, 491–511.

Rasero, F. S., Monteoliva, M. and Mayor, F. (1968). Enzymes related to 4-aminobutyrate metabolism in intestinal parasites. *Comp. Biochem. Physiol.* **25**, 693–701.

Read, C. P. (1951). Studies on the enzymes and intermediate products of carbohydrate degradation in the cestode *Hymenolepis diminuta. Expl. Parasit.* **1**, 1–18.

Read, C. P. (1952). Contributions to cestode enzymology. I. The cytochrome system and succinic dehydrogenase in *Hymenolepis diminuta. Expl. Parasit.* **1**, 353–362.

Read, C. P. (1953). Contributions to cestode enzymology. II. Some anaerobic dehydrogenases in *Hymenolepis diminuta. Expl. Parasit.* **2**, 341–347.

Read, C. P. (1956). Carbohydrate metabolism of *Hymenolepis diminuta. Expl. Parasit.* **5**, 325–344.

Read, C. P. (1957). The role of carbohydrates in the biology of cestodes. III. Studies on two species from dogfish. *Expl. Parasit.* **6**, 288–293.

Read, C. P. (1959). The role of carbohydrates in the biology of cestodes. VIII. Some conclusions and hypotheses. *Expl. Parasit.* **8**, 365–382.

Read, C. P. (1967). Carbohydrate metabolism in *Hymenolepis* (Cestoda). *J. Parasit.* **53**, 1023–1029.

Read, C. P. and Rothman, A. H. (1957a). The role of carbohydrates in the biology of cestodes. I. The effect of dietary carbohydrate quality on the size of *Hymenolepis diminuta. Expl. Parasit.* **6**, 1–7.

Read, C. P. and Rothman, A. H. (1957b). The role of carbohydrates in the biology of cestodes. II. The effect of starvation on glycogenesis and glucose consumption in *Hymenolepis. Expl. Parasit.* **6**, 280–287.

Read, C. P. and Rothman, A. H. (1957c). The role of carbohydrates in the biology of cestodes. IV. Some effects of host dietary carbohydrate on growth and reproduction of *Hymenolepis. Expl. Parasit.* **6**, 294–305.

Read, C. P. and Simmons, J. E., Jr (1963). Biochemistry and physiology of tapeworms. *Physiol. Rev.* **43**, 263–305.

Read, C. P., Schiller, E. L. and Phifer, K. (1958). The role of carbohydrates in the biology of cestodes. V. Comparative studies on the effects of host dietary carbohydrate on *Hymenolepis* spp. *Expl. Parasit.* **7**, 198–216.

Reid, W. M. (1942). Certain nutritional requirements of the fowl cestode *Raillietina cesticillus* (Molin) as demonstrated by short periods of starvation of the host. *J. Parasit.* **28**, 319–340.

Reissig, M. and Colucci, A. V. (1968). Localization of glycogen in the cestode, *Hymenolepis diminuta. J. Cell Biol.* **39**, 754–763.

Reuter, J. (1967). Studies on plerocercoids of *Diphyllobothrium dendriticum.* I. The dry weight/fresh weight ratio of the tissues under various physiological conditions. *Acta Acad. åbo., Ser. B*, **27**, 1–15.

Reuter, J. (1971). Studies on plerocercoids of *Diphyllobothrium dendriticum.* IV. The influence of oxygen and of a tissue homogenate on the glycogen contents and weight changes of the larvae. *Acta Acad. åbo., Ser. B*, **31**, 1–6.

Reynolds, C. H. (1980). Phosphoenolpyruvate carboxykinase from the rat and from the tapeworm *Hymenolepis diminuta*. Effects of inhibitors and transition metals on the carboxylation reaction. *Comp. Biochem. Physiol.* **65B**, 481–487.

Poach, P. J. (1981). Glycogen synthase and glycogen synthase kinases. *Curr. Topics Cell. Regul.* **20**, 45–105.

Roberts, L. S. (1961). The influence of population density on patterns and physiology of growth in *Hymenolepis diminuta* (Cestoda: Cyclophyllidea) in the definitive host. *Expl. Parasit.* **11**, 332–371.

Roberts, L. S. (1966). Developmental physiology of cestodes. I. Host dietary carbohydrate and the "crowding effect" in *Hymenolepis diminuta*. *Expl. Parasit.* **18**, 305–310.

Roberts, L. S. (1980). Development of *Hymenolepis diminuta* in its definitive host. *In* "Biology of the Tapeworm *Hymenolepis diminuta*", (H. Arai, ed.), pp. 357–423. Academic Press, New York and London.

Roberts, L. S. and Platzer, E. G. (1967). Developmental physiology of cestodes. II. Effects of changes in host dietary carbohydrate and roughage on previously established *Hymenolepis diminuta*. *J. Parasit.* **53**, 85–93.

Roberts, L. S., Bueding, E. and Orrell, S. A. (1972). Developmental physiology of cestodes—XI. Synthetic activity in various molecular weight fractions of glycogen during development of *Hymenolepis diminuta*. *Comp. Biochem. Physiol.* **43B**, 825–836.

Rolleston, F. S. (1972). A theoretical background to the use of measured concentrations of intermediates in study of the control of intermediary metabolism. *In* "Current Topics in Cellular Regulation", (B. L. Horecker and E. R. Stadtman, eds), Vol. 5, pp. 47–75. Academic Press, New York and London.

Rothstein, M. and Mayoh, H. (1964). Glycine synthesis and isocitrate lyase in the nematode, *Caenorhabditis briggsae*. *Biochem. biophys. Res. Commun.* **14**, 43–47.

Rybicka, K. (1967). Embryogenesis in *Hymenolepis diminuta* II. Glycogen distribution in the embryos. *Expl. Parasit.* **20**, 98–105.

Sakamoto, T. and Sugimura, M. (1969). Studies on echinococcosis. XXI. Electron microscopical observations on general structure of larval tissue of multilocular *Echinococcus*. *Jap. J. vet. Res.* **17**, 67–101.

Salminen, K. (1974). Succinate dehydrogenase and cytochrome oxidase in adult and plerocercoid *Diphyllobothrium latum*. *Comp. Biochem. Physiol.* **49B**, 87–92.

Saz, H. J. (1971). Anaerobic phosphorylation in *Ascaris* mitochondria and the effects of anthelmintics. *Comp. Biochem. Physiol.* **39B**, 627–637.

Saz, H. J. (1972). Comparative biochemistry of carbohydrates in nematodes and cestodes. *In* "Comparative Biochemistry of Parasites", (H. Van den Bossche, ed.), pp. 33–47. Academic Press, New York and London.

Saz, H. J. (1981a). Energy metabolism of parasitic helminths: adaptations to parasitism. *A. Rev. Physiol.* **43**, 323–341.

Saz, H. J. (1981b). Energy generation in parasitic helminths. *In* "The Biochemistry of Parasites", (G. M. Slutzky, ed.), pp. 177–189. Pergamon Press, Oxford.

Saz, H. J. and Lescure, O. L. (1969). The functions of phosphoenolpyruvate carboxykinase and malic enzyme in the anaerobic formation of succinate by *Ascaris lumbricoides. Comp. Biochem. Physiol.* **30**, 49–60.

Scheibel, L. W. and Saz, H. J. (1966). The pathway for anaerobic carbohydrate dissimilation in *Hymenolepis diminuta. Comp. Biochem. Physiol.* **18**, 151–162.

Scheibel, L. W., Saz, H. J. and Bueding, E. (1968). The anaerobic incorporation of ^{32}P into adenosine triphosphate by *Hymenolepis diminuta. J. Biol. Chem.* **243**, 2229–2235.

Schiller, E. L., Bueding, E., Turner, V. M. and Fisher, J. (1975). Aerobic and anaerobic carbohydrate metabolism and egg production of *Schistosoma mansoni in vitro. J. Parasit.* **61**, 385–389.

Singh, B. B., Singh, K. S. and Dwaraknath, P. K. (1977). Levels of glycogen and total lipids in *Thysaniezia giardi* (Cestoda: Anoplocephalidae). *Abstrs. 1st Natl. Congress Parasit., Baroda, 24–26 Feb. 77, Indian Soc. Parasit.*

Sinha, D. P. and Hopkins, C. A. (1967). Studies on *Schistocephalus solidus.* IV. The effect of temperature on growth and maturation *in vitro. Parasitology*, **57**, 555–566.

Smyth, J. D. (1969). "The Physiology of Cestodes" W. H. Freeman and Co., San Francisco.

Soderling, T. R. (1979). Regulatory functions of protein multisite phosphorylation. *Mol. Cell. Endocrinol.* **16**, 157–179.

Taylor, A. E. R., McCabe, M. and Longmuir, I. S. (1966). Studies on the metabolism of larval tapeworms (Cyclophyllidea: *Taenia crassiceps*) II. Respiration, glycogen utilization, and lactic acid production during culture in a chemically defined medium. *Expl. Parasit.* **19**, 269–275.

Tkachuck, R. D., Saz, H. J., Weinstein, P. P., Finnegan, K. and Mueller, J. F. (1977). The presence and possible function of methylmalonyl-CoA-mutase and propionyl-CoA-carboxylase in *Spirometra mansonoides. J. Parasit.* **63**, 769–774.

Wack, M., Komuniecki, R. and Roberts, L. S. (1983). Amino acid metabolism in the rat tapeworm, *Hymenolepis diminuta. Comp. Biochem. Physiol.* **74B**, 399–402.

Walkey, M. and Fairbairn, D. (1973). L(+)-lactate dehydrogenases from *Hymenolepis diminuta* (Cestoda). *J. expl. Zool.* **183**, 365–373.

Ward, C. W. and Fairbairn, D. (1970). Enzymes of beta-oxidation and the tricarboxylic acid cycle in adult *Hymenolepis diminuta* (Cestoda) and *Ascaris lumbricoides* (Nematoda). *J. Parasit.* **56**, 1009–1012.

Wardle, R. A. (1937). The physiology of the sheep tapeworm, *Moniezia expansa* Blanchard. *Can. J. Res., Sect. D.* **15**, 117–126.

Watts, S. D. M. and Fairbairn, D. (1974). Anaerobic excretion of fermentation acids by *Hymenolepis diminuta* during development in the definitive host. *J. Parasit.* **60**, 621–625.

Wertheim, G., Zeledon, R. and Read, C. P. (1960). Transaminases of tapeworms. *J. Parasit.* **46**, 497–499.

Chapter 10

Lipid Metabolism

J. Barrett

I. INTRODUCTION

Lipids are a heterogenous group of compounds, all of which are soluble in organic solvents, but relatively insoluble in water. The varied chemical structure of lipids means that there is no general method of classification (Barrett, 1981).

In the cell, lipids have a variety of functions. They are major structural components in cell membranes, and in most organisms triacylglycerols are an important energy reserve. Other lipids are involved in metabolic regulation, cell surface interactions and glycosyltransferase reactions. Lipid components also form part of the cytochrome chain and membrane transport mechanisms, and occur in association with proteins and carbohydrates in lipoproteins and glycolipids.

II. COMPOSITION AND DISTRIBUTION

Lipid usually forms about 20% of the dry weight of cestodes (range

BIOLOGY OF THE EUCESTODA Vol. 2
ISBN 0–12–062102–9

Table I. Lipid composition of some Cestodes.

Species	% Lipid		% Phospholipid	References
	wet wt	dry wt		
Calliobothrium verticillatum	7	25	39	Beach *et al.*, 1973
Cotugnia digonopora	—	7.6	—	Nigam, 1979
Diphyllobothrium latum	1.6	—	31	von Brand, 1952
Dipylidium caninum	2.4	—	21	Chopra *et al.*, 1978
Echinococcus granulosus (larva)	2.0	12.7–14.3	57	Agosin *et al.*, 1957; Fraya *et al.*, 1980; Fraya and Haddad, 1980; McManus, 1981
(adult)	—	7.5–12.2	—	McManus, 1981
Eubothrium crassum	—	37.3	64.5	Vysotzkaja and Sidorov, 1973
Hymenolepis citelli	3.9	16.1	25	Harrington, 1965
Hymenolepis diminuta	5.8	30.9	24	Ginger and Fairbairn, 1966*a*; Mettrick and Cannon, 1970
Lacistorhynchus tenuis	1.2–10	—	50	Buteau *et al.*, 1971
Ligula intestinalis (plerocercoid)	—	14.2	27.4	Vysotzkaja and Sidorov, 1973
Moniezia denticulata	1.3	16.2	—	von Brand, 1952
Moniezia expansa	3.4	30.1	15	von Brand, 1952
Orgymatobothrium musteli	2.5–10	—	25	Buteau *et al.*, 1971
Proteocephalus exiguus	—	35.6	49	Vysotzkaja and Sidorov, 1973
Raillietina fuhrmanni	—	10.8	—	Nigam, 1979

Raillietina cesticillus	3.2	20	—	Botero and Reid, 1969; von Brand, 1952
Schistocephalus solidus (plerocercoid)	2.6	9.8	52.1	Körting and Barrett, 1977; Vysotzkaja and Sidorov, 1973
Spirometra mansonoides				
(adult)	—	20	58	Beach *et al.*, 1980a
(coracidium)	—	25	64	Beach *et al.*, 1980a
(procercoid)	—	10	58	Beach *et al.*, 1980a
(plerocercoid)	—	17	64	Beach *et al.*, 1980a
Taenia hydatigena (larva)	—	18.2	30	Kassis and Frayha, 1973
Taenia marginata	1.1	4.9	—	von Brand, 1952
Taenia plicata	9.1	33.1	—	von Brand, 1952
Taenia saginata	3.8	31.1	4.3	Cmelik and Bartl, 1956
Taenia solium	14	—	—	von Brand, 1952
Taenia taeniaeformis				
(adult)	3.8	10.6	50	McMahon, 1961
(larva)	2.3	6.9	47	McMahon, 1961
Triaenophorus nodulosus	—	22.5	51.8	Vysotzkaja and Sidorov, 1973

5–35%) and of the total lipid around 35% is phospholipid (range 15–64%). Data for the lipid content of cestodes is given in Table I.

The total lipid content of a tapeworm can be influenced by a variety of extrinsic and intrinsic factors. The lipid content of *Hymenolepis diminuta*, for example, depends on the lipid content of the host diet, the presence of roughage in the diet, the age of the worm, the density of infection, and the host species (Overturf and Dryer, 1968; Roberts and Platzer, 1967; Roberts, 1961; Warren and Daugherty, 1957). The level of lipid in the host diet has also been shown to affect the lipid content in several other cestodes including *Raillietina cesticillus* (Botero and Reid, 1969), *R. fuhrmanni* and *Cotugnia digonopora* (Nigam, 1979). In the shark tapeworms *Lacistorhynchus tenuis* and *Orgymatobothrium musteli*, the lipid content varies both with location of the parasite and host species (Buteau *et al.*, 1971), while Vysotzkaja and Sidorov (1973) have reported that the age of the worm, the host species and the season all affected the lipid content of a number of cestodes from freshwater fish (*Eubothrium crassum*, *Proteocephalus exiguus*, *Triaenophorus nodulosus* and the plerocercoids of *Ligula intestinalis* and *Schistocephalus solidus*). Finally, in *Echinococcus granulosus*, the lipid content of both the adult and larva vary with the host species (McManus, 1981). In adult tapeworms, lipid is found primarily in the parenchyma, with little lipid detectable histochemically in the tegument or other organs. Large lipid droplets have been reported from the uterus of *Dipylidium caninum*, lipid droplets have also been described from the lumen of the excretory canals in *E. granulosus* and *Moniezia expansa*, and lipid is said to be concentrated around the excretory canals in *H. diminuta* (von Brand, 1973). Finally, what appear to be lipid droplets accumulate intercellularly in the tissues of plerocercoids of *L. intestinalis* during *in vitro* culture (Smyth, 1949). Small amounts of lipid (triacylglycerols, cholesterol, phospholipids and carotenoids) have also been reported from cestode calcareous corpuscles (von Brand, 1973).

In *H. diminuta*, the lipid content varies along the length of the strobila; the lowest content at the anterior end, the highest in the gravid proglottides (Fairbairn *et al.*, 1961; Overturf and Dryer, 1968; Mettrick and Cannon, 1970). A similar gradient exists in *R. cesticillus* (Hedrick, 1958), but in *Taenia saginata*, the anterior end is said to have the highest lipid content and the gravid segments the least (von Brand, 1973). When ripe proglottides are shed, lipid in the parenchyma is lost since it is not used by the adult or by the eggs. In some cases, it is possible that the lipids in the proglottides might attract the intermediate host to eat the eggs.

An interesting feature of lipid distribution in cestodes (and parasitic helminths generally), is that much of the lipid in the eggs is extra-embryonic, occurring in the space between the embryo and the egg shell. When the eggs hatch, the lipid is discarded (Beach *et al.*, 1980*a*). Similarly, in cysticercoids, much of the lipid is left in the cyst tissue when the scolex emerges. Most of the lipid in cysticerci, however, is in the protoscoleces, with very little in the cyst fluid or cyst wall (Cmelik, 1952; Kassis and Frayha, 1973; Kilejian *et al.*, 1962).

Detailed lipid analyses are available for a variety of cestodes (Tables II and III). In general, the different lipid classes and their relative proportions are similar to those of free-living organisms. However, the phospholipid content of cestodes, around 35%, is rather higher than that found in most free-living invertebrates.

A. Fatty acids

A great variety of saturated and unsaturated fatty acids ranging from C_{10} to C_{28}, including branched and odd-numbered fatty acids, have been found in cestode lipids. Fatty acids occur both as free fatty acids and esterified as components of complex lipids.

(1) Free fatty acids

The levels of free fatty acids in tissues are normally low (less than 1% of total lipids), even in tissues with an active lipid metabolism. Free fatty acid levels in cestodes are also generally low, but a value of 5% of the total lipids has been reported for the cysticerci of *T. hydatigena* and *E. granulosus* (Kassis and Frayha, 1973; Frayha *et al.*, 1980) and 7.7% for adult *T. saginata* (Cmelik and Bartl, 1956). High levels of free fatty acids occur in the larval stages of a variety of parasitic helminths, and this may be related to the lipid changes which accompany infection (Barrett, 1981). High levels of free fatty acids could also be the result of hydrolysis of complex lipids during extraction and analysis.

Prostaglandins are formed by cyclization of C_{20} unsaturated acids and have profound physiological effects in mammals. Prostaglandins have not been described in cestodes although they do occur in other invertebrates (nematodes and arthropods).

(2) Esterified fatty acids

There is a large literature on the fatty acid composition of whole lipid extracts and of isolated lipid classes from a range of cestode species. As in other organisms, different lipid classes have different and often characteristic fatty acid compositions. Moreover, the same lipid class from different tissues of the same parasite may show variations. In *H. diminuta*, the fatty acid composition of polar lipids from isolated tegumental components

Table II. Neutral lipid fractions identified in Cestodes.

Fraction	Species
Free fatty acids	*C. verticillatum*[1], *Eubothrium crassum*[2], *E. granulosus larvae*[3,4,5], *H. citelli*[6], *H. diminuta*[6,7,8,9,10], *L. tenuis*[11], *L. intestinalis* plerocercoid[2], *O. musteli*[11], *Proteocephalus exiguus*[2], *S. solidus* plerocercoid[2], *S. mansonoides* adult and larval stages[12,13], *T. hydatigena* larvae[14], *T. saginata*[15], *Triaenophorus nodulosus*[2]
Monoacylglycerols	*C. verticillatum*[1], *E. granulosus* larvae[3,4,5], *H. diminuta*[7,8,9,10,16], *L. tenuis*[11], *O. musteli*[11], *S. mansonoides* adult and larval stages[13], *T. hydatigena* larvae[14]
Diacylglycerols	*C. verticillatum*[1], *E. granulosus* larvae[4,5], *H. diminuta*[7,8,9,10,16], *L. tenuis*[11], *O. musteli*[11], *S. mansonoides* adult and larval stages[13], *T. hydatigena* larvae[14]
Triacylglycerols	*C. verticillatum*[1], *E. crassum*[2], *E. granulosus* larvae[3,4,5], *H. diminuta*[7,8,9,10,16], *L. tenuis*[11], *L. intestinalis* plerocercoid[2], *O. musteli*[11], *P. exiguus*[2], *S. solidus* plerocercoid[2], *S. mansonoides* adult and larval stages[12,13], *T. hydatigena* larvae[14], *T. nodulosus*[2]
Diacylglycerol alkyl ethers	*C. verticillatum*[1]
Sterols and sterol esters	*Avitellina centripunctata*[17], *C. verticillatum*[1,11], *Cotugnia digonopora*[18], *D. latum*[19], *E. crassum*[2], *E. granulosus* larvae[3,4,5,14,20,21,22,23,24], *E. multilocularis* larvae[21], *H. diminuta*[7,8,9,10,25], *L. tenuis*[11], *Moniezia expansa*[17], *Moniezia* sp.[26], *O. musteli*[11], *P. exiguus*[2], *Raillietina fuhrmanni*[18], *S. solidus* plerocercoid[2], *S. mansonoides* adult and larval stages[12,13], *Stilesia globipunctata*[17], *T. hydatigena* larvae[14,21], *T. saginta*[27], *T. taeniaeformis*[26], *T. nodulosus*[2]
Wax esters	*C. verticillatum*[1]
Hydrocarbons	*E. granulosus* larvae[21,22], *E. multilocularis* larvae[21], *H. diminuta*[7], *T. hydatigena* larvae[21,22]

(brush borders and vesicles) is different from that of polar lipids prepared from whole worms (Cain *et al.*, 1977).

In general, a similar range of fatty acids is found in the lipids of parasites and free-living organisms. The major fatty acids in both are C_{16} and C_{18}; in mammals C_{18} acids account for 50–60% of the total acids. Some cestodes (*H. diminuta*, *R. cesticillus*, larval *T. hydatigena*) have a considerably higher percentage of C_{18} acids in their lipids (70–80%), while some of the cestodes from sharks (*Calliobothrium verticillatum*, *O. musteli*, *Thysanocephalum cephalum*, *Poecilancistrium caryophyllum*, *Dasyrhynchus giganteus*, *L. tenuis*) have a much lower percentage of C_{18} acids (20–30%). In most mammals and free-living invertebrates the major C_{18} acid is oleic ($C_{18:1}$), and this is also true in the majority of cestodes. There are exceptions: linoleic ($C_{18:2}$) is the dominant acid in *H. diminuta* and *R. cesticillus*, while stearic ($C_{18:0}$) is the predominant C_{18} acid in the procercoids of *Spirometra mansonoides*, larval *T. hydatigena* and adult *C. digonopora* and *R. fuhrmanni*. In cestodes, some 80–90% of the total C_{18} acids are unsaturated, compared with about 60–80% in mammals.

The degree of unsaturation of the fatty acids of whole lipid extracts can be expressed either as the percentage of acids with double bonds or as the average number of double bonds per molecule (unsaturation index). In mammals, 50–60% of the fatty acids are unsaturated (unsaturation index 0.7–0.8). Many cestodes also fall within this range, but several have markedly higher levels of unsaturated fatty acids: 70–80% (unsaturation index 1.3–3.6) in *H. diminuta*, *T. cephalum*, *C. verticillatum*, *O. musteli*, *R. cesticillus*, *L. tenuis*, *E. crassum* and *Diphyllobothrium dendriticum*). The very high unsaturation indices found in tapeworms from sharks reflect the presence in these cestodes of long chain polyunsaturated acids characteristic of shark lipids. The lipids of *E. crassum* also contain large amounts of C_{22} polyunsaturated fatty acids, and again these fatty acids are abundant in the fish host (Smirnov and Sidorov, 1979).

Cestode lipids usually contain small amounts (less than 1% of odd numbered (C_{15}, C_{17}) and branched chain acids (br-C_{15}, br-C_{17}). The procercoids of *S. mansonoides* contain 5.3% odd-numbered branched chain

---

(1) Beach *et al.*, 1973. (2) Vysotzkaja and Sidorov, 1973. (3) Vessel *et al.*, 1972. (4) Frayha *et al.*, 1980. (5) Frayha and Haddad, 1980. (6) Harrington, 1965. (7) Ginger and Fairbairn, 1966*a*. (8) Overturf and Dryer, 1968. (9) Cain *et al.*, 1977. (10) Webb and Mettrick, 1975. (11) Buteau *et al.*, 1971. (12) Meyer *et al.*, 1966. (13) Beach *et al.*, 1980*a*. (14) Kassis and Frayha, 1973. (15) Cmelik, 1952. (16) Buteau and Fairbairn, 1969. (17) Nigam and Premvati, 1980*a*. (18) Nigam and Premvati, 1980*b*. (19) Smyth, 1969. (20) Frayha, 1968. (21) Frayha, 1971. (22) Frayha, 1974. (23) Thorson *et al.*, 1968. (24) Digenis *et al.*, 1970. (25) Fairbairn *et al.*, 1961. (26) Thompson *et al.*, 1960. (27) Cmelik and Bartl, 1956. (28) Webb and Mettrick, 1971. (29) Webb and Mettrick, 1973. (30) Chopra *et al.*, 1978. (31) von Brand, 1952. (32) McMahon, 1961. (33) Lesuk and Anderson, 1941.

acids, and again this reflects the high levels of these unusual fatty acids in the copepod host (Beach *et al.*, 1980*a*). The overall fatty acid composition of an organism is strongly influenced by two factors: diet and environmental temperature.

(3) Effect of diet

There is often a close correlation between the fatty acid composition of an organism and the fatty acid content of its diet. The similarity between host and cestode fatty acids is especially noticeable in tapeworms from sharks. Sharks characteristically contain large amounts of C_{20} and C_{22} polyunsaturated acids, and these acids are also found in high concentrations in shark tapeworms, occurring in all of the lipid classes (Buteau *et al.*, 1969). High levels of branched C_{15} and C_{17} acids occur in the lipids of copepods and similarly high levels of these fatty acids are found in procercoids of *Spirometra*.

There is, unfortunately, no simple method for comparing fatty acid compositions. Fatty acid analyses are usually given as moles %, and this is not suitable for most statistical methods since the values are not independently variable; also the identification of the minor components in fatty acid analyses are often unreliable. Mean square distance and rank correlation coefficients have both been suggested as ways of comparing fatty acid compositions (Cain and French, 1975; Barrett, 1981); a better quantitative method may be the cosine measure (Nei, 1972). These statistical techniques enable quantitative comparisons to be made between different fatty acid compositions. If this is done, it is found that there is a high correlation (correlation coefficients in parentheses) between the total fatty acid composition of *H. diminuta* and rat intestinal contents (0.85), *R. cesticillus* and chicken intestinal contents (0.86), *R. fuhrmanni* and *C. digonopora* and chicken intestinal contents (0.84 and 0.75), larval *E. granulosus* and infected host liver (0.85), and between the tapeworms of sharks and their hosts as follows: *C. verticillatum* (0.93); *D. giganteus* (0.92); *Grillotia simmonsi* (0.72); *P. caryophyllum* (0.9); *T. cephalum* (0.9). Finally, the different life-cycle stages of *S. mansonoides* all show a fairly good correlation between their total fatty acid compositions and that of their respective hosts as follows: procercoid (0.86); plerocercoid (0.85); adult (0.94). It has also been demonstrated in a variety of cestodes (*H. diminuta, R. cesticillus, R. fuhrmanni* and *C. digonopora*) that altering the fatty acid composition of the host's diet results in corresponding changes in the fatty acids of the tapeworm lipids (Overturf and Dryer, 1968; Botero and Reid, 1969; Nigam, 1979).

Although there is a high correlation between the fatty acid composition of the total lipid extracts of a number of tapeworms and their hosts lipids, the same is not necessarily true for the different individual lipid classes. Taking those species for which fatty acid analyses are available for the various lipid classes (*H. diminuta*, *C. verticillatum*, *S. mansonoides*), it is found that there is a high correlation for the acylglycerol fractions, but rather less for the sterol ester, phosphatidylcholine and phosphatidylethanolamine fraction. The fatty acid composition of the phosphatidylinositol, phosphatidylserine, cardiolipin and cerebroside fractions, however, show a much lower correlation with the host fatty acid composition (Barrett, unpublished). It is, therefore, the triacylglycerol fraction that most closely mirrors the dietary fatty acids, while the structural lipids have a more conservative composition. So, although there is a close correspondence between the fatty acid composition of cestodes and the fatty acids of their hosts, the incorporation of dietary fatty acids into the different lipid fractions is none the less a regulated process.

(4) Effect of temperature

Animals which live at higher temperatures usually have lipids with a higher melting point than those living at lower temperatures, and this is reflected in the degree of unsaturation of their fatty acids. Unsaturated fatty acids have lower melting points than the corresponding saturated fatty acids, so the lipids of animals which live at higher temperatures, for example birds and mammals, have a lower unsaturation index than the lipids of animals which live at lower temperatures, such as fish. The same principle seems to apply to cestodes; from published data, the average unsaturation index for cestodes of birds and mammals is 1.53 (mean for 5 species), while that of cestodes from fish is 2.4 (mean for 8 species).

Organisms modulate the melting point of their lipids in response to temperature change to maintain the physical state of their membranes within narrow limits. The fluidity of cell membranes affects the activity of membrane-bound enzymes and the functioning of transport mechanisms. Depot lipids must similarly be maintained in a suitable physical state. In ectotherms, hosts and parasites presumably modulate their lipids in unison, in response to environmental temperature change. The parasites of endotherms present a particularly interesting problem with respect to temperature change. During their life-cycles, the different stages of such parasites are faced with rapid changes in ambient temperature, the free-living stages or the larval stages in an ectotherm passing into the tissues of a mammal or bird (at 35–43°C), and the infective stages passing from the endothermic host back to the free-living environment. Very little is known of the adaptations which occur in the lipids of parasites in response to these rapid changes in environmental temperature. Free-living organisms have been shown to respond to a drop in temperature by increasing the proportion of long-chain polyunsaturated acids in their lipids. In the goldfish, for example, temperature acclimation from 30° to 5°C results in an increase in the mean chain-length of fatty acids from 17.7 to 20.1 and an increase in unsaturation index from 2.2 to 2.5 (Van den Thillart and de Bruin, 1981). The melting point of a fatty acid also depends on chain length: the longer the chain, the higher the melting point. However, the effect on melting point of altering chain length is relatively minor compared with the effect of introducing double bonds. Straightforward changes in average chain length do not seem to be a factor in temperature adaptation in animals. Indeed, in the cold-adapted goldfish the average chain length increases rather than decreases (due to an increase in the proportion of long-chain polyunsaturated acids), so the change in the unsaturation index is apparently more important. Branched chain fatty acids have a lower melting point than their straight chain analogues, and in bacteria, the proportion of branched chain acids in the lipids decreases as the temperature increases. As well as changes in the overall fatty acid composition of membrane lipids, temperature change may also be accompanied by alterations in the relative proportions of the different lipid classes. In the goldfish, acclimation at 5°C and 30°C results in a drop in the phosphatidylcholine/phosphatidylethanolamine ratio in the membranes from 1.1 to 2.3. In hibernating mammals, the amount of lysophosphoglycerides in the membranes increases during hibernation, and changes in sterol content and the existence of protein modifiers of membrane fluidity have also been suggested. The occurrence of relatively large amounts of lysophosphoglycerides in certain cestodes, the presence of branched chain acids in other species, and the high levels of free-fatty acids in some larval stages may all possibly be related to temperature adaptation (Barrett, 1981). However, the only cestode for which lipid analyses are available for all the larval stages is *S. mansonoides* (Meyer *et al.*, 1966; Beach *et al.*, 1980*a*). In *S. mansonoides*, the coracidium has the highest unsaturation index (1.63), followed by the adult (1.35) and the plerocercoid and procercoid (both 1.13). Average chain length of the fatty acids is 17.2 for the adult and coracidium, 16.7 in the plerocercoid and 16.2 in the procercoid. So, in the different stages of *S. mansonoides*, neither changes in the unsaturation index nor changes in the average chain length can be correlated with changes in environmental temperature.

During infection or during the release of free-living stages from an infected endothermic host, the different parasite stages are subjected to extremely rapid changes in ambient temperature. Possibly the lipids of intermediate stages are adapted to function over relatively wide temperature ranges, so that the transition from one environment to the next need not necessarily be accompanied by simultaneous changes in lipid composition.

B. Acylglycerols

Triacylglycerol is the major neutral lipid in most cestodes, comprising around 75% of the total neutral lipids. Exceptions are the larvae of *E. granulosus* and the adult, procercoid and plerocercoid stages of *S. mansonoides* where the cholesterol content exceeds or equals the triacylglycerol levels (Vessel *et al.*, 1972; Frayha *et al.*, 1980; Beach *et al.*, 1980*a*). In *H. diminuta*, the triacylglycerols show a high degree of stereospecificity with 92% of the acids esterified at the C_2 position being unsaturated (Buteau and Fairbairn, 1969).

Monoacylglycerols and 1,2 and 1,3 diacylglycerols are usually present in small amounts (less than 1% of total lipids) in lipid extracts from cestodes. Small amounts of diacylglycerol alkyl ethers containing C_{16}, C_{18} and $C_{18:1}$ ether-linked residues have been found in the shark tapeworm *C. verticillatum* (Beach *et al.*, 1973).

C. Phosphoglycerides

The parent compound of phosphoglycerides is phosphatidic acid. However, it is usually only present in very small amounts in animal tissues, including cestodes, and phosphatidic acid has, so far, only been identified in *H. diminuta* and *S. mansonoides* (Table III).

Phosphatidylcholine and phosphatidylethanolamine are usually the major phospholipids present in animal tissues, together making up about 70% of the total. These two phospholipids appear to be universally distributed in cestodes; phosphatidylcholine is usually the major one, and the PC/PE ratio of cestodes ranges from 1 to 3.8. However in *T. saginata*, phosphatidylcholine and phosphatidylinositol, rather than phosphatidylethanolamine, appear to account for most, if not all, of the phospholipid fraction (Cmelik and Bartl, 1956). Fatty acid analyses of the phosphatidylcholine fraction of *H. diminuta*, *C. verticillatum*, *L. tenuis* and *S. mansonoides* suggest that, as is usually found in mammalian tissues, there is one saturated fatty acid and one unsaturated fatty acid per mole (Ginger and Fairbairn, 1966*a*; Buteau *et al.*, 1971; Beach *et al.*, 1973; Meyer *et al.*, 1966; Beach *et al.*, 1980*a*). In contrast, the phosphatidylcholine fraction from larval *T. taeniaeformis* seems to consist largely of the dipalmitate compound (Lesuk and Anderson, 1941).

Phosphatidylserine and phosphatidylinositol are normally present as minor components of cestode lipids (accounting for 5–10% of the total phospholipids each). In *T. hydatigena* larvae, however, phosphatidylinositol accounts for 40% of the phospholipids (Kassis and Frayha, 1973), and phosphatidylinositol is also a major component of the lipids of *T. saginata*

Table III. Phospholipid fractions identified in Cestodes.

Fraction	Species
Phosphatidic acid	*H. diminuta*[7,10,16,28,29], *S. mansonoides* adult and larval stages[13]
Phosphatidylcholine	*C. verticillatum*[1], *D. caninum*[30], *D. latum*[31], *E. granulosus* larvae[3,4], *H. citelli*[6], *H. diminuta*[6,7,8,9,10,16,28,29], *L. tenuis*[11], *O. musteli*[11], *S. mansonoides* adult and larval stages[12,13], *T. hydatigena* larvae[14], *T. saginata*[27], *T. taeniaeformis* adult and larvae[32,33]
Phosphatidylethanolamine	*C. verticillatum*[1], *D. caninum*[30], *E. granulosus* larvae[3,4], *H. citelli*[6], *H. diminuta*[6,7,8,9,10,16,25,28,29], *L. tenuis*[11], *O. musteli*[11], *S. mansonoides* adult and larval stages[12,13], *T. hydatigena* larvae[14], *T. taeniaeformis* adult and larvae[32]
Phosphatidylserine	*C. verticillatum*[1], *D. caninum*[30], *E. granulosus* larvae[4], *H. citelli*[6], *H. diminuta*[6,7,8,9,10,25,28,29], *L. tenuis*[11], *O. musteli*[11], *S. mansonoides* adult and larval stages[12,13], *T. hydatigena* larvae[14], *T. taeniaeformis* adult and larvae[32]
Phosphatidylinositol	*D. caninum*[30], *E. granulosus* larvae[3,4], *H. citelli*[6], *H. diminuta*[6,7,8,9,10,25,28,29], *S. mansonoides* adult and larval stages[12,13], *T. hydatigena* larvae[14], *T. saginata*[27]
Phosphatidylglycerol	*H. diminuta*[10]
Cardiolipin	*C. verticillatum*[1], *H. citelli*[6], *H. diminuta*[6,7,8,9,10,28,29], *S. mansonoides* adult and larval stages[12,13]
Plasmalogens	*H. citelli*[6], *H. diminuta*[6,7,8,25]
Lysophosphoglycerides	*D. caninum*[30], *E. granulosus* larvae[3,4], *H. diminuta*[7,9,10,25,28,29], *S. mansonoides* adult and larval stages[13], *T. hydatigena* larvae[14]
Sphingomyelin	*E. granulosus* larvae[3,4], *H. citelli*[6], *H. diminuta*[6,9], *S. mansonoides* adult and larval stages[13], *T. hydatigena* larvae[14], *T. taeniaeformis* adult and larvae[32]
Cerebrosides	*D. caninum*[30], *D. latum*[31], *E. granulosus* larvae[3], *H. citelli*[6], *H. diminuta*[6,7,9,25,28,29], *M. expansa*[31], *S. mansonoides* adult and larval stages[12], *T. taeniaeformis* adult and larvae[32,33]

(Cmelik and Bartl, 1956). In contrast, McMahon (1961) could detect no phosphatidylinositol in adult or larval *T. taeniaeformis*.

Cardiolipin is a major constituent of the inner mitochondrial membrane, and small amounts have been detected in lipid extracts from several cestodes (Table III). Plasmalogens are phosphoglycerides in which the 1-hydroxyl group is joined via an ether link to a long alk-1-enyl chain, instead of being esterified to a fatty acid. Plasmalogens occur widely as minor components in the tissues of vertebrates and invertebrates; ethanolamine plasmalogen is usually the major plasmalogen, with lesser amounts of choline plasmalogen. In cestodes, plasmalogens have been identified in *H. diminuta* and *H. citelli* (Table III), and in *H. diminuta*, as is usual, ethanolamine plasmalogen is the dominant plasmalogen, with lesser amounts of choline plasmalogen (Ginger and Fairbairn, 1966*a*).

Lysophosphoglycerides are formed from the corresponding phosphoglyceride by the removal of one of the fatty acids, and usually occur in tissues in relatively low amounts (less than 1% of total lipids). In lipid extracts from cestodes, lysophosphatidylcholine is usually the major phosphoglyceride, with lesser amounts of lysophosphatidylethanolamine. However, a number of helminths including *T. hydatigena* larvae and adult *D. caninum* contain relatively large amounts of lysophosphoglycerides (6–8% of total lipids; Kassis and Frayha, 1973; Chopra *et al.*, 1978). The biochemical significance of these high levels of lysophosphoglycerides in some cestodes is not known (Barrett, 1981). However, it should be remembered that lysophosphoglycerides, like free fatty acids and mono- and diacylglycerols, can be the result of hydrolysis of complex lipids during extraction and analysis.

D. Sphingolipids

The sphingolipids are important components of cell membranes and are especially abundant in nervous tissue. As a backbone, sphingolipids have the complex amino alcohol sphingosine or a related base. Some thirty different amino alcohols have been isolated from the sphingolipids of different organisms. However, there has been no specific work on the nature of the base in cestode sphingolipids. Sphingosine has been identified

---

(1) Beach *et al.*, 1973. (2) Vysotzkaja and Sidorov, 1973. (3) Vessel *et al.*, 1972. (4) Frayha *et al.*, 1980. (5) Frayha and Haddad, 1980. (6) Harrington, 1965. (7) Ginger and Fairbairn, 1966*a*. (8) Overturf and Dryer, 1968. (9) Cain *et al.*, 1977. (10) Webb and Mettrick, 1975. (11) Buteau *et al.*, 1971. (12) Meyer *et al.*, 1966. (13) Beach *et al.*, 1980*a*. (14) Kassis and Frayha, 1973. (15) Cmelik, 1952. (16) Buteau and Fairbairn, 1969. (17) Nigam and Premvati, 1980*a*. (18) Nigam and Premvati, 1980*b*. (19) Smyth, 1969. (20) Frayha, 1968. (21) Frayha, 1971. (22) Frayha, 1974. (23) Thorson *et al.*, 1968. (24) Digenis *et al.*, 1970. (25) Fairbairn *et al.*, 1961. (26) Thompson *et al.*, 1960. (27) Cmelik and Bartl, 1956. (28) Webb and Mettrick, 1971. (29) Webb and Mettrick, 1973. (30) Chopra *et al.*, 1978. (31) von Brand, 1952. (32) McMahon, 1961. (33) Lesuk and Anderson, 1941.

in *M. expansa* and sphingosine and/or sphinganine have been found in the adults and larvae of *T. taeniaeformis* (McMahon, 1961; von Brand, 1973). In the latter case it is not known if sphinganine is actually part of the sphingolipids, as it is also an intermediate in sphingosine synthesis.

The most abundant of the sphingolipids are the sphingomyelins, and cestode lipids normally contain measurable amounts of sphingomyelins (2 to 5% of total lipids). Cerebrosides, which are glycosphingolipids, have also been demonstrated in several species of cestodes (Table III). The cerebrosides from *H. diminuta, M. expansa* and adult and larval *T. taeniaeformis* contain galactose, although Harrington (1965) identified glucose as the hydrolysis product of the glycolipid fraction from *H. diminuta* and *H. citelli*. In mammals, galactose cerebrosides are characteristic of nervous tissue, and glucose cerebrosides are characteristic of non-nervous tissue. Characteristically, cerebrosides contain long chain saturated and mono-unsaturated fatty acids, and in mammals hydroxy-acids are common. The cerebrosides of *H. diminuta* contain the characteristic $C_{24:0}$ and $C_{24:1}$ fatty acids, and hydroxy-acids have been tentatively identified (Webb and Mettrick, 1971, 1973). Hydroxy-acids have also been described from *M. expansa* (von Brand, 1933).

Glycosphingolipids containing sulphated sugars are known as sulphatides. Sulphatides have been tentatively identified in the lipids of *D. caninum* and larval *E. granulosus* (Chopra *et al.*, 1978; Vessel *et al.*, 1972). In addition to the data in Table III, unspecified glycolipids have also been noted in larval and adult *S. mansonoides* (Meyer *et al.*, 1966).

E. Waxes

Wax esters, that is long chain alcohols esterified with long chain acids, have been described in *C. verticillatum* (Beach *et al.*, 1973). These wax esters contain unsaturated C_{14}, C_{16} and C_{18} alcohols esterified with $C_{20:5}$ and $C_{22:6}$ acids with lesser amounts of $C_{16:0}$, $C_{16:1}$, $C_{18:1}$, $C_{18:3}$ and $C_{20:4}$ acids. Nothing is known of the biosynthesis of wax esters in helminths. However, the distribution of chain lengths in the alcohol moiety suggests that, as in other organisms, the alcohols are formed by reduction of the corresponding fatty acid. Long-chain alcohols have also been found in larval *T. hydatigena* (Frayha, 1971).

A hydrocarbon fraction has been identified in the lipids of *H. diminuta* and in the larvae of *T. hydatigena* and *E. granulosus* (Frayha, 1971, 1974).

F. Terpenes

Measurable amounts of 2-*trans*, 6-*trans* farnesol have been found in *H. diminuta* (Frayha and Fairbairn, 1969; Fioravanti and MacInnis, 1977). Farnesol is a precursor of higher prenoids and is also a juvenile hormone mimic. A compound, similar to and possible identical with 2-*trans*, 6-*trans* farnesol, has been found in the protoscoleces of *E. granulosus* and the cysticerci of *T. hydatigena* (Frayha, 1971, 1974). Farnesal, the aldehyde of farnesol, has been tentatively identified from *E. granulosus* protoscoleces (Thorson *et al.*, 1968), and a number of additional unidentified isoprenoid compounds have been isolated from this tissue (Digenis *et al.*, 1970). The possible hormonal role of farnesol in cestodes is discussed later.

The triterpene squalene has been tentatively identified in lipids from larval *E. granulosus* and *T. hydatigena* (Frayha, 1971, 1974). Squalene, a precursor of cholesterol is normally present in only small amounts in tissues.

Carotenoids have been demonstrated in a number of cestodes (Table IV), but none have been found in *H. diminuta* (Barrett, unpublished). Nothing is known of carotenoid metabolism in cestodes, although it is probable that the carotenoid content of the parasites reflects the carotenoids available from the host (Barrett, 1981).

The ubiquinones consist of a quinone ring with a long isoprenoid side chain, and these compounds act as hydrogen carriers in biological systems. Ubiquinone-10 and rhodoquinone-10 have been found in the different life-cycle stages of *S. mansonoides* (Beach *et al.*, 1980*b*); rhodoquinone was the only quinone found in *M. expansa* (Allen, 1973), and this quinone has also been isolated from *H. diminuta.*

Table IV. Carotenoids identified in Cestodes.

Species	Carotenoid
Ligula intestinalis (plerocercoid)	Astaxanthin, astaxanthin ester, cryptoxanthin, taraxanthin, 4-keto-α-carotene
Schistocephalus solidus (plerocercoid)	γ-Carotene derivatives (celaxanthin, gazanioxanthin), canthaxanthin, unidentified ketocarotenoid, lutein-like carotenoid, violaxanthin-like carotenoid
Taenia saginata (adult)	Esterified astaxanthin, isozeaxanthin, α-cryptoxanthin, γ-carotene oxide, possibly ξ-carotene oxide, 4-hydroxy-4-ketocarotene

Data from Czeczuga, 1972; 1974; 1976.

G. Steroids

The major steroid found in cestodes, and helminths generally, is cholesterol. Indeed, with few exceptions, cholesterol seems to be the only steroid present in the cestode species so far investigated (Table II). In *C. verticillatum* and *L. tenuis*, in addition to cholesterol, two unidentified sterols account for 2–15% of the sterol fraction (Buteau *et al.*, 1971) and small amounts of unidentified sterols have been reported from larval *E. granulosus* (Thorson *et al.*, 1968) and *H. diminuta* (Fioravanti and MacInnis, 1977), although, again, cholesterol is the major component in both cases.

A variable proportion of the tissue cholesterol is esterified with long chain fatty acids. In cestodes, free sterols predominate with about one quarter of the steroid fraction being esterified.

III. SYNTHESIS

Cestodes readily incorporate labelled precursors into their complex lipids, although the actual pathways of lipid synthesis and the enzyme systems involved have not been studied in detail.

A. Fatty acid synthesis

None of the cestodes, so far studied, (*H. diminuta*, *C. verticillatum*, *S. mansonoides*) can synthesize long chain fatty acids *de novo*, nor can these helminths de-saturate preformed fatty acids (Jacobsen and Fairbairn, 1967; Beach *et al.*, 1973; Meyer *et al.*, 1966). Fatty acid synthesis in cestodes seems to be limited to simple chain lengthening by the sequential addition of acetyl-CoA: *H. diminuta* can elongate $C_{16:0}$ and $C_{18:0}$ (but not $C_{18:1}$ or $C_{18:2}$) to C_{20}, C_{22}, C_{24} and C_{26} acids, while *S. mansonoides* has been shown to elongate C_{16}, C_{18}, $C_{18:1}$, $C_{18:2}$ and $C_{18:3}$ to yield small amounts of C_{20} and C_{22} acids. The mechanism for chain lengthening in these tapeworms is not known, but it may be similar to the mammalian mitochondrial fatty acid elongation system. Acyl-CoA carboxylase, the enzyme catalysing the conversion of acetyl-CoA to malonyl-CoA, has been isolated from *S. mansonoides* (Meyer *et al.*, 1978). This reaction is the first committed step in fatty acid synthesis. The enzyme from *S. mansonoides* is, however, capable of carboxylating both propionyl- and butyryl-CoA as well as acetyl-CoA and it seems likely that the acyl-CoA carboxylase of *S. mansonoides* is involved primarily in propionate production from carbohydrate and not in fatty acid synthesis. In eukaryotes fatty acid de-saturation requires molecular oxygen and involves microsomal cytochromes. Most adult cestodes live in oxygen-poor environments and cestodes, as well as other helminths, seem to lack a number of oxygen-dependent pathways (Fairbairn, 1970). Microsomal cytochrome systems also seem to be missing in parasitic helminths.

Pathways absent from adult parasites are often present in the larval stages (Barrett, 1981). However, in *S. mansonoides* both the adult and plerocercoid are unable to synthesize fatty acids, but whether the other larval stages of cestodes similarly lack the ability to make fatty acids has not been studied. The inability to synthesize long-chain fatty acids *de novo* or to de-saturate preformed fatty acids may in fact be characteristic of the Platyhelminths as a whole and not just a feature of the cestodes. No fatty acid synthesis has been demonstrated in trematodes, and the free-living turbellarians *Dugesia dorotocephala* and *Convoluta roscoffensis* also can neither synthesize nor de-saturate long chain fatty acids (Meyer *et al.*, 1970; Meyer *et al.*, 1979). A wide range of saturated and unsaturated fatty acids occur in the lipids of cestodes and presumably these must all be obtained from the diet as "essential" fatty acids.

The inability of cestodes to synthesize or de-saturate long chain fatty acids may explain in part why the fatty acid composition of tapeworms is similar to that of their host. Nevertheless, tapeworms are able to control the fatty acid composition of their lipids. They could do this either by regulating the uptake of different fatty acids from the environment or by preferential acylation of free fatty acids in their own free fatty acid pool, or both.

The limited synthetic abilities of cestodes presents problems with regard to changes in environmental temperature. One of the responses to low temperature in free-living animals is an increase in the activity of the fatty acid desaturase system. Cestodes, however, can neither synthesize saturated fatty acids nor de-saturate preformed fatty acids. Cestodes and eukaryotes generally are also unable to reduce unsaturated fatty acids to saturated ones. The only mechanisms available to cestodes for modifying the melting point of their lipids is chain lengthening and altering the gross lipid composition of their membranes.

B. Acylglycerol synthesis

Several cestodes have been shown to incorporate labelled fatty acids and labelled glycerol or glucose into their mono-, di- and triacylglycerol fractions (Meyer *et al.*, 1966; Ginger and Fairbairn, 1966*b*; Jacobsen and Fairbairn, 1967; Bailey and Fairbairn, 1968; Overturf and Dryer, 1968; Kilejian *et al.*, 1968; Buteau and Fairbairn, 1969; Beach *et al.*, 1973; Webb and Mettrick, 1975). In *H. diminuta*, these studies indicate that triacylglycerols are synthesized via the phosphatidic acid/diacylglycerol pathway and that the monoacylglycerol route is at best only a minor pathway.

Lysolecithin is readily absorbed and metabolized by *H. diminuta* (Buteau and Fairbairn, 1969). It is possible that *H. diminuta* (and other intestinal parasites) may hydrolyse lysolecithin to lysophosphatidic acid which can then be synthesized into triacylglycerol.

In addition to incorporation into acylglycerols, *C. verticillatum* also incorporates labelled docosahexaenoic acid ($C_{22:6}$) into diacylglyceryl ethers (Beach *et al.*, 1973).

C. Phospholipid and sphingolipid synthesis

The incorporation of label from inorganic ^{32}P and ^{14}C labelled fatty acids, amino acids, glycerol and glucose into phospholipids has been demonstrated in several species of cestodes. Thus, *H. diminuta* can incorporate precursors into phosphatidic acid, phosphatidylcholine, phosphatidylethanolamine, phosphatidylserine, phosphatidylinositol, cardiolipin, cerebrosides and lysophosphoglycerides (Ginger and Fairbairn, 1966*b*; Jacobsen and Fairbairn, 1967; Bailey and Fairbairn, 1968; Overturf and Dryer, 1968; Buteau and Fairbairn, 1969; Webb and Mettrick, 1971, 1973, 1975), *C. verticillatum* has been shown to synthesize phosphatidylcholine, phosphatidylethanolamine, phosphatidylserine and cardiolipin (Beach *et al.*, 1973), and *S. mansonoides* to synthesize phosphatidylcholine, phosphatidylethanolamine, phosphatidylserine, phosphatidylinositol and cardiolipin (Meyer *et al.*, 1966). In all of these parasites, different phospholipid classes become labelled at different rates and this presumably reflects different rates of phospholipid turnover.

There are two possible pathways for phosphatidylethanolamine synthesis, the direct decarboxylation of phosphatidylserine and synthesis from ethanolamine via CDP-ethanolamine. In *H. diminuta*, ^{14}C-serine is readily incorporated into ethanolamine. Since eukaryotes cannot decarboxylate free serine to give ethanolamine, synthesis is probably via phosphatidylserine. CDP-ethanolamine has been identified in extracts of *T. saginata* (Antoniewicz *et al.*, 1977), so there is evidence for both routes of phosphatidylethanolamine synthesis in cestodes.

Phosphatidylserine synthesis in *H. diminuta* appears to be via phosphatidic acid (Webb and Mettrick, 1973):

phosphatidic acid + CTP → CDP-diacylglycerol + pyrophosphate
CDP-diacylglycerol + serine → phosphatidylserine + CMP

This pathway is characteristic of micro-organisms rather than mammals where phosphatidylserine synthesis is by base exchange as follows:

phosphatidylethanolamine + serine ⇌ phosphatidylserine + ethanolamine

Phosphatidylinositol synthesis in *H. diminuta* again involves phosphatidic acid, the other precursor, inositol, being synthesized from glucose, presumably via glucose-6-phosphate (Webb and Mettrick, 1971, 1975). This suggests that the pathway in *H. diminuta* is the same as in mammals:

phosphatidic acid + CTP → CDP-diacylglycerol + pyrophosphate
CDP-diacylglycerol + inositol → phosphatidylinositol + CMP

Sphingolipid synthesis has not been investigated in *H. diminuta*, but ^{14}C-serine is rapidly incorporated into sphingosine by this parasite, which suggests that the pathway is probably the same as in vertebrates:

palmitoyl-CoA + serine → 3-dehydrosphinganine → sphinganine
→ sphingosine

The intermediate sphinganine has been isolated from larval *T. taeniaeformis* (McMahon, 1961). Apart from these few studies on *H. diminuta* nothing is known of the pathways of phospholipid synthesis in cestodes.

D. Steroid and steroid ester synthesis

None of the cestodes so far studied appear to be able to synthesize steroids *de novo* (*H. diminuta*, Jacobsen and Fairbairn, 1967; Frayha and Fairbairn, 1969; protoscoleces of *E. granulosus*, Frayha, 1968, 1971, 1974; Digenis *et al.*, 1970; cysticerci of *E. multilocularis*, Frayha, 1971; cysticerci of *T. hydatigena*, Frayha, 1971, 1974; adults and larvae of *S. mansonoides*, Meyer *et al.*, 1966). However, many invertebrate groups appear to lack the ability to synthesize steroids; trematodes and turbellarians are unable to make steroids and so the absence of *de novo* steroid synthesis may be a peculiarity of platyhelminths in general and not just of the cestodes (Meyer *et al.*, 1970; Meyer *et al.*, 1979; Barrett *et al.*, 1970). Like the desaturation of fatty acids, steroid synthesis has an absolute requirement for molecular oxygen; for example the cyclization of squalene via squalene epoxide to give lanosterol being catalysed by a microsomal mono-oxygenase. As discussed above, cestodes seem to lack a number of these oxygen dependent pathways. However, the block to steroid synthesis in cestodes appears to be in squalene formation, rather than in its cyclization and neither *H. diminuta* nor the larvae of *E. granulosus* nor *T. hydatigena* can synthesize squalene *de novo*. The origin of the small amounts of squalene isolated from cestodes is not known, but *in vitro* tapeworms have been shown to be able to absorb squalene from the incubation medium (Frayha and Fairbairn, 1969).

Although they are unable to synthesize steroids, cestodes none the less readily incorporate exogenous steroids and fatty acids into their steroid ester fraction. The pathway of steroid ester synthesis, either direct acylation of cholesterol or via a transferase reaction, has not been studied.

E. Synthesis and physiological role of prenoids

Despite the absence of steroid synthesis, cestodes are able to synthesize polyisoprenoids. The rat tapeworm *H. diminuta* can synthesize the sesquiterpenoid farnesol (three isoprenoid units) from acetate or mevalonate (Frayha and Fairbairn, 1969). Originally it was thought that *H. diminuta* produced the 2-*cis*, 6-*trans* isomer of farnesol, but it only synthesizes the 2-*trans*, 6-*trans* isomer (Fioravanti and MacInnis, 1977). The absence of any of the other isomers of farnesol in *H. diminuta* suggests that this parasite lacks the enzyme sesquiterpenoid isomerase. Farnesol is also an intermediate in steroid synthesis; the natural isomer for steroid synthesis is 2-*trans*, 6-*trans* farnesol, so the failure of *H. diminuta* to synthesize squalene is not due to steric hindrance. The larvae of *E. granulosus* and *T. hydatigena* convert acetate to hydroxymethylglutarate, mevalonate and a compound similar to 2-*trans*, 6-*trans* farnesol. A number of unidentified isoprenoids, including possibly the aldehyde farnesal, have also been isolated from *E. granulosus* protoscoleces (Thorson *et al.*, 1968; Digenis *et al.*, 1970).

Farnesol and related compounds are juvenile hormone mimics, and it seemed possible that farnesol might have a similar hormonal function in cestodes. Thorson *et al.* (1968) isolated a steroid fraction from *E. granulosus* larvae which depressed the *in vitro* growth of *H. diminuta* cysticercoids, while a "farnesal-like" fraction was reported to stimulate growth. However, Fioravanti and MacInnis (1976) found no stimulatory effect on the *in vitro* growth of *H. diminuta* by 2-*cis*, 6-*trans* farnesol, 2-*trans*, 6-*trans* farnesol, farnesal or farnesyl methyl ether. At high concentrations these compounds were all toxic, confirming an earlier report that farnesyl methyl ether was toxic to *H. diminuta in vitro* (Tofts and Meerovitch, 1974). So, at present, there is no evidence for farnesol or a related prenoid having any hormonal role in tapeworms.

IV. CATABOLISM

In mammals, complex lipids are broken down by lipases to yield fatty acids, glycerol and the various polar groups from the phospholipids. Non-specific esterases which may be capable of hydrolysing lipids have been demonstrated histochemically in cestodes (for review see Lee *et al.*, 1963), and the protoscoleces of *E. granulosus* appear to be able to hydrolyse cholesterol esters (Digenis *et al.*, 1970). Phospholipase activity has been demonstrated in *H. diminuta* (Webb and Mettrick, 1973), and a monoacylglycerol lipase was demonstrated in the tegument of *H. diminuta* by Bailey and Fairbairn (1968), but this was not confirmed by Overturf and Dryer (1968).

In aerobic organisms, the fatty acids released by the action of the various lipases are oxidized by the beta-oxidation sequence.

A. Beta-oxidation

There is no evidence for an active beta-oxidation sequence in cestodes. No radioactive carbon dioxide was released from ^{14}C-U-palmitate by the adults, cysticercoids (activated and unactivated) or hexacanths (hatched) of *H. diminuta*, the adults of *Khawia sinensis*, *Bothriocephalus gowkongensis* and *B. scorpii* or the plerocercoids of *S. solidus*, *L. intestinalis* and *T. nodulosus*, indicating the lack of an active beta-oxidation sequence (Ward and Fairbairn, 1970; Barrett and Körting, 1977; Körting and Barrett, 1978; Barrett, unpublished). There is, however, some histochemical and analytical evidence that the free-swimming coracidia of pseudophyllidean cestodes may utilize lipid (Barrett, 1977; Beach *et al.*, 1980*a*).

Despite the apparent absence of a functional beta-oxidation sequence in adult and larval cestodes, beta-oxidation enzymes have been demonstrated in several species (Table V). The intracellular distribution of these enzymes has been studied in the plerocercoids of *S. solidus*; acyl-CoA synthetase occurred both in the mitochondria and in the cytoplasm, the remainder of the beta-oxidation enzymes being mitochondrial (Barrett and Körting, 1977). The occurrence of acyl-CoA synthetase in the cytoplasm is characteristic of tissues with a low fatty acid oxidation capacity; in animals with a high beta-oxidation activity the enzymes are all mitochondrial. In mammals, fatty acids are transported into the mitochondria via a carnitine-dependent mechanism and a similar system may exist in cestodes (Barrett and Körting, 1977).

Unlike the inability to synthesize steroids and fatty acids, the absence of a beta-oxidation sequence is not a general feature of the Platyhelminths, and although trematodes as well as cestodes may lack the ability to oxidize fatty acids, free-living turbellarians have active beta-oxidation pathways. The absence of a functional beta-oxidation sequence in cestodes, despite the presence of the enzymes, could be related either to the relative unimportance of the tricarboxylic acid cycle in these parasites or to the low environmental pO_2. The absence of an active tricarboxylic acid cycle would limit the further catabolism of acetyl-CoA produced by beta-oxidation. The large amounts of reducing power (NADH, reduced

Table V. The activities of the beta-oxidation enzymes in Cestodes.

Enzyme	Activity (nmoles/min/mg protein at 30°C)		
	Hymenolepis diminuta (adult)	*Schistocephalus solidus* (plerocercoid)	*Ligula intestinalis* (plerocercoid)
Acyl-CoA synthetase (short chain)	—	5.3	3.0
Acyl-CoA synthetase (long chain)	1.9	0.5	0.4
Acyl-CoA dehydrogenase	1.5	0.7	1.2
Enoyl-CoA hydratase	0	11.6	0
3-Hydroxyacyl-CoA dehydrogenase	0	2.2	0
Acetyl-CoA acyltransferase	12	9.6	4.7

Data from Ward and Fairbairn, 1970; Barrett and Körting, 1977; Körting and Barrett, 1978.

flavoprotein) produced during beta-oxidation require reoxidation via an oxidase system. The beta-oxidation sequence cannot, therefore, function anaerobically, and is only found in aerobic organisms.

It is possible to devise a scheme for the co-fermentation of lipid and carbohydrate, such that the reducing equivalents formed during beta-oxidation are reoxidized by coupling with the reduction of fumarate to succinate and the acetyl-CoA could be cleaved and acetate excreted. There is, however, no evidence from carbon balance studies for the co-fermentation of lipid and carbohydrate (Barrett and Körting, 1977).

The role of the beta-oxidation enzymes in cestodes is not clear. In the adult they could be associated primarily with the developing eggs in readiness for the larval stages. However, in cestodes the larval stages also seem to lack an active beta-oxidation sequence. The beta-oxidation enzymes may be involved in the chain lengthening of fatty acids by a malonyl-CoA independent pathway. In mammals this is a mitochondrial system which catalyses the condensation of an acetyl-CoA molecule with a long chain fatty acyl-CoA molecule. The system requires two NADPH for each C_2 added and appears to take place by the reversal of the steps of beta-oxidation. In mammals, the synthetic route differs from the degradative one in that both the 3-(OH)-acyl-CoA dehydrogenase and the acyl-CoA dehydrogenase of the synthetic route are NADP-linked. In beta-oxidation, the 3-(OH)-acyl-CoA dehydrogenase is NAD-linked, and the acyl-CoA dehydrogenase is flavoprotein-linked. However, neither NADP-linked 3-(OH)-acyl-CoA dehydrogenase nor NADP-linked acyl-CoA dehydrogenase have yet been found in cestodes.

The mitochondrial chain lengthening system can also, in theory, be coupled to carbohydrate catabolism. The breakdown of one mole of glucose to two moles of acetyl-CoA gives a net production of 2ATP and four NADH; the condensation of two acetyl-CoA with a long chain fatty acyl-CoA would utilize four NADH so there would be a net production of 2ATP per glucose. The initial activation of the long chain fatty acid would require the equivalent of two ATP (fatty acyl-CoA synthetase uses both high energy bonds of ATP, yielding AMP rather than ADP). This scheme would also require reducing equivalents to be transported into the mitochondria, since glycolysis is cytoplasmic while the chain lengthening system is mitochondrial. Energetically this system has no advantage over glycolysis, but could be useful if, for some reason, the excretion of organic acids was restricted. It has frequently been suggested that lipid could be an end-product of carbohydrate catabolism in parasitic helminths (for review see von Brand, 1973). However, there is no evidence, either from isotope studies or carbon balance studies that long chain fatty acids can act as terminal acceptors for NADH or acetyl-CoA in parasitic helminths, nor are there any reports of lipid excretion by tapeworms, despite the presence of lipid droplets in the excretory system.

The catabolism of the aliphatic amino acids valine, leucine and isoleucine involve steps analogous to those of beta-oxidation. At least one tapeworm is able to catabolize aliphatic amino acids to a minor extent (Fisher and Starling, 1970). However, in other organisms, different enzymes are involved in amino acid catabolism and fatty acid oxidation, and the same is likely to be true for cestodes.

B. Alpha- and omega-oxidation

When cestodes are incubated with universally labelled palmitate, a small amount of radioactive carbon dioxide is produced and there is some evidence that this arises by decarboxylation of the fatty acid (alpha-oxidation). In omega-oxidation carbon atoms are removed from the C terminal end of the fatty acid; there is no evidence for this in cestodes.

C. Steroid catabolism

Nothing is known of steroid catabolism in cestodes. In mammals, steroids are partly degraded to bile acids and partly excreted unchanged. A significant amount of the cholesterol absorbed by *H. diminuta* was reported to be metabolized to a more polar, unidentified compound (Frayha and Fairbairn, 1968). Bile acids have not been isolated from tapeworms, and *H. diminuta* appears to be relatively impermeable to them (Surgan and Roberts, 1976).

V. SUMMARY

The lipid composition of cestodes appears to be similar to that of other organisms. Cestodes are, however, unable to synthesize steroids or long chain fatty acids *de novo*, nor can they de-saturate pre-formed fatty acids; fatty acid synthesis in tapeworms is limited to simple chain lengthening by the sequential addition of acetyl-CoA. Nevertheless, cestodes, given the precursors, synthesize their own complex neutral and phospholipid classes and appear able to regulate the fatty acid composition of their lipids.

Although they probably possess lipases, cestodes are unable to catabolize fatty acids and lack an active beta-oxidation sequence. The absence of lipid catabolism in cestodes poses the problem as to why these parasites accumulate lipid if they are unable to catabolize it. Some of the lipid may be destined for incorporation into the eggs, but there is no evidence that the larval stages of cestodes can catabolize lipids either. Possibly cestodes have to absorb large amounts of lipid in order to get enough of a particular fatty acid or fat-soluble vitamin, and the excess lipid is stored rather than being excreted again. Alternatively, lipid accumulation in adult tapeworms may be related to the retention of potentially immunogenic compounds.

VI. REFERENCES

Agosin, M., von Brand, T., Rivera, G. F. and McMahon, P. (1957). Studies on the metabolism of *Echinococcus granulosus* I. General chemical composition and respiratory reactions. *Expl. Parasit.* **6**, 37–51.

Allen, P. C. (1973). Helminths: comparison of their rhodoquinone. *Expl. Parasit.* **34**, 211–219.

Antoniewicz, K., Baer, W. and Chmiel, J. (1977). Phosphate compounds in the strobila of *Taenia saginata* Goeze, 1782 (Cestoda, Taeniidae). *Acta parasit. pol.* **24**, 283–293.

Bailey, H. H. and Fairbairn, D. (1968). Lipid metabolism in helminth parasites-V. Absorbtion of fatty acids and monoglycerides from micellar solution by *Hymenolepis diminuta* (Cestoda). *Comp. Biochem. Physiol.* **26**, 819–836.

Barrett, J. (1977). Energy metabolism and infection in helminths. *In* "Parasite Invasion", (A. E. R. Taylor and R. Muller, eds), *Symp. Brit. Soc. Parasit.* **15**, 121–144.

Barrett, J. (1981). "Biochemistry of Parasitic Helminths", Macmillan, London.

Barrett, J. and Körting, W. (1977). Lipid catabolism in the plerocercoids of *Schistocephalus solidus* (Cestoda: Pseudophyllidea). *Int. J. Parasit.* **7**, 419–422.

Barrett, J., Cain, G. D. and Fairbairn, D. (1970). Sterols in *Ascaris lumbricoides* (Nematoda), *Macracanthorhynchus hirudinaceus* and *Moniliformis dubius* (Acanthocephala) and *Echinostoma revolutum* (Trematoda). *J. Parasit.* **56**, 1004–1008.

Beach, D. H., Mueller, J. F. and Holz, G. G. (1980*a*). Lipids of stages in the life cycle of the cestode *Spirometra mansonoides. Molec. Biochem. Parasit.* **1**, 249–268.

Beach, D. H., Mueller, J. F. and Holz, G. G. (1980*b*). Benzoquinones in stages of the life cycle of the cestode *Spirometra mansonoides. Molec. Biochem. Parasit.* **1**, 269–278.

Beach, D. H., Sherman, I. W. and Holz, G. G. (1973). Incorporation of docosahexaenoic fatty acid into the lipids of a cestode of marine elasmobranchs. *J. Parasit.* **59**, 655–666.

Botero, H. and Reid, W. M. (1969). *Raillietina cesticillus*: Fatty acid composition. *Expl. Parasit.* **25**, 93–100.

von Brand, T. (1933). Untersuchungen über den Stoffbestand einiger Cestoden und den Stoffwechsel von *Moniezia expansa. Z. vergl. Physiol.* **18**, 562–596.

von Brand, T. (1952). "Chemical physiology of Endoparasitic Animals", Academic Press, New York and London.

von Brand, T. (1973). "Biochemistry of Parasites", Academic Press, New York and London.

Buteau, G. H. and Fairbairn, D. (1969). Lipid metabolism in helminth parasites. VIII. Triglyceride synthesis in *Hymenolepis diminuta* (Cestoda). *Expl. Parasit.* **25**, 265–275.

Buteau, G. H., Simmons, J. E. and Fairbairn, D. (1969). Lipid metabolism in helminth parasites IX. Fatty acid composition of shark tapeworms and of their hosts. *Expl. Parasit.* **26**, 209–213.

Buteau, G. H., Simmons, J. E., Beach, D. H., Holz, G. G. and Sherman, I. W. (1971). The lipids of cestodes from pacific and atlantic coast triakid sharks. *J. Parasit.* **57**, 1272–1278.

Cain, G. D. and French, J. A. (1975). Effects of parasitism by the lung fluke, *Haematoloechus medioplexus*, on the lung fatty acid and sterol composition in the bullfrog, *Rana catesbiana. Int. J. Parasit.* **5**, 159–164.

Cain, G. D., Johnson, W. J. and Oaks, J. A. (1977). Lipids from subcellular fractions of the tegument of *Hymenolepis diminuta. J. Parasit.* **63**, 486–491.

Chopra, A. K., Jain, S. K., Vinayak, V. F. and Khuller, G. K. (1978). Phospholipid composition of *Dipylidium caninum. Experientia*, **34**, 1457–1458.

Cmelik, S. (1952). Zur Kenntnis der Lipoide aus den Cystenmembranen von *Taenia echinococcus. Hoppe-Sayler's Z. Physiol. Chem.* **289**, 78–79.

Cmelik, S. and Bartl, Z. (1956). Zusammensetzung der Lipide von *Taenia saginata. Hoppe-Seyler's Z. Physiol. Chem.* **305**, 170–176.

Czeczuga, B. (1972). The occurrence of γ-carotene derivatives in *Schistocephalus solidus* (Müller, 1776) plerocercoids (Cestoda, Ligulidae). *Acta parasit. pol.* **20**, 533–537.

Czeczuga, B. (1974). Carotenoid pigments in plerocercoids of *Ligula intestinalis* (L. 1758), (Cestoda, Ligulidae). *Bull. Acad. pol. Sci. Cl. II Sér Sci. biol.* **22**, 499–502.

Czeczuga, B. (1976). Astaxanthin ester, a dominating carotenoid in *Taenia saginata* (Cestoidea). *Bull. Acad. pol. Sci. Cl. II Sér Sci. biol.* **23**, 829–832.

Digenis, G. A., Thorson, R. E. and Konyalian, A. (1970). Cholesterol biosynthesis and lipid biochemistry in the scolex of *Echinococcus granulosus. J. Pharm. Sci.* **59**, 676–679.

Fairbairn, D. (1970). Biochemical adaptation and loss of genetic capacity in helminth parasites. *Biol. Rev.* **45**, 29–72.

Fairbairn, D., Wertheim, G., Harpur, R. P. and Schiller, E. L. (1961). Biochemistry of normal and irradiated strains of *Hymenolepis diminuta. Expl. Parasit.* **11**, 248–263.

Fioravanti, C. F. and MacInnis, A. J. (1976). The *in vitro* effects of farnesol and derivatives on *Hymenolepis diminuta. J. Parasit.* **62**, 749–755.

Fioravanti, C. F. and MacInnis, A. J. (1977). The identification and characterization of a prenoid constituent (farnesol) of *Hymenolepis diminuta* (Cestoda). *Comp. Biochem. Physiol.* **57B**, 227–233.

Fisher, F. M. and Starling, J. A. (1970). The metabolism of L-Valine by *Calliobothrium verticillatum* (Cestoda: Tetraphyllidea): identification of α-ketoisovaleric acid. *J. Parasit.* **56**, 103–107.

Frayha, G. J. (1968). A study on the synthesis and absorbtion of cholesterol in hydatid cysts (*Echinococcus granulosus*). *Comp. Biochem. Physiol.* **27**, 875–878.

Frayha, G. J. (1971). Comparative metabolism of acetate in the taeniid tapeworms *Echinococcus granulosus*, *E. multilocularis* and *Taenia hydatigena. Comp. Biochem. Physiol.* **39B**, 167–170.

Frayha, G. J. (1974). Synthesis of certain cholesterol precursors by hydatid protoscolices of *Echinococcus granulosus* and cysticerci of *Taenia hydatigena. Comp. Biochem. Physiol.* **49B**, 93–98.

Frayha, G. J. and Fairbairn, D. (1968). Lipid metabolism in helminth parasites VII. Absorbtion of cholestrol by *Hymenolepis diminuta* (Cestoda). *J. Parasit.* **54**, 1144–1146.

Frayha, G. J. and Fairbairn, D. (1969). Lipid metabolism in helminth parasites-VI. Synthesis of 2-cis, 6-trans farnesol by *Hymenolepis diminuta* (Cestoda). *Comp. Biochem. Physiol.* **28**, 1115–1124.

Frayha, G. J. and Haddad, R. (1980). Comparative chemical composition of protoscolices and hydatid cyst fluid of *Echinococcus granulosus* (Cestoda). *Int. J. Parasit.* **10**, 359–364.

Frayha, G. J., Bahr, G. M. and Haddad, R. (1980). The lipids and phospholipids of hydatid protoscolices of *Echinococcus granulosus* (Cestoda). *Int. J. Parasit.* **10**, 213–216.

Ginger, C. D. and Fairbairn, D. (1966*a*). Lipid metabolism in helminth parasites. I. The lipids of *Hymenolepis diminuta* (Cestoda). *J. Parasit.* **52**, 1086–1096.

Ginger, C. D. and Fairbairn, D. (1966*b*). Lipid metabolism in helminth parasites. II. The major origins of the lipids of *Hymenolepis diminuta* (Cestoda). *J. Parasit.* **52**, 1097–1107.

Harrington, G. W. (1965). The lipid content of *Hymenolepis diminuta* and *Hymenolepis citelli*. *Expl. Parasit.* **17**, 287–295.

Hedrick, R. M. (1958). Comparative histochemical studies on cestodes II. The distribution of fat substances in *Hymenolepis diminuta* and *Raillietina cesticillus*. *J. Parasit.* **44**, 75–84.

Jacobsen, N. S. and Fairbairn, D. (1967). Lipid metabolism in helminth parasites III. Biosynthesis and interconversion of fatty acids by *Hymenolepis diminuta* (Cestoda). *J. Parasit.* **53**, 355–361.

Kassis, A. I. and Frayha, G. J. (1973). Lipids of the cysticerci of *Taenia hydatigena* (Cestoda). *Comp. Biochem. Physiol.* **46B**, 435–443.

Kilejian, A., Sauer, K. and Schwabe, C. W. (1962). Host–parasite relationships in echinococcosis VIII. Infrared spectra and chemical composition of the hydatid cyst. *Expl. Parasit.* **12**, 377–392.

Kilejian, A., Ginger, C. D. and Fairbairn, D. (1968). Lipid metabolism in helminth parasites. IV. Origins of the intestinal lipids available for absorption by *Hymenolepis diminuta* (Cestoda). *J. Parasit.* **54**, 63–68.

Körting, W. and Barrett, J. (1977). Carbohydrate catabolism in the plerocercoids of *Schistocephalus solidus* (Cestoda:Pseudophyllidea). *Int. J. Parasit.* **7**, 411–417.

Körting, W. and Barrett, J. (1978). Studies on beta-oxidation in the plerocercoids of *Ligula intestinalis* (Cestoda:Pseudophyllidea). *Z. ParasitKde*, **57**, 243–246.

Lee, D. L., Rothman, A. H. and Senturia, J. B. (1963). Esterases in *Hymenolepis* and in *Hydatigera*. *Expl. Parasit.* **14**, 285–295.

Lesuk, A. and Anderson, R. J. (1941). Concerning the chemical composition of *Cysticercus fasciolaris*. II. The occurrence of a cerebroside containing dihydrosphingosine and of hydrolecithin in cysticercus larvae. *J. biol. Chem.* **139**, 457–469.

McMahon, P. (1961). Phospholipids of larval and adult *Taenia taeniaeformis*. *Expl. Parasit.* **11**, 156–160.

McManus, D. P. (1981). A biochemical study of adult and cystic stages of *Echinococcus granulosus* of human and animal origin from Kenya. *J. Helminth.* **55**, 21–27.

Mettrick, D. F. and Cannon, C. E. (1970). Changes in the chemical composition of *Hymenolepis diminuta* (Cestoda:Cyclophyllidea) during prepatent development within the rat intestine. *Parasitology*, **61**, 229–243.

Meyers, F., Kimura, S. and Mueller, J. F. (1966). Lipid metabolism in the larval and adult forms of the tapeworm *Spirometra mansonoides. J. biol. Chem.* **241**, 4224–4232.

Meyer, F., Meyer, H. and Bueding, E. (1970). Lipid metabolism in the parasitic and free-living flatworms, *Schistosoma mansoni* and *Dugesia dorotocephala. Biochim. biophys. Acta*, **210**, 257–266.

Meyer, H., Mueller, J. F. and Meyer, F. (1978). Isolation of an acylCoA carboxylase from the tapeworm *Spirometra mansonoides. Biochem. biophys. Res. Commun.* **82**, 834–839.

Meyer, H., Provasoli, L. and Meyer, F. (1979). Lipid biosynthesis in the marine flatworm *Convoluta roscoffensis* and its algal symbiont *Platymonas convoluta. Biochim. biophys. Acta*, **573**, 464–480.

Nei, M. (1972). Genetic distance between populations. *Am. Nat.* **106**, 283–292.

Nigam, S. C. (1979). Effect of host diet on fatty acid composition of *Cotugnia digonopora* (Pasquelle 1891) and *Raillietina fuhrmanni* (Southwell, 1922). *Ind. J. Parasit.* **3**, 67–69.

Nigam, S. C. and Premvati, G. (1980*a*). Presence of cholesterol in the neutral lipids of three sheep cestodes. *J. Helminth.* **54**, 215–218.

Nigam, S. C. and Premvati, G. (1980*b*). Unsaponifiable lipids of *Cotugnia digonopora* and *Raillietina fuhrmanni* (Cestoda:Cyclophyllidea). *Folia Parasit.* **27**, 59–61.

Overturf, M. and Dryer, R. L. (1968). Lipid metabolism in the adult cestode *Hymenolepis diminuta. Comp. Biochem. Physiol.* **27**, 145–175.

Roberts, L. S. (1961). The influence of population density on patterns and physiology of growth in *Hymenolepis diminuta* (Cestoda:Cyclophyllidea) in the definitive host. *Expl. Parasit.* **11**, 332–371.

Roberts, L. S. and Platzer, E. G. (1967). Developmental physiology of cestodes. II. Effects of changes in host dietary carbohydrate and roughage on previously established *Hymenolepis diminuta. J. Parasit.* **53**, 85–93.

Smirnov, L. P. and Sidorov, V. S. (1979). Fatty acid composition of the cestodes *Eubothrium crassum* and *Diphyllobothrium dendriticum. Parazitologiya*, **13**, 522–529.

Smyth, J. D. (1949). Studies on tapeworm physiology IV. Further observations on the development of *Ligula intestinalis in vitro. J. exp. Biol.* **26**, 1–14.

Smyth, J. D. (1969). "The Physiology of Cestodes", Oliver & Boyd, Edinburgh.

Surgan, M. H. and Roberts, L. S. (1976). Adsorbtion of bile salts by the cestodes *Hymenolepis diminuta* and *H. microstoma. J. Parasit.* **62**, 78–86.

Thompson, M. J., Mosettig, E. and von Brand, T. (1960). Unsaponifiable lipids of *Taenia taeniaeformis* and *Moniezia* sp. *Expl. Parasit.* **9**, 127–130.

Thorson, R. E., Digenis, G. A., Berntzen, A. and Konyalian, A. (1968). Biological activities of various lipid fractions from *Echinococcus granulosus* scolices on *in vitro* cultures of *Hymenolepis diminuta. J. Parasit.* **54**, 970–973.

Tofts, J. and Meerovitch, E. (1974). The effect of farnesyl methyl ether, a mimic of insect juvenile hormone, on *Hymenolepis diminuta in vitro. Int. J. Parasit.* **4**, 211–218.

Van den Thillart, G. and de Bruin, G. (1981). Influence of environmental temperature on mitochondrial membranes. *Biochim. biophys. Acta*, **640**, 439–447.

Vessal, M., Zekavat, S. Y. and Mohammadzadeh-K, A. A. (1972). Lipids of *Echinococcus granulosus* protoscolices. *Lipids*, **7**, 289–296.

Vysotzkaja, R. U. and Sidorov, V. S. (1973). On lipid contents of some helminths from freshwater fishes. *Parazitologiya*, **7**, 51–57.

Ward, C. W. and Fairbairn, D. (1970). Enzymes of beta-oxidation and the tricarboxylic acid cycle in adult *Hymenolepis diminuta* (Cestoda) and *Ascaris lumbricoides* (Nematoda). *J. Parasit.* **56**, 1009–1012.

Warren, M. and Daugherty, J. (1957). Host effects on the lipid fraction of *Hymenolepis diminuta*. *J. Parasit.* **43**, 521–526.

Webb, R. A. and Mettrick, D. F. (1971). Pattern of incorporation of ^{32}P into the phospholipids of the rat tapeworm *Hymenolepis diminuta*. *Can. J. Biochem.* **49**, 1209–1212.

Webb, R. A. and Mettrick, D. F. (1973). The role of serine in the lipid metabolism of the rat tapeworm *Hymenolepis diminuta*. *Int. J. Parasit.* **3**, 47–58.

Webb, R. A. and Mettrick, D. F. (1975). The role of glucose in the lipid metabolism of the rat tapeworm *Hymenolepis diminuta*. *Int. J. Parasit.* **5**, 107–112.

Chapter 11

Electron transport systems

K. S. Cheah

I. INTRODUCTION

The electron transport systems of helminths have not been investigated as intensively as those of mammalian, bacterial and plant tissues. The main reasons are probably due firstly, to the difficulty in getting sufficient material, secondly, to the lack of adequate facilities and thirdly, to the fact that biochemists in general tend to avoid working with parasites. The outcome is that only very few detailed studies have been reported, and to date only the electron transport systems of large and readily available cestodes have been investigated. In this review the major emphasis will be on mitochondrial structure and respiration, electron transport carriers, end-products of mitochondrial oxidation, peroxidase and cytochrome components.

BIOLOGY OF THE EUCESTODA Vol. 2
ISBN 0-12-062102-9

II. MITOCHONDRIA

A. Structure

The ultrastructural configuration of mitochondria in cestodes both *in situ* and *in vitro* has been investigated using positively and negatively stained thin section preparations. Mitochondria isolated from *Moniezia expansa* (Cheah, 1971a) and *Hymenolepis diminuta* (Harlow and Byram, 1971), fixed with glutaraldehyde and then stained with uranyl acetate and lead citrate, showed clearly defined outer and inner membranes, outer compartments and intracristal and matrix spaces. Unfixed preparations negatively stained with potassium phosphotungstate, showed the presence of elementary particles attached to the mitochondrial cristae of *M. expansa* (Cheah, 1971a) and *H. diminuta* (Harlow and Byram, 1971). The structural characteristics of cestode mitochondria are very similar to those of mammalian aerobic mitochondria. A more comprehensive study of the localization and configuration of mitochondria in different types of cells of *Lacistorhynchus tenuis* has also been carried out (Lumsden, 1967). Considerable variation in number, form and size of mitochondria was observed, with most of the mitochondria being localized in the tegument and a moderate number in the medullary parenchymal cells. Mitochondria of the tegument, tegumentary cytons, muscle and nerve cells contained only a few short cristae while the cristae of the medullary parenchymal cell mitochondria were more extensively developed with regard to the number and length of the cristae. Mitochondria were also relatively well developed in cell types associated with the reproductive tissues.

B. Oxygen consumption

All parasitic cestodes so far investigated are facultative aerobes, exhibiting varying degrees of tolerance to lack of oxygen, but consuming oxygen when available (von Brand, 1973). It is well established that oxygen is readily available in the intestinal environment (Rogers, 1949; Podesta and Mettrick, 1974). The lumenal bulk aqueous phase of rat is aerobic with an oxygen tension of 40–50 mm Hg and experimental evidence has demonstrated sufficient oxygen in the intestine of rat to support *H. diminuta* having an aerobic type of metabolism (Podesta and Mettrick, 1974). Oxygen consumption by cestodes is generally low in the absence of exogenous substrate. In the presence of glucose, the oxygen consumption of *H. diminuta* was stimulated about threefold (Read, 1956). *H. diminuta* can also be cultured to adulthood in a gas mixture of N_2 (97%) and CO_2 (3%) (Schiller, 1965), and oxygen, even up to 20%, did not interfere significantly with its development (Roberts and Mong, 1969). Some tapeworms are known to thrive best in atmospheric air, for example *H. microstoma* (de Rycke and Berntzen, 1967), larval *Taenia crassiceps* (Taylor, 1963) and *Echinococcus granulosus* (Smyth *et al.*, 1967). With *H. microstoma*, oxygen was required for the development of larval to adult stages (de Rycke and Berntzen, 1967). Oxygen consumption by cestodes, however, can be affected by variation in the oxygen tension. With *H. diminuta*, the rate of oxygen consumption in 5% oxygen was about two-thirds that in 21% oxygen, and this was further reduced in 2.5% oxygen (Read, 1956). A sharp decline in oxygen consumption was also observed with *E. granulosus* when the oxygen tension was less than 5% in the gas phase (Read and Simmons, 1963), but the scoleces of this cestode could carry out aerobic metabolism under an oxygen tension known to exist *in vivo* (Farhan *et al.*, 1959). Comparative studies of the rate of oxygen consumption between the larval and adult stages of *T. taeniaeformis* conducted under 21% oxygen showed that the adult consumed oxygen at a rate twice that of the larval form (von Brand and Bowman, 1961).

The observations that cestodes consume oxygen (Read, 1956; Farhan *et al.*, 1959; Read and Simmons, 1963; de Rycke and Berntzen, 1967; Smyth *et al.*, 1967; von Brand, 1973) and that oxygen is available *in vivo* (Rogers, 1949; Podesta and Mettrick, 1974) have directed research towards characterizing the properties of cestode mitochondria. Most of the work on oxygen consumption by mitochondria of cestodes has been carried out using large intestinal tapeworms because of the availability of sufficient material for mitochondrial isolation. The first detailed study on oxygen uptake by cestode mitochondria was on *M. expansa* (Cheah, 1967a) using manometric techniques. The oxidation of α-glycerophosphate by the mitochondrial fraction of *M. expansa* was three times faster than that of succinate, and five times greater than NADH (Cheah, 1967a). The oxidation of NADH by these mitochondrial preparations suggested that some of the isolated mitochondria were probably damaged, possibly due to the homogenization procedure employed. The isolation of cestode mitochondria was subsequently improved (Cheah, 1971a) to yield mitochondria from *M. expansa* that did not oxidize exogenous NADH (indicating that the isolated organelles were intact) and which exhibited clearly the classical State 3 to State 4 transition induced by addition of exogenous ADP during the oxidation of α-glycerophosphate and succinate. The coupled mitochondria of *M. expansa* could oxidize, in decreasing order of activity, α-glycerophosphate, succinate, pyruvate plus malate, L-glutamate and pyruvate. Ascorbate *plus* tetramethyl-p-phenylenediamine (TMPD) was oxidized at the same rate as succinate, and L-glutamate *plus* malate at the same rate as pyruvate *plus* malate. Similar substrates were also oxidized by mitochondria isolated from *H. diminuta*, but these mitochondria were unable to oxidize β-hydroxybutyrate, citrate, or isocitrate *plus* NADP even in the presence of malate (Rahman and Meisner, 1973). Palmitoyl-CoA was not oxidized by *H. diminuta* mitochondria even in the presence of carnitine (Rahman and Meisner, 1973). In contrast to the findings of Rahman and Meisner (1973), mitochondria isolated from the same cestode were demonstrated to oxidize isocitrate (Yorke and Turton, 1974). With mitochondria from *T. taeniaeformis*, only the oxidation of α-glycerophosphate and malate was observed (Weinbach and von Brand, 1970), and the oxidation of both of these substrates was stimulated by exogenous vitamin K_3. Mitochondria from *T. hydatigena* (Cheah, 1967b) and *Diphyllobothrium latum* (Salminen, 1974) could also oxidize succinate. Isolated mitochondria from cestodes so far investigated have one distinct feature in common: α-glycerophosphate was the most readily oxidized of all substrates tested (Cheah, 1967b, 1971a; Weinbach and von Brand, 1971; Rahman and Meisner, 1973; Yorke and Turton, 1974).

C. Oxidative phosphorylation

One of the main features of isolated mammalian mitochondria is their ability to exhibit oxidative phosphorylation and respiratory control which may readily be shown using an oxygen electrode. The first experiments with isolated cestode mitochondria were carried out with *T. taeniaeformis* (Weinbach and von Brand, 1970). Unfortunately, the isolated mitochondrial preparations were completely uncoupled, as demonstrated by the lack of response to ADP, 2,4-dinitrophenol or oligomycin, even though bovine serum albumin was added to remove fatty acids which are known to uncouple isolated mitochondria (Weinbach and Garbus, 1966; Vazquez-Colon *et al.*, 1966). Some portion of the oxidative phosphorylation process was, however, shown to remain intact since oligomycin, an inhibitor of mitochondrial ATPase (Slater, 1967), inhibited the ATPase activity of the mitochondrial preparation of *T. taeniaeformis*. Weinbach and von Brand (1970) concluded that the failure to demonstrate either respiratory control or oxidative phosphorylation in isolated mitochondria of *T. taeniaeformis* was a technical rather than a physiological problem. These technical difficulties were attributed to the calcium content of the cestode, and its ability to stimulate endogenous ATPase activity of the isolated mitochondria. Mitochondrial ATPase is a latent enzyme, and its activity can be increased through damage of the isolated mitochondrial membrane. The first convincing evidence that cestode mitochondria, like classical aerobic mammalian mitochondria, could carry out oxidative phosphorylation was obtained with *M. expansa* (Cheah, 1971a). The oxidation of α-

glycerophosphate, succinate and ascorbate *plus* TMPD by isolated mitochondria of *M. expansa* was stimulated by exogenous ADP when measured with an oxygen electrode. The ADP-stimulated oxidation of these substrates was inhibited by oligomycin, an inhibitor of oxidative phosphorylation, and the inhibited respiration was relieved by the uncoupler, carbonyl cyanide p-trifluoromethoxy-phenylhydrazone (FCCP), as also occurs in mammalian skeletal muscle mitochondria (Cheah, 1970, 1972). Mitochondria isolated from *M. expansa* oxidized both α-glycerophosphate and succinate with an ADP/O ratio of about 1.5, and respiratory control indices of 1.4 and 1.6 respectively, indicating that the oxidation of both of these substrates involved two coupling sites, as in mammalian mitochondria. Furthermore, the amount of ADP required to give half-maximal acceleration of succinate oxidation (K_m) for *M. expansa* was found to be similar to the succinoxidase system of pigeon cardiac and liver mitochondria. Atractyloside, a specific inhibitor of the adenine nucleotide translocator for the formation of mitochondrial ATP (Klingenberg *et al.*, 1970), inhibited 92% of the oxidative phosphorylation of *M. expansa* mitochondria at a concentration of 0.5 μg atractyloside per mg protein in the presence of 1.2 mM ADP. The existence of oxidative phosphorylation was also demonstrated in mitochondria of another cestode by two independent groups of workers. By means of the oxygen electrode, coupled respiration in *H. diminuta* was shown with α-glycerophosphate, succinate, pyruvate *plus* malate, glutamate *plus* malate and α-ketoglutarate *plus* malate (Rahman and Meisner, 1973), and also with isocitrate (York and Turton, 1974). The respiratory control index for the mitochondrial oxidation of succinate in *H. diminuta* (Rahman and Meisner, 1973) was similar to that of *M. expansa* mitochondria (Cheah, 1971a), but the value of the respiratory control index for α-glycerophosphate oxidation of *H. diminuta* was much lower than that of *M. expansa* (Cheah, 1971a). The value of the ADP/O ratio of 0.6 for succinate oxidation in *H. diminuta* was also very much lower than 1.5 observed for *M. expansa*. The lower value of the respiratory control index and the ADP/O ratio in *H. diminuta* mitochondria might be due to the loss of cytochrome *c* during the isolation of the mitochondria since the addition of exogenous cytochrome *c* improved the values of both the respiratory control index and the ADP/O ratio (Rahman and Meisner, 1973). The loss of cytochrome *c* could be accounted for by leakiness in the mitochondrial outer membrane (Cheah, 1971b) during the process of isolation of the mitochondria. Another interesting feature of *H. diminuta* mitochondria was the value for the ADP/O ratio, which was dependent on the amount of exogenous ADP used in the experiment. For the oxidation of pyruvate *plus* malate for example, the value of ADP/O ratio was increased from 0.08 to 1.7 following an increase of ADP from 7 to 290 μM; the maximal value calculated was 2.8 for the oxidation of this NAD-linked substrate (Rahman and Meisner, 1973). As with *M. expansa*, atractyloside completely inhibited the ADP-stimulated respiration in mitochondria of *H. diminuta* (Rahman and Meisner, 1973).

III. ELECTRON TRANSPORT CARRIERS

The electron carriers most intensively investigated in cestode mitochondria are the flavoproteins, non-haem iron and cytochromes. The first evidence that cytochromes might be present in cestodes was reported for *D. latum*. A well defined absorption band occurring between 520–530 nm was detected in this cestode with a hand spectroscope, but no absorption bands characteristic of reduced α-peaks of cytochrome components were observed (Friedheim and Baer, 1933). A dithionite-reduced component with an absorption maximum between 550 and 560 nm was reported for the whole homogenate preparation of *M. benedeni*, and this was suggested to be contributed by a mixture of *b*- and *c*-type cytochromes (van Grembergen, 1944). However, the dithionite-reduced pyridine haemochromogen showed an α-peak at 558 nm, implying the presence of a predominant *b*-type cytochrome. If a *c*-type cytochrome were present, it comprised only a minor component. In a homogenized preparation of *M. benedeni*, oxygen uptake was stimulated by *p*-phenylenediamine, and was inhibited by cyanide. These observations led van Grembergen (1944) to conclude that the electron transport carriers in *M. benedeni* resembled those of mammalian tissue. With *H. diminuta*, there was some indication that cytochrome *c* oxidase was present, since the ascorbate-reduced exogenous cytochrome *c* stimulated oxygen uptake in homogenates of this cestode (Read, 1952). A lapse of 14 years occurred before a detailed investigation on the oxidase systems of *M. expansa* was carried out using manometric and spectroscopic techniques (Cheah, 1967a, 1968). The spectroscopic techniques specially adapted to monitor absorption changes of cytochromes and flavoproteins in highly turbid suspensions at room and liquid-nitrogen temperatures were originally devised by Chance (1957), and they were widely adopted for detecting electron carriers in plants, bacteria and mammalian tissues. The oxidation of α-glycerophosphate, succinate and NADH was determined in the presence of various specific inhibitors, and the reduction of various cytochromes by these substrates was followed at room and liquid-nitrogen temperatures using a Cary (Model 14R) spectrophotometer specially modified to record spectral changes of turbid suspensions at both these temperatures. This detailed investigation led to the proposal of the electron transport system of *M. expansa* illustrated in Fig. 1 (Cheah, 1967a), a conclusion based on manometric and spectroscopic

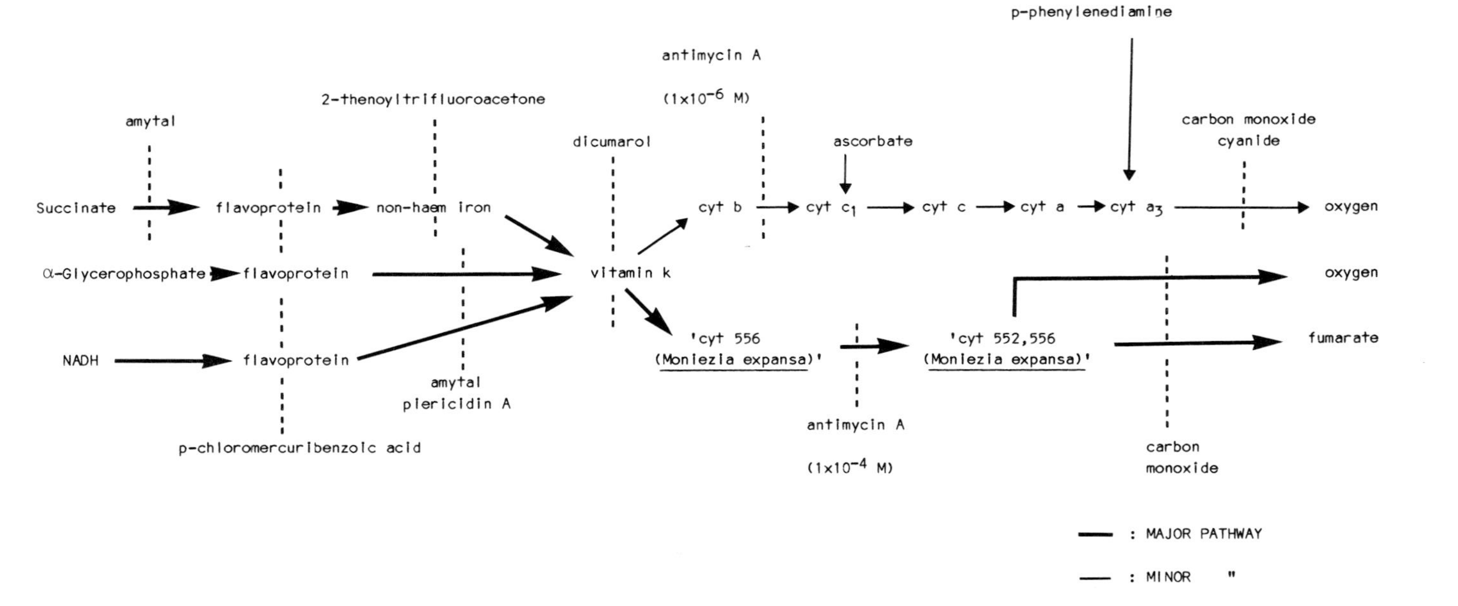

Fig. 1. Proposed electron transport system of *Moniezia expansa.*

studies using respiratory inhibitors and electron donors. The oxidation of α-glycerophosphate and NADH showed the same degree of sensitivity to amytal, but the oxidation of succinate was more resistant to this inhibitor in that 60% inhibition was observed with 20 mM amytal compared with 84% and 89% inhibition for α-glycerophosphate and NADH, respectively. The concentration of amytal required to produce a significant inhibition of these three separate systems was about ten times greater than that normally required in rat liver mitochondria oxidizing glutamate (Hochstein *et al.*, 1965), or for succinate and NADH oxidation in dark aerobically grown *Rhodospirillum rubrum* (Taniguchi and Kamen, 1965), or for α-glycerophosphate oxidation by *Trypanosomaa rhodesiense* (Grant and Sargent, 1960). The results obtained with *M. expansa*, however, compared well with *Staphylococcus aureus* in that elevated levels of amytal were required to produce a significant inhibition of oxidation of both succinate and NADH (Taber and Morrison, 1964). Of interest was the observation that amytal inhibited the oxidation of succinate in *M. expansa*, since it was generally accepted that amytal blocked electron transfer only to NAD-linked flavoprotein and also inhibited oxidation when the preparation was capable of phosphorylation (Chance and Hollunger, 1963).

That mitochondria of *M. expansa* could carry out oxidative phosphorylation was shown subsequently with an oxygen electrode (Cheah, 1971a). The involvement of non-haem iron in the oxidative pathway of succinate, and not of NADH, in *M. expansa* was demonstrated by inhibition with 2-thenoyltrifluoroacetone, a potent inhibitor of succinate oxidation through its ability to chelate non-haem iron (Singer *et al.*, 1956; Tappel, 1960). Unfortunately, the effect of this inhibitor on the oxidation of α-glycerophosphate by *M. expansa* mitochondria was not tested (Cheah, 1967a), but it is likely that non-haem iron is also involved in this pathway since it has been demonstrated in the mitochondria of *T. taeniaeformis*, *H. diminuta* and *H. microstoma* by electron paramagnetic resonance (EPR) measurements (Weinbach, 1972). The characteristic signal of non-haem iron at g = 1.94 was clearly observed when mitochondria of *T. taeniaeformis*, *H. diminuta* and *H. microstoma* were oxidizing α-glycerophosphate, and also when the non-haem iron was chemically reduced with dithionite. Once reduced with α-glycerophosphate, the non-haem iron could be oxidized by air resulting in the disappearance of the EPR signal, implying that the non-haem iron was involved in the oxidation of α-glycerophosphate in the cestode mitochondria. Furthermore, the characteristic EPR signal of reduced non-haem iron was prevented by treating mitochondria with 2-thenoyltrifluoroacetone prior to α-glycerophosphate.

p-Chloromercuribenzoic acid, a highly specific inhibitor for functional sulphydryl groups (Crane *et al.*, 1956; Tyler *et al.*, 1965) effectively blocked the oxidation of succinate, NADH and α-glycerophosphate in *M. expansa* mitochondria, but the oxidation of α-glycerophosphate was found to be less sensitive to this inhibitor than that of succinate and NADH (Cheah, 1967a). The participation of specific flavoproteins in the oxidation of these three separate substrates was also shown spectroscopically (Cheah, 1967a), and by the different stimulating effects on oxygen consumption produced by methylene blue and phenazine methosulphate in the presence and absence of *p*-chloromercuribenzoic acid. Methylene blue did not stimulate oxygen uptake in the presence of NADH, but stimulation was observed with both α-glycerophosphate and succinate (Cheah, 1967c). The α-glycerophosphate pathway could be differentiated from the succinate pathway by the pronounced differences in sensitivity to 0.2 mM *p*-chloromercuribenzoic acid, which inhibited only 39% of the oxidation of α-glycerophosphate but 89% of succinate oxidation (Cheah, 1967a). Vitamin K was suggested as a probable component linking the flavoproteins to the cytochrome components in the branched electron transport system of *M. expansa* (see Fig. 1). This conclusion was based on the inhibition obtained with dicoumarol (Cheah, 1967a), a vitamin-K antagonist (Collentine and Quick, 1951; Griminger and Donis, 1960) commonly employed as an inhibitor of vitamin K-linked enzymes in the electron transport systems of tissues from various sources (Brodie and Gray, 1956; Brodie and Ballantine, 1960; Benziman and Pérez, 1965).

The classical inhibitors of the respiratory chain system of mammalian mitochondria, antimycin A and cyanide, were also employed to demonstrate their inhibitory effect on the oxidation of various substrates in a *M. expansa* mitochondrial particulate fraction (Cheah, 1967a). The effect of both antimycin A and cyanide, in addition to piericidin A, on the oxidation of NAD^+-linked substrate, pyruvate *plus* malate, were further investigated with coupled mitochondria of *M. expansa* using an oxygen electrode (Cheah, 1971a) instead of a Warburg apparatus (Cheah, 1967a). Complete inhibition of the ADP-stimulated respiration of pyruvate *plus* malate oxidation by *M. expansa* mitochondria was observed with 50 pmol piericidin A per mg protein, an amount similar to that required for inhibition in mammalian skeletal muscle mitochondria (Cheah, 1970). Incomplete inhibition of the succinoxidase system by antimycin A, and incomplete inhibition of the succinoxidase, α-glycerophosphate oxidase and cytochrome oxidase activities by cyanide in the presence of ADP was observed with coupled *M. expansa* mitochondria. These data confirmed previous findings of the existence of an antimycin A and cyanide-insensitive, and an antimycin A and cyanide-sensitive pathway in the mitochondria of *M. expansa* (Cheah, 1967a). The participation of cytochromes in the oxidation of succinate, α-glycerophosphate and NADH was detected and resolved mainly by recording difference spectra at liquid nitrogen temperature. Under these conditions, the reduced α-peaks of the *a*, *b* and *c*-type cytochromes were clearly defined and further differentiated by the use of specific inhibitors in conjunction with appropriate electron donors (see Fig. 1).

These studies established that adult *M. expansa* mitochondria possessed all the cytochromes of mammalian mitochondria in addition to another *b*-type cytochrome, sensitive only to high concentration of antimycin A, and another terminal oxidase, cytochrome *o* (Cheah, 1967a, 1968). The additional cytochrome *b*, designated "Cytochrome 556 (*Moniezia expansa*)" (77°K), and cytochrome *o*, designated "Cytochrome 552, 556 (*Moniezia expansa*)" (77°K), following the recommendations of the commission on Enzyme Nomenclature (see Florkin and Stotz, 1965), constituted the major pathway for the electron transport system of *M. expansa* mitochondria, and this pathway was responsible for the observed antimycin A and cyanide-insensitivity in the cestode mitochondria. The two different *b*-type cytochromes were differentiated spectroscopically by using different concentrations of antimycin A. As in mammalian mitochondria, the *b*-type cytochrome was detected in the presence of 1.0 μM antimycin A (Cheah, 1968), and the new *b*-type cytochrome in the major pathway by 0.1 mM antimycin A (Cheah, 1967a). The mammalian *b*-type cytochrome was also observed following reduction with either α-glycerophosphate, succinate or NADH (Cheah, 1967a, 1968). Cytochrome c_1 and cytochrome *c* in *M. expansa* mitochrondria were observed spectroscopically after reduction with appropriate substrates. Thus, cytochrome c_1 and cytochrome *c*, in addition to cytochrome *a* and cytochrome a_3, were clearly resolved at liquid-nitrogen temperature following reduction with ascorbate in the presence of 1.0 mM cyanide (Cheah, 1967a), and also by reduction with succinate in the presence of 1.0 mM cyanide (Cheah, 1968). The two terminal oxidases were characterized by the formation of CO-complexes at room temperature using either dithionite as a chemical reducing agent (Cheah, 1967a) or α-glycerophosphate as a substrate reducing agent (Cheah, 1968). The overlapping absorption band of the predominant cytochrome *o*-CO complex on that of cytochrome a_3-CO complex in the Soret region was clearly resolved by recording spectra at various time intervals at room temperature. As in the bacterial system (Broberg and Smith, 1967b), CO bound to cytochrome a_3 more readily than to cytochrome *o* in *M. expansa* mitochondria (Cheah, 1968). A branched electron transport chain system with two terminal oxidases, cytochrome *o* and cytochrome a_3, was also detected in mitochondria from adult *T. hydatigena* (Cheah, 1967b). The oxidation of succinate by *T. hydatigena* mitochondria involved the same type of cytochromes as those observed for mitochondria from adult *M. expansa* in which the major pathway involves a *b*-type cytochrome sensitive only to high concentration of antimycin A, and linking to cytochrome *o* as the major terminal oxidase. As with *M. expansa* mitochondria, all the substrate-reducible cytochromes in the branched electron transport chain of *T. hydatigena* were resolved at liquid-nitrogen temperature, and the terminal oxidases by the formation of CO-complexes.

In addition to *M. expansa* and *T. hydatigena*, branched electron transport systems might also exist in mitochondria of *T. taeniaeformis* (Weinbach and von Brand, 1970). With mitochondria from this cestode, the oxidation of α-glycerophosphate by larvae and adults showed both antimycin and cyanide-sensitive and insensitive pathways, both of which were inhibited about 80% with 0.05 mM dicumarol. Thus, it appeared that vitamin K might also be involved, as in *M. expansa* mitochondria (Cheah, 1967a). Mitochondria of *T. taeniaeformis* contained amounts of pyridine nucleotide similar to those found in rat liver mitochondria, and like *M. expansa* mitochondria, these catalysed the oxidation of NAD^+-linked substrates as shown by inhibitor studies using an oxygen electrode (Weinbach and von Brand, 1970). Preliminary spectroscopic studies of mitochondria of *T. taeniaeformis* suggested that no cytochrome *c* or c_1 and little or no cytochrome a/a_3 were present (Weinbach and von Brand, 1979). An unknown CO-binding pigment, probably acting as a terminal oxidase, and a modified cytochrome *b* were detected, but no details were given as to whether these cytochrome components in the mitochondria of *T. taeniaeformis* were substrate-reducible. The electron transport system of *H. diminuta* was initially investigated with a Warburg apparatus using whole-worm homogenates (Read, 1952). Succinate dehydrogenase activity was clearly shown to be present using methylene blue as the electron acceptor, and the enzyme was inhibited by malonate, a specific inhibitor of succinic dehydrogenase. Cytochrome oxidase activity was monitored using ascorbate-reduced cytochrome *c* as a substrate, and the enzyme activity was observed only if the reducing agent ascorbate was present. Experimental evidence from whole-muscle homogenate preparations suggested that cytochrome oxidase was not a limiting factor for the transfer of electrons from succinic dehydrogenase to oxygen (Read, 1952).

A lapse of twenty-one years occurred before the electron transport system of this cestode was re-investigated using coupled mitochondria. By means of an oxygen electrode in the presence of suitable electron donors, antimycin A-sensitive and insensitive pathways were shown to be present in *H. diminuta* mitochondria (Rahman and Meisner, 1973). Pyridine nucleotides were involved in the pathway for the oxidation of NAD^+-linked substrate since 2.8 mM rotenone inhibited the oxidation of pyruvate *plus* malate, of L-glutamate *plus* malate and of NADH. The oxidation of α-glycerophosphate by *H. diminuta* mitochondria was shown to be more sensitive to cyanide than the mitochondria of *T. taeniaeformis* (Weinbach and von Brand, 1970) in that 84% inhibition was observed with 0.2 mM cyanide for *H. diminuta* compared with 41% inhibition for *T. taeniaeformis* (Weinbach and von Brand, 1970) by 1.0 mM cyanide. The α-glycerophosphate oxidase system of coupled mitochondria of *M. expansa* showed about the same sensitivity to cyanide (Cheah, 1971a) as *H. diminuta* mitochondria in that 75% inhibition of α-glycerophosphate oxidation was observed with 0.1 mM cyanide. An active cytochrome oxidase activity, initially observed with a whole homogenate preparation (Read, 1952), was also detected polarographically in *H. diminuta* mitochondria using the electron donors ascorbate *plus* TMPD (Rahman and Meisner, 1973), which donate electrons to cytochrome a_3 via *c*-type cytochrome in mammalian mitochondria. The cytochrome oxidase activity was inhibited 82% with 0.2 mM cyanide, a situation similar to that observed for coupled *M. expansa* mitochondria (Cheah, 1971a).

Unlike mitochondria isolated from *M. expansa* (Cheah, 1971a) and *H. diminuta* (Rahman and Meisner, 1973), no cytochrome oxidase activity was detected polarographically with ascorbate *plus* TMPD in mitochondria from *T. taeniaeformis* (Weinbach and von Brand, 1970). Succinic dehydrogenase and cytochrome oxidase activities in plerocercoid and adult *D. latum* were investigated spectroscopically using a mitochondrial particulate fraction (Salminen, 1974). No significant difference in the kinetic constants of plerocercoid and adult succinate dehydrogenase activities was observed using phenazine methosulphate as an electron acceptor. Succinate cytochrome *c* oxidoreductase activity was present only in adult *D. latum*, and the specific activity of cytochrome oxidase and its K_m, measured by the reoxidation of ascorbate-reduced cytochrome *c*, in adults was much higher than in the plerocercoid. The lack of succinate cytochrome oxidoreductase and the lower specific activities of succinate CoQ oxidoreductase and cytochrome oxidase suggested the aerobic electron transport pathway to be of minor importance in the plerocercoid. The high cytochrome oxidase activity in adult *D. latum* could account for the observed rate of succinate oxidation, and this led to the suggestion that the electron transport chain of this cestode resembled that of the mammalian electron transport system, although an alternative pathway capable of operating anaerobically was not totally excluded (Salminen, 1974).

Oxygen and fumarate were postulated to be the terminal acceptors for the branched electron transport chain system in *M. expansa* (Cheah, 1967a,b). Oxygen was the sole terminal hydrogen acceptor for the minor pathway involving all the mammalian-type cytochromes with cytochrome a_3 as its terminal oxidase. The major pathway, consisting of the new *b*-type cytochrome, sensitive only to high concentrations of antimycin A, with an *o*-type cytochrome as its terminal oxidase could interact directly either with oxygen or fumarate (see Fig. 1). In the absence of fumarate, and under low or high oxygen tension, water and hydrogen peroxide were the end-products of substrate oxidation (Cheah, 1967a). In the presence of fumarate, under the same conditions, conversion to succinate occurred (Cheah and Bryant, 1966; Cheah, 1967a). The interaction of fumarate with the *o*-type cytochrome was demonstrated either spectroscopically by following the re-oxidation of the reduced cytochrome by fumarate (Cheah, 1967c) or by following the formation of radioactive succinate from radioactive fumarate (Cheah and Bryant, 1966; Cheah, 1967c). The reaction between the *o*-type cytochrome and fumarate was found to be highly specific since neither maleate nor crotonate could reoxidize the reduced cytochrome (Cheah, 1967c). Both maleate, a cis-isomer of fumarate, and crotonate are C_4-acids with an unsaturated bond in the same position as fumarate. The same terminal acceptors could be operating in the mitochondria of *T. hydatigena* (Cheah, 1967b), since the electron transport chain of this cestode is similar to that of *M. expansa*. Unfortunately, the direct reoxidation of reduced cytochrome *o* by fumarate and the formation of succinate from fumarate has not been investigated with *T. hydatigena*. With the exception of *M. expansa* and *T. hydatigena*, no other published reports are available concerning the involvement of an *o*-type cytochrome in the reduction of fumarate to succinate. It is highly probable that the mechanism for the reduction of fumarate to succinate in mitochondria of *M. expansa* is a common feature in all cestodes living in environments similar to *M. expansa*, particularly those cestodes possessing an electron transport system known to be both sensitive and insensitive to antimycin A and cyanide, and having succinate as one of the major end-products of carbohydrate metabolism.

IV. HYDROGEN PEROXIDE FORMATION

Several intestinal parasites have been reported to form hydrogen peroxide during the oxidation of various substrates. With a mitochondria-containing fraction from *M. expansa*, the oxidation of succinate and NADH (Cheah and Bryant, 1966) and α-glycerophosphate (Cheah, 1967a) resulted in hydrogen peroxide formation, detected spectroscopically by the formation of tetraguaiacol from guaiacol. Using α-glycerophosphate as a typical substrate, hydrogen peroxide production was inhibited by 40 mM amytal, which also completely inhibited the oxidation of this substrate by the mitochondrial fraction of *M. expansa* (Cheah, 1967a). Hydrogen peroxide formation was also inhibited by fumarate and the 12 000 g supernatant fraction, but not by concentrations of antimycin A up to 0.1 mM, a concentration which inhibited α-glycerophosphate oxidation by

M. expansa by only 62% (Cheah, 1967a). With *M. expansa*, the production of hydrogen peroxide was shown to be closely linked with the *o*-type terminal oxidase, and was an end-product of substrate oxidation by the electron transport chain system (Cheah, 1967a). Mitochondria of *T. hydatigena* (Cheah, 1967b) would also be expected to produce hydrogen peroxide since this cestode also possesses a branched electron transport system with an *o*-type cytochrome as its major terminal oxidase. Unfortunately, hydrogen peroxide formation from substrate oxidation by mitochondria from *T. hydatigena* was not analysed spectroscopically as in *M. expansa*. Attempts to detect hydrogen peroxide formation in mitochondria of *T. taeniaeformis* as an end-product of α-glycerophosphate oxidation were unsuccessful (Weinbach and von Brand, 1970), and no attempts were made to detect hydrogen peroxide in the mitochondrial oxidation of *H. diminuta* (Rahman and Meisner, 1973; Yorke and Turton, 1974) and *D. latum* (Salminen, 1974).

V. PEROXIDASE

The 12 000 g supernatant fraction from *M. expansa*, which prevented hydrogen peroxide formation during substrate oxidation (Cheah, 1967a) contained a haemoprotein identified spectroscopically and histochemically to be peroxidase (Cheah, 1967d). The peroxidase was shown histochemically, using benzidine, to be localised just below the tegument in immature and mature proglottides of *M. expansa* (Cheah, 1967c), with the mature proglottis containing at least three times more enzyme than in the immature proglottis. Peroxidase was also reported in *H. diminuta* (Threadgold *et al.*, 1968) and *H. citelli* (Rothman, 1968). The enzyme was demonstrated to be present in the mitochondria of the surface syncytium of the tegument, the tegumentary cells and the parenchyma, but appeared to be absent in the muscle mitochondria of *H. diminuta* (Threadgold *et al.*, 1968). Similar observations were also reported for *H. citelli* (Rothman, 1968). The presence of peroxidase in the mitochondria of *H. diminuta* and *H. citelli* indicated that the enzyme was involved in respiration by destroying the toxic hydrogen peroxide formed during substrate oxidation as suggested previously for *T. hydatigena* (Cheah, 1967b) and *M. expansa* (Cheah, 1967a). Furthermore, peroxidase activity was localized only just below the tegument (Cheah, 1967c) where the cestode mitochondria were located (Rothman, 1968).

IV. CYTOCHROMES

M. expansa is the only cestode with its cytochromes characterized at various stages of growth. The types of cytochromes in the unembryonated eggs, scoleces and immature and mature proglottides were determined in whole homogenate preparations at liquid-nitrogen temperature (Cheah, 1967c) using the same technique as for isolated cestode mitochondria (Cheah, 1967a,b, 1968); similar types of cytochromes were observed at all stages of growth when reduced with dithionite (Cheah, 1967c). All preparations of *M. expansa* showed similar reduced absorption bands contributed by *a*, *b* and *c*-type cytochromes as in mitochondria isolated from the adult cestode (Cheah, 1967a,b). Cytochromes of *M. expansa* could be extracted with 0.1% sodium deoxycholate. Spectroscopic analyses of the deoxycholate extract and residue (Cheah, 1967c) accounted for the different types of cytochromes observed in the mitochondrial preparation (Cheah, 1967b, 1968). The *o*-type cytochrome linked to fumarate reduction could also be extracted with 0.1% deoxycholate (Cheah, 1967c), but this haemoprotein was not subjected to further purification. Enzymatically active cytochrome c_{550} was extracted from *M. expansa* with 0.3% aluminium sulphate, purified by chromatography with carboxymethyl (CM-52) cellulose and Sephadex G-25, and its properties investigated (Cheah, 1975). *Moniezia* cytochrome *c* was found to be a basic haemoprotein of molecular weight 12 000 with an isoelectric point of about pH 9.4. At room temperature, the cytochrome showed absorption bands at 550, 521 and 415 nm in the reduced form, and at 560, 526 and 410 nm in the oxidized form. At liquid nitrogen temperature, the reduced form showed only two α-bands at 547 and 538 nm, and these spectral characteristics were different from mammalian cytochrome c_{550} which possesses three reduced α-bands (Estabrook, 1966). The spectra of reduced purified horse heart cytochrome c_{550}, recorded under the same conditions as *Moniezia* cytochrome c_{550}, showed α-bands at 548, 546 and 538 nm (Cheah, 1975). It would appear that the absorption band of either cytochrome $c_{\alpha 1}$ (548 nm) or $c_{\alpha 2}$ 546 nm) observed in horse heart cytochrome c_{550} might be missing in *Moniezia* cytochrome c_{550}, or that they were replaced by a single absorption band at 547 nm. Purified *Moniezia* cytochrome c_{550} could also reconstitute the cytochrome oxidase activity of cytochrome *c*-deficient mammalian skeletal muscle mitochondria, and its restorative ability was 90% that of purified horse heart cytochrome c_{550} (Cheah, 1975).

VII. CONCLUSION

Mitochondria of cestodes have similar structural characteristics to mammalian skeletal muscle mitochondria, and are capable of carrying out oxidative phosphorylation. In contrast to mammalian mitochondria, the electron transport system of cestode mitochondria so far studied consists both of antimycin and cyanide-sensitive and insensitive pathways. Detailed spectroscopic studies from mitochondria of *M. expansa* and *T. hydatigena* suggest that the electron transport system has two terminal oxidases. The major pathway consists of a *b*-type cytochrome which is sensitive only to high concentrations of antimycin A and linked directly to an *o*-type cytochrome. The minor pathway resembles the classical mammalian electron transport system with cytochrome a_3 as its terminal oxidase. The branched-chain electron transport system of *M. expansa* has two terminal oxidases, with oxygen and fumarate as terminal acceptors, and the presence of peroxidase presents a remarkable example of adaptation of cestodes living in an environment where the oxygen tension can be low and variable but not necessarily strictly zero. In the presence of O_2 (high or low tension), but in the absence of fumarate, water and hydrogen peroxide are the end-products of mitochondrial metabolism. Because it is toxic, hydrogen peroxide is probably destroyed by the peroxidase. Under aerobic conditions, but in the presence of fumarate, no hydrogen peroxide is formed. In a strictly anaerobic environment, the minor terminal oxidase, cytochrome a_3, does not function. Electrons derived from the oxidation of various substrates are then transferred via a fumarate reductase to fumarate resulting in the formation of succinate. This is the most likely explanation for succinate production and its secretion by *M. expansa* and other helminths under anaerobic conditions (von Brand, 1933, 1966). Branched-chain electron transport systems also appear to be a common feature in other helminths (Kikuchi *et al.*, 1959; Cheah and Chance, 1970; Bryant, 1970; Hayashi and Terada, 1973; Cheah and Pritchard, 1975; Cheah, 1976; Kohler and Bachmann, 1980), and this is probably linked directly to their occupation of similar habitats which have a low oxygen tension.

VIII. REFERENCES

Benziman, M. and Pérez, L. (1965). The participation of vitamin K in malate oxidation by *Acetobacter xylinum*. *Biochem. biophys. Res. Commun.* **19**, 127–132.

Broberg, P. L. and Smith, L. (1967). The cytochrome system of *Bacillus megaterium KM*. The presence and properties of two CO-binding cytochromes. *Biochim. biophys. Acta*, **131**, 479–489.

Brodie, A. F. and Ballantine, J. (1960). Oxidative phosphorylation in fractionated bacterial systems. III. Specificity of vitamin K reactivation. *J. biol. Chem.* **235**, 232–237.

Brodie, A. F. and Gray, C. T. (1956). Phosphorylation coupled to oxidation in bacterial extracts. *J. biol. Chem.* **219**, 853–862.

Bryant, C. (1970). Electron transport in parasitic helminths and protozoa. *In* "Advances in Parasitology", (B. Dawes, ed.), Vol. 8, pp. 139–172, Academic Press, New York and London.

Chance, B. (1957). Techniques for the assay of the respiratory enzymes. *In* "Methods in Enzymology", (S. P. Colowick and N. O. Kaplan, eds), Vol. 4, pp. 273–329, Academic Press, New York and London.

Chance, B. and Hollunger, G. (1963). Inhibition of electron and energy transfer in mitochondria. I. Effects of amytal, rotenone, progesterone and methylene glycol. *J. biol. Chem.* **238**, 418–431.

Cheah, K. S. (1967a). The oxidase systems of *Moniezia expansa* (Cestoda). *Comp. Biochem. Physiol.* **23**, 277–302.

Cheah, K. S. (1967b). Spectrophotometric studies on the succinate oxidase system of *Taenia hydatigena. Comp. Biochem. Physiol.* **20**, 867–875.

Cheah, K. S. (1967c). "Studies on the Oxidative Metabolism of *Moniezia expansa* (Cestoda)", Ph.D. thesis, Australian National University, Canberra, Australia.

Cheah, K. S. (1967d). Histochemical and spectrophotometric demonstration of peroxidase in *Moniezia expansa* (Cestoda). *Comp. Biochem. Physiol.* **21**, 351–356.

Cheah, K. S. (1968). The respiratory components of *Moniezia expansa* (Cestoda). *Biochim. biophys. Acta*, **153**, 718–720.

Cheah, K. S. (1970). Evidence for the existence of an α-glycerophosphate oxidase system with three phosphorylation sites and sensitive to rotenone and piericidin A. *FEBS Lett.* **10**, 109–112.

Cheah, K. S. (1971a). Oxidative phosphorylation in *Moniezia* muscle mitochondria. *Biochim. biophys. Acta*, **253**, 1–11.

Cheah, K. S. (1971b). Effect of loss of cytochrome *c* following storage of mitochondria *in situ. FEBS Lett.* **19**, 105–108.

Cheah, K. S. (1972). Fuscin, an inhibitor of respiration and oxidative phosphorylation in ox-neck muscle mitochondria. *Biochim. biophys. Acta*, **275**, 1–9.

Cheah, K. S. (1975). Purification and properties of *Moniezia* cytochrome *c*-550. *Comp. Biochem. Physiol.* **51B**, 41–45.

Cheah, K. S. (1976). Electron transport system of *Ascaris*-muscle mitochondria. *In* "Biochemistry of Parasites and Host-Parasite Relationships", (H. Van den Bossche, ed.), pp. 133–143, North-Holland Publishing Company.

Cheah, K. S. and Bryant, C. (1966). Studies on the electron transport system of *Moniezia expansa* (Cestoda). *Comp. Biochem. Physiol.* **19**, 197–223.

Cheah, K. S. and Chance, B. (1970). The oxidase systems of *Ascaris*-muscle mitochondria. *Biochim. biophys. Acta*, **223**, 55–60.

Cheah, K. S. and Pritchard, R. K. (1975). The electron transport system of *Fasciola hepatica* mitochondria. *Int. J. Parasit.* **5**, 183–186.

Collentine, G. E. and Quick, A. J. (1951). The interrelationship of vitamin K and dicumarin. *Amer. J. med. Sci.* **222**, 7–12.

Crane, F. L., Glenn, J. L. and Greene, D. E. (1956). Studies on the electron transfer particle. IV. The electron transfer particle. *Biochim. biophys. Acta*, **22**, 475–487.

de Rycke, P. H. and Berntzen, A. K. (1967). Maintenance and growth of *Hymenolepis microstoma* (Cestoda: Cyclophyllidea) *in vitro*. *J. Parasit.* **53**, 352–354.

Estabrook, R. W. (1966). Enzymatic and spectral properties of various types of cytochrome *c*. *In* "Heme and Hemoproteins", (B. Chance, R. W. Estabrook and T. Yonetani, eds), pp. 405–409, Academic Press, New York and London.

Farhan, I., Schwabe, C. W. and Zobel, C. R. (1959). Host–parasite relationships in echinococcosis. III. Relations of environmental oxygen tension to the metabolism of hydatid scolices. *Am. J. trop. Med. Hyg.* **8**, 473–478.

Florkin, M. and Stotz, E. H. (1965). The classification and nomenclature of cytochromes. *In* "Comprehensive Biochemistry", Vol. 13, pp. 21–25, Elsevier, Amsterdam.

Friedheim, E. A. H. and Baer, J. C. (1933). Untersuchungen über die Atmung von *Diphyllobothrium latum* (L). Ein Beitrag zur Kenntnis der Atmungsfermente. *Biochem. Z.* **265**, 329–337.

Grant, P. T. and Sargent, J. R. (1960). Properties of L-glycerophosphate oxidase and its role in the respiration of *Tryanosoma rhodesiense*. *Biochem. J.* **76**, 229–237.

Griminger, P. and Donis, O. (1960). Potency of vitamin K_1 and two analogues in counteracting the effects of dicumarol and sulfaquinoxaline in the chick. *J. Nutr.* **70**, 361–368.

Harlow, D. R. and Byram, J. E. (1971). Isolation and morphology of the mitochondrion of the cestode, *Hymenolepis diminuta*. *J. Parasit.* **57**, 559–565.

Hayashi, E. and Terada, M. (1973). The influence of oxygen pressure on the survival time of *Ascaris lumbricoides suum*. (4). On the electron transport system containing cytochrome *c* peroxidase in *Ascaris muscle*. *Jap. J. Parasit.* **22**, 1–12.

Hochstein, P., Laszlo, J. and Miller, D. (1965). A unique, dicumarol-sensitive, non-phosphorylating oxidation of DPNH and TPNH catalyzed by streptonigrin. *Biochem. biophys. Res. Commun.* **19**, 289–295.

Kikuchi, G., Ramirez, J. and Gusman Barron, E. S. (1959). Electron transport system in *Ascaris lumbricoides*. *Biochim. biophys. Acta*, **36**, 335–342.

Klingenberg, M., Grebe, K. and Heldt, H. W. (1970). On the inhibition of the adenine nucleotide translocation by bongkrekic acid. *Biochem. biophys. Res. Commun.* **39**, 344–351.

Kohler, P. and Bachmann, R. (1980). Mechanisms of respiration and phosphorylation in *Ascaris* muscle mitochondria. *Mol. Biochem. Parasit.* **1**, 75–90.

Lumsden, R. D. (1967). Ultrastructure of mitochondria in a cestode, *Lacistorhynchus tenius* (V. Beneden, 1858) *J. Parasit.* **53**, 65–77.

Podesta, R. B. and Mettrick, D. F. (1974). Pathophysiology of cestode infections: Effect of *Hymenolepis diminuta* on oxygen tensions, pH and gastrointestinal function. *Int. J. Parasit.* **4**, 277–292.

Rahman, R. and Meisner, H. (1973). Respiratory studies with mitochondria from the rat tapeworm, *Hymenolepis diminuta*. *Int. J. Biochem.* **4**, 153–162.

Read, C. P. (1952). Contributions to cestode enzymology. I. The cytochrome system and succinic dehydrogenase in *Hymenolepis diminuta*. *Expl. Parasit.* **1**, 353–362.

Read, C. P. (1956). Carbohydrate metabolism of *Hymenolepis diminuta*. *Expl. Parasit.* **5**, 325–344.

Read, C. P. and Simmons, J. E., Jr (1963). Biochemistry and physiology of tapeworms. *Physiol. Rev.* **43**, 263–305.

Roberts, L. S. and Mong, F. N. (1969). Developmental physiology of cestodes. IV. *In vitro* development of *Hymenolepis diminuta* in the presence and absence of oxygen. *Expl. Parasit.* **26**, 166–174.

Rogers, W. P. (1949). On the relative importance of aerobic metabolism in small nematode parasites of the alimentary tract. I. Oxygen tensions in the normal environment of the parasites. *Aust. J. Scient. Res. (Ser. B)*, **2**, 157–165.

Rothman, A. H. (1968). Peroxidase in platyhelminth cuticular mitochondria. *Expl. Parasit.* **23**, 51–55.

Salminen, K. (1974). Succinate dehydrogenase and cytochrome oxidase in adult and plerocercoid *Diphyllobothrium latum. Comp. Biochem. Physiol.* **47B**, 87–92.

Schiller, E. L. (1965). A simplified method for the *in vitro* cultivation of the rat tapeworm, *Hymenolepis diminuta. J. Parasit.* **51**, 516–518.

Singer, T. P., Kearney, E. B. and Bernath, P. (1956). Studies on succinic dehydrogenase. II. Isolation and properties of the dehydrogenase from beef heart. *J. biol. Chem.* **233**, 599–613.

Slater, E. C. (1967). Applications of inhibitors and uncouplers for a study of oxidative phosphorylation. *In* "Methods in Enzymology: Oxidation and Phosphorylation", (R. W. Estabrook and M. E. Pullman, eds), Vol. II, pp. 207–225, Academic Press, London and New York.

Smyth, J. D., Miller, H. J. and Howkins, A. B. (1967). Further analysis of the factors controlling strobilization, differentiation and maturation of *Echinococcus granulosus in vitro. Expl. Parasit.* **21**, 31–41.

Taber, H. W. and Morrison, M. (1964). Electron transport in Staphylococci. Properties of a particle preparation from exponential phase of *Staphylococcus aureus. Archs. Biochem. Biophys.* **105**, 367–379.

Taniguchi, S. and Kamen, M. D. (1965). The oxidase system of heterotropically-grown *Rhodospirillum rubrum. Biochim. biophys. Acta*, **96**, 395–428.

Tappel, A. L. (1960). Inhibition of electron transport by antimycin A, alkyl hydroxy napthoquinones and metal coordination compounds. *Biochem. Pharmac.* **3**, 289–296.

Taylor, A. E. R. (1963). Maintenance of larval *Taenia crassiceps* (Cestoda: Cyclophyllidea) in a chemically defined medium. *Expl. Parasit.* **14**, 304–310.

Threadgold, L. T., Arme, C. and Read, C. P. (1968). Ultrastructural localization of a peroxidase in the tapeworm, *Hymenolepis diminuta. J. Parasit.* **54**, 802–807.

Tyler, D. D., Butow, R. A., Gonze, J. and Estabrook, R. W. (1965). Evidence for the existence and function of an occult, reactive sulphydryl group in the respiratory chain DPNH dehydrogenase. *Biochem. biophys. Res. Commun.* **19**, 551–557.

Van Grembergen, G. (1944). Le métabolisme respiratoire du cestode, *Moniezia benedeni* (Moniez, 1879). *Enzymologia*, **11**, 268–281.

Vázquez-Colón, L., Ziegler, F. D. and Elliot, W. B. (1966). On the mechanism of fatty acid inhibition of mitochondrial metabolism. *Biochemistry*, **5**, 1134–1139.

von Brand, T. (1933). Untersuchungen über den Stoffbestand einiger Cestoden und den Stoffwechsel von *Moniezia expansa. Z. vergl. Physiol.* **18**, 562–596.

von Brand, T. (1966). "Biochemistry of Parasites", Academic Press, New York and London.

von Brand, T. (1973). "Biochemistry of Parasites", 2nd edition, pp. 368–425, Academic Press, New York and London.

von Brand, T. and Bowman, I. B. R. (1961). Studies on the aerobic and anerobic metabolism of larval and adult *Taenia taeniaeformis*. *Expl. Parasit.* **11**, 276–297.

Weinbach, E. C. (1972). Role of non-heme iron in cestode respiration. *In* "Comparative Biochemistry of Parasites", (H. van den Bossche, ed.), pp. 433–444, Academic Press, New York and London.

Weinbach, E. C. and Garbus, J. (1966). Restoration by albumin of oxidative phosphorylation and related reactions. *J. biol. Chem.* **241**, 169–175.

Weinbach, E. C. and von Brand, T. (1970). The biochemistry of cestode mitochondria. I. Aerobic metabolism of mitochondria from *Taenia taeniaeformis*. *Int. J. Biochem.* **1**, 39–56.

Yorke, R. E. and Turton, J. A. (1974). Effects of fasciolicidal and anti-cestode agents on the respiration of isolated *Hymenolepis diminuta* mitochondria. *Z. ParasitKde.* **45**, 1–10.

Chapter 12

Pathology of the Invertebrate Host–Metacestode Relationship

R. S. Freeman

I. INTRODUCTION

For an orderly discussion of the pathology of invertebrate host–metacestode relationships one must understand the taxonomy, life histories and the terminology used to describe the ontogeny of species in the various taxa. Wardle *et al.* (1974) proposed that only tetrafossate (four-suckered) cestodes are "true cestodes" belonging, therefore, in their new class Eucestoda. They proposed the new class Cotyloda to receive the following: the orders Caryophyllidea, Spathebothriidea, a revised order Diphyllidea (to receive the family Diphyllobothriidae), Pseudophyllidea (which includes the remaining families of the old order), and, lastly, Gyrocotylidea

BIOLOGY OF THE EUCESTODA Vol. 2
ISBN 0–12–062102–9

and Amphilinidea. All cestodes in the class Cotyloda were considered to be "pseudotapeworms". Unfortunately, such a jumble of characters was used to define the Cotyloda that various species now included in the order Pseudophyllidea could equally well belong, by their concept, in the class Eucestoda. Their system completely overlooked what probably is the single, ubiquitious and, therefore, presumably most primitive feature of all members of the subclass Eucestoda as now accepted, namely the presence of the oncosphere or hexacanth larva as the first stage in the life-cycle (Freeman, 1982a,b). Until there is better evidence to the contrary, the classification of Schmidt (1970), which is essentially based on this fact, will be followed here. This classification has all orders with species having hexacanth larvae belonging in the subclass Eucestoda Southwell, 1930, and assigns Gyrocotylidea and Amphilinidea, with lycophore larvae, to the subclass Cestodaria Monticelli, 1891. The order Pseudophyllidea includes the family Diphyllobothriidae as well as the other families that Wardle *et al.* (1974) accept in this order.

A universal feature of all eucestode life histories is that the oncosphere, enclosed partly or completely within membranes or other coverings associated with the egg, must be eaten by a suitable first host (Freeman, 1973). In the gut lumen, the oncosphere is released from these coverings, activates, penetrates the gut wall, and enters a parenteral site (e.g. for cyclophyllideans see Lethbridge, 1980). Here it metamorphoses and initiates growth to the metacestode (Freeman, 1973). The term "metacestode", first proposed by Wardle and McLeod (1952) and as used here, includes all cestode forms or stages following oncospheral metamorphosis and before the initiation of adulthood shown by incipient gonad formation. In aquatic eucestode life-cycles the first host is always an invertebrate, most frequently a crustacean or annelid, or possibly another arthropod or invertebrate (Dollfus, 1923b, 1931; Hall, 1929; Wardle and McLeod, 1952; Mackiewicz, 1972). Similarly, in most terrestrial eucestode life-cycles, the first host is some insect or other arthropod, annelid, mollusc, or possibly another invertebrate (Hall, 1929; Wardle and McLeod, 1952; Joyeaux and Baer, 1961). However, the oncospheres of all species of the family Taeniidae, some Dilepididae, *Hymenolepis nana* and perhaps a few other species (e.g. possibly *Oochoristica* spp. *sensu lato*) may develop parenterally in various vertebrate first hosts (Wardle and McLeod, 1952; Abuladze, 1964).

Many aquatic cestodes may require, or can undergo, further metacestode growth either enterally or parenterally in a second or third host before attaining sexual maturity in the gut lumen of the final host. In marine cycles, the second or even third hosts, as well as the first, may be invertebrates. Transfer from host to host is passive with the parasite or the host containing the parasite being eaten by the next host (Freeman, 1973).

Generally, in freshwater and marine cestode life-cycles, the second and third intermediate or paratenic hosts, when necessary, are various fish or, more rarely, higher vertebrates. Final metacestode development to the plenametacestode (see Freeman, 1981) preceding the adult is typically completed in the gut lumen of the final vertebrate host (Wardle, 1935; Freeman, 1964; Fischer and Freeman, 1973). Thus, via passive transfer from host to host, the cestode moves up the food chain or web in each biotope. As a rule, however, only a limited number of species of hosts in a particular biotope are suitable hosts for metacestode development, i.e. there usually is some level of host restriction. It follows that pathology induced by metacestodes may occur parenterally in invertebrates or vertebrates, and potentially in the gut lumen of both.

Notwithstanding the fact that most commonly the first host for most cestodes is some invertebrate, surprisingly little is known about the pathogenicity of oncospheres or developing metacestodes in invertebrates. Much more is known about the pathology associated with metacestode–vertebrate interactions, particularly those in humans or domestic animals. Pathology in wild fish, birds and mammals also has received some attention. This will be omitted here.

Pathology associated with migrating oncospheres and developing metacestodes may result from mechanical, physiological or immunological changes. It may depend on the number of metacestodes present and where they locate. Reactions of invertebrate or vertebrate hosts to the parasites may differ. Both may, however, mount non-specific cellular, e.g. phagocytic, responses as has been reported from various invertebrates (Salt, 1963; Poinar, 1969; Tripp, 1969; Cheng and Rifkin, 1970). That vertebrates may utilize a battery of specific immunogenic cellular and humoral responses is well known (Anderson, 1974; Roberts, 1978; Leid and Williams, 1979). Conversely, there is a paucity of experimental evidence concerning invertebrate immunity (Tripp, 1969) and little has been published concerning cellular and humoral responses of invertebrates to invading oncospheres and growing metacestodes.

Presumably, successful host reactions result in killing, isolating, or otherwise restricting the metacestodes in order to prevent or minimize the harm they may cause to the host. Occasionally, the reaction of the host, and the resulting damage to itself, seems out of proportion to the direct damage caused by the parasite. As Read (1972) pointed out, to understand the host–parasite relationship, the host and parasite must not be considered as separate entities, but rather they should be considered as a biological unit. He likened it to "hostparasite", which in systems terms would be an expression of a host–parasite interaction system which differs from its parts. Similarly, Anderson (1978) showed that, from the standpoint of population dynamics, the host and parasite form an interacting unit. Thus, pathological effects may vary from one host–parasite interaction to another even when the same species are involved. Davies *et al.* (1980) put forth the novel idea that perhaps the ultimate overall purpose of the evolution of cellular and humoral response systems by hosts, and various avoidance mechanisms by parasites, may be Nature's way of establishing stable co-existing populations or colonies. They suggested that man, along with other vertebrates, may be a complex colony of organisms, with the vertebrate being the dominant, but not the only, species of the association. Brown and Threlfall (1968) were of like mind when, as a consequence of their study on parasites of the squid *Illex illecebrosus*, they stated: "Too often parasitism is confused with pathogenicity".

This is what parasitologists frequently see, namely, that the majority of animals in most natural populations, including humans in many parts of the world (le Riche, 1967), have on or in them various combinations of animal organisms, including cestodes, frequently causing little obvious harm to the hosts. An obvious answer to the query of whether a parasitic infection ever is beneficial to a host is, however, that if an initial apparently light infection can immunize a host against a potential subsequent, massive, clinically threatening infection, then undeniably such initial infection is beneficial. At the other extreme, Bauer (1958), who worked with fish, considered that every parasite exerts some harm on its host. He went on to enumerate various circumstances and changes under which an apparently harmless organism may become pathogenic. In short, he considers that every parasite has a noxious effect on its host, no matter how small.

Pathogenesis and pathology are determined usually by examination of gross specimens and, frequently, histological material as well. Abnormal behaviour may have a pathological basis, and mortality attributable to metacestodes also indicates a reduction of host well-being, even if it is not always evident exactly how death comes about. All such parameters will be considered here depending on the types of data available. Also, where it is known, some observations on what the metacestode may do to "resist" actions of the host will be included (e.g. Ubelaker *et al.*, 1970; Lackie, 1976). Unfortunately, very few studies have considered the pathology resulting from metacestodes infecting invertebrate hosts. Such information frequently arises as incidental observations in studies of cestode life histories or ecology. Because of this, some valuable information probably was inadvertently omitted from this chapter. The guides to early literature on metacestodes in invertebrates by Dollfus (1923b–1976) and Hall (1929) were useful in writing this text. For convenience the data will be considered by orders, essentially following the classification of the Eucestoda proposed by Schmidt (1970). Discussion of cestodarians, i.e. the amphilinids and gyrocotylids, will be omitted.

II. THE CARYOPHYLLIDEA Beneden *in* Olsson, 1893, AND SPATHEBOTHRIIDEA Wardle and McLeod, 1952

The caryophyllids and spathebothriids are considered together for convenience. There are those (e.g. Joyeux and Baer, 1961; Sandeman and Burt, 1972) who consider that they belong to the same family, or as separate families but in the order Pseudophyllidea. Most recent workers accept that they constitute separate orders (e.g. Schmidt, 1970; Mackiewicz, 1972; Wardle *et al.*, 1974). Both orders are unique among eucestodes in that as adults they do not undergo apolysis. The caryophyllids are monozoic. [Mehlhorn *et al.* (1981) develop arguments why the terms segment, monozoic and polyzoic should not be used in reference to cestodes. There are, however, well established counter arguments (e.g. see Beklemishev, 1964, pp. 197–201). Accordingly, these well known and established terms will be used here.] The spathebothriids lack segmentation, but are potentially polyzoic since they are multiproglottate, i.e. each strobila contains a series of reproductive units. Eggs are discharged via permanent uterine pores. At least some species of both orders become sexually mature and may produce viable eggs while in a parenteral site in the presumed first host (Calentine, 1964; Kennedy, 1965; Stark, 1965; Mackiewicz, 1972; Sandeman and Burt, 1972). As far as is known, the caryophyllids utilize oligochaete annelids as first hosts, and the spathebothriids utilize gammarid amphipods and, possibly, mysids. The life-cycles for a number of caryophyllids have been worked out experimentally (for review, Mackiewicz, 1972), but not for spathebothriids (Sandeman and Burt, 1972).

Normally, less than 3% of suitable, wild annelids are infected with caryophyllids, and usually with fewer than 5 to 10 parasites per annelid. Experimentally, much higher prevalences and intensities of infection are possible, but the heavily infected oligochaetes die quickly (Mackiewicz, 1972). Calentine (1965a) found only single infections with *Biacetabulum* spp. in nature, and showed experimentally that more than one metacestode per annelid usually killed the host. Earlier, Calentine (1964) showed that metacestodes of *Archigetes iowensis* usually localized in the posterior seminal vesicle of the host. Hosts with more than four metacestodes died within 100 days. Other annelids with one or two metacestodes survived up to 2 years. Later, Calentine and DeLong (1966), working with *A. sieboldi*, found that mortality varied from 71% to 83% within 24 days in four laboratory cultures of naturally infected annelids. There was an average of 4.6 (1–15) cestodes per host. Exactly what killed the annelids was not stated. Kennedy (1965) maintained, on the other hand, that up to 20 developing metacestodes of *A. limnodrili* could infect suitable annelids without injury to the host because most metacestodes in such infections died. The cause of death of supernumerary metacestodes was not established experimentally. *Archigetes* spp. may develop to gravid adults in annelids and escape from the host by rupturing the body wall (Calentine, 1964; Kennedy, 1965). Kennedy (1972) stated that almost all caryophyllids have been noted to produce severe effects upon their oligochaete hosts and that hosts may die. Calentine (1965b) found that a maximum of three fully developed metacestodes of *Hunterella nodulosa*, which are comparatively small, could be tolerated by oligochaete hosts. Obviously, relatively few caryophyllids can come to full metacestode size or sexual maturity in oligochaete hosts without causing serious harm. Generally, death is associated with rupture of the host's body wall (Calentine, 1964, 1965a,b; Kennedy, 1965).

Death of the annelid host by rupture of its body wall by the caryophyllid, while the most obvious result, is not the only effect observed. The oligochaetes *Tubifex tempeltoni* (2–4 weeks post-hatching) were readily infected with *Biacetabulum* spp., but the hosts did not mature sexually. In contrast, sexually mature oligochaetes of the same species were infected experimentally, but oncospheres seldom survived more than 5 days (Calentine, 1965a). Inability of infected juvenile oligochaetes to mature sexually apparently is true of many, but not all, caryophyllid-oligochaete interactions, since some juvenile *Limnodrilus hoffmeisteri* infected with *H. nodulosa* did mature sexually (Calentine, 1965b). Calentine (1965b) observed that, although *Monobothrium hunteri* may not kill *L. hoffmeisteri*, the parasites caused localized tissue damage in the host, such as destroying septal walls and damaging or destroying the body-wall musculature. In this same study, the annelid *Tubifex tubifex* was re-infected 30 days after initial infection with *Glaridacris catostomi*, indicating that a protective immunity did not follow the initial infection.

Calentine *et al.* (1970) fed eggs of six species of caryophyllids to up to 11 species of oligochaetes. They found that in most cases, inability to infect an annelid was caused by failure of eggs to hatch. If the oncosphere hatched, became activated and entered the oligochaete coelom, cellular and chemical responses were evident in some hosts. Some young cestodes were killed by "... certain phagocytic cells which line the septa and body wall (and possibly coelomocytes) ..." (Calentine *et al.*, 1970). The authors suggested that occasionally there may have been "chemical incompatibility", since in some cases young parasites invariably died even after they had escaped from the host cells. In other cases, however, young metacestodes continued developing after escaping from the phagocytic cells. Size of the annelid host was a critical factor in some cases because in smaller hosts, the growing metacestode mechanically ruptured the host's body wall causing subsequent death. Calentine *et al.* (1970) showed that hosts in which metacestodes developed fully survived only approximately half as long as uninfected hosts under similar test conditions. These authors concluded: "Termination of infection by host death was three times more common than was termination by death of the parasite within the annelid". In contrast, Kennedy (1972), working with natural populations of *Psammoryctes barbatus* infected with *Caryophyllaeus laticeps*, concluded that, overall, the parasite had only a slight effect upon the host population although growth progressively decreased and reproduction was inhibited in infected hosts. The host's life-span was not affected. This study did not ascertain either mortality during early stages of infection or the real effect of single or multiple infections on young tubificids, for infected tubificids could not be identified until October, although parasites probably were acquired in July or August. From October on, most infections had a single parasite per tubificid, although some hosts contained two or three parasites in all months.

Spathebothriidean adults of *Cyathocephalus truncatus* and *Diplocotyle olrikii* [Wardle *et al.* (1974), unlike Sandeman and Burt (1972), retain *Diplocotyle olrikii* as a valid species separate from *Bothrimonus sturionis*; this is followed here], like adult caryophyllids, are relatively small. The adults may be found in the gut of salmonids, coregonids and a variety of other fresh and estuarine cold-water fish. The juvenile and early adult stages are known from the haemocoel of *Gammarus* spp., *Pontoporeia* sp. and *Mysis relicta* (see Amin, 1978). The method by which gammarids become infected has not been demonstrated experimentally, but field data suggest it may be by ingestion of the larvated eggs (Wiśniewski, 1932; Sandeman and Burt, 1972). Amin (1978) suggests that *M. relicta* may be a transport host that is infected by eating an infected *Pontoporeia affinis*. There is some controversy whether the oncosphere of *C. truncatus*, as now described, is fully differentiated (see Freeman, 1982b). Fish become infected by eating infected gammarids.

Wiśniewski (1932) maintained that it is the small, young gammarids which become infected with *Cyathocephalus* sp. and that the host and parasite grow in parallel. He suggested, furthermore, that gammarids become immune to further infection, and that this accounts for the fact that large, infected gammarids usually have a single, sexually mature cestode. Apparently, gammarids may form a cyst around the developing worm, probably with the aid of hypodermal cells, which finally take on a cuticular character and may kill the worm. According to Wiśniewski (1932), these capsule-like cysts may also protect the host from toxic effects even if the worms are not killed. The parasites are often able to free themselves from these cysts. As Wiśniewski (1932) stressed, his conclusions were based solely on observations of natural infections; alternate conclusions may be derived just as easily from such data. For example, could it be that more than one *Cyathocephalus* sp. is usually fatal to gammarids, hence large gammarids with more than one worm are scarce?

Or conversely, could it be that even if several worms commenced their development in a single gammarid, that only one usually succeeds due to intraspecific competition? Since the complete life-cycle is not known, the crucial experiments necessary to verify these alternate explanations remain to be done.

The complete life-cycle of *D. olrikii* has not been verified experimentally, although it is certain that fish become infected following ingestion of parasitized gammarids (Sandeman and Burt, 1972). Numerous attempts by these authors experimentally to infect gammarids failed. All of their experiments were based on the assumption that gammarids become infected following ingestion of larvated eggs, and that such larvae subsequently develop into sexually mature cestodes in gammarids as well as in fish. Sandeman and Burt (1972) were presumably influenced by Stark (1965) who suggested that a developmental stage in a host preceding the gammarid is unlikely since a procercoid, because of its size, would probably be killed by the masticatory process of an amphipod. That there might be a first host preceding the gammarid is suggested, however, by observations of both Stark (1965) and Sandeman and Burt (1972) who found no naturally infected gammarids with early developmental stages of *Diplocotyle* sp. The smallest worm collected by the latter authors was 7 mm long and had a recognizable holdfast. This contrasts with the size of an oncosphere which is approximately 0.025 mm in length (Burt and Sandeman, 1969). The worms appeared quite suddenly in gammarids by late winter or early spring, and by late spring, depending on where collections were made, most worms were sexually mature reaching a maximum length of 110 mm (Stark, 1965; Burt and Sandeman, 1969; Sandeman and Burt, 1972). Theoretically, therefore, a pregammarid host cannot be ruled out.

Sandeman and Burt (1972) found that, "Infected gammarids lack pigmentation to the extent that the outline of the larvae could be distinguished through the body wall" (Fig. 1). They found one, and occasionally two, worms in the haemocoel (Fig. 2), but only once did they find three worms. They stated that the parasite burden could weigh more than the host (Fig. 2). Both sexes of gammarids were equally susceptible to infection (Stark, 1965). There was a more marked physical effect on females than on males, with infected females being significantly larger than uninfected ones. The worms, which because of their length are folded several times, usually grow just under the heart where the ovary is normally located. Stark (1965) found no trace of the ovary in infected female gammarids, and assumed this was due to mechanical influence of the parasite. Furthermore, the infected females had bristleless oostegites characteristic of the non-breeding female (Stark, 1965). Sandeman and Burt (1972) also found that female gonads were smaller in infected than in

Fig. 1 Photograph of the worm *Bothrimonus* (=*Diplocotyle*) sp. visible within the haemocoel of its gammarid host (from Sandeman and Burt (1972), *Journal of the Fisheries Research Board of Canada*; reproduced by permission).

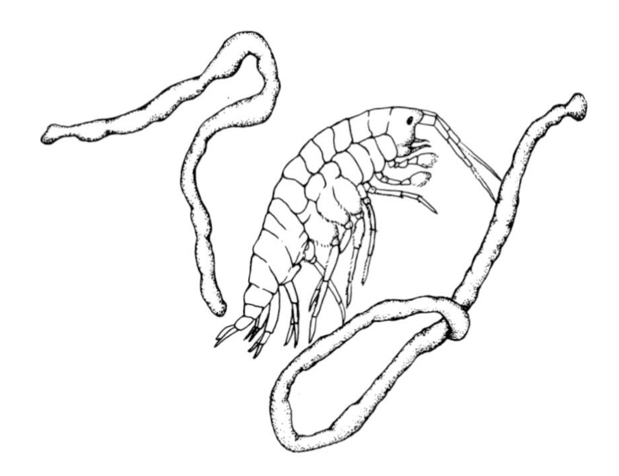

Fig. 2. Drawing of two *Bothrimonus* (=*Diplocotyle*) sp. freshly dissected from their gammarid host (after Sandeman and Burt (1972), *Journal of the Fisheries Research Board of Canada*; by permission).

uninfected hosts. Scott and Bullock (1974) found another species of amphipod, *Psammonyx nobilis*, infected with *Bothrimonus sturionis*. Again the females apparently had no gonads, whereas infected males appeared normal. Stark (1965) made the interesting observation that gammarids with large parasites became sluggish, tending to float upwards and sticking to the water surface, apparently too weak to free themselves from the surface tension. Obviously, infection with even a single *Diplocotyle* sp. may be markedly pathogenic to a gammarid host.

III. LECANICEPHALIDEA Baylis, 1920

Cestodes in this order were originally placed as a superfamily in the order Tetraphyllidea. The composition of the Lecanicephalidea was defined more clearly by Wardle and McLeod (1952) and Yamaguti (1959), but the order is still somewhat confused (Euzet and Combes, 1965). All adults infect the spiral valve of elasmobranchs, but no complete life-cycle is known. Metacestodes considered to belong to the genus *Tylocephalum* Linton, 1890 were originally described from the Ceylon pearl oyster (Shipley and Hornell, 1904, 1905; Herdman and Hornell, 1906; Jameson, 1912). Subsequently, similar cestodes were reported from other bivalve molluscs (Willey, 1907; Dollfus, 1923a; Sparks, 1963; Cheng, 1966; Cheng and Rifkin, 1968; Wolf, 1976; Stephen, 1978). The specific identity of none of these metacestodes has been confirmed by experimental feedings, but most frequently they are referred to either as being in the genus *Tylocephalum* or *Tetragonocephalum* Shipley and Hornell, 1905. Schmidt (1970) apparently follows Baer (1948) and Euzet and Combes (1965), who consider that the genus *Tylocephalum* is a junior synonym of *Tetragonocephalum*; this is accepted here. Most references to these metacestodes, nevertheless, refer them to the genus *Tylocephalum*. To avoid confusion, therefore, they will be referred to as *Tetragonocephalum* (= *Tylocephalum*) sp. in the following discussion.

Cheng (1966) proposed that the large ciliated organisms he found associated with the American (=eastern) oyster, *Crassostrea virginica* (Gmelin), were "hookless coracidia" that invaded the oyster and developed into metacestodes of *Tetragonocephalum* (= *Tylocephalum*) sp. Subsequent statements by Cheng (1967, 1973), Cheng and Rifkin (1968), Wolf (1976) and others tended to support this assumption. However, Freeman (1982b) indicated that such ciliated animals are unlikely to be coracidia of *Tetragonocephalum* (= *Tylocephalum*) sp. What these organisms are, and whether they can penetrate into molluscs, remains to be determined experimentally. Here it is accepted that how molluscs become infected with *Tetragonocephalum* (= *Tylocephalum*) sp. still remains uncertain, although Freeman (1982b) suggested that the mollusc is possibly a second host which becomes infected by eating a procercoid that had developed in a copepod. It is assumed here, nevertheless, that the tissue parasites studied by numerous workers (Sparks, 1963; Cheng, 1966, 1967, 1973; Cheng and Rifkin, 1968, 1970; Rifkin and Cheng, 1968; Rifkin *et al.*, 1969; Cake, 1979; and possibly Wolf, 1976, and Stephen, 1978) all belong to the same lecanicephalid genus, although this remains to be verified experimentally. The pathology associated with these metacestodes has been studied, since it was first suggested that they may be responsible for pearl formation in the Ceylon pearl oyster (Shipley and Hornell, 1904, 1905; Herdman and Hornell, 1906).

Sparks (1963) first reported encysted metacestodes of *Tetragonocephalum* (= *Tylocephalum*) sp. from the eastern oyster from Pearl Harbor, Hawaii. This apparently was the first record of such metacestodes from molluscs from outside the Orient. Each metacestode was enclosed in a thick fibrous cyst measuring approximately 100 by 120 μm. They occurred in the region of the digestive diverticula or the wall of the intestine, but were primarily in the connective tissue between the digestive diverticula and below the intestine. Sparks (1963) assumed that these extensive cysts were due to a fibrosis of the host tissue. Condition Indices, estimates of the oysters' fitness, were lower than normal for the season, but it was not clear whether this was due to crowding on the oyster beds or due to heavy metacestode infection.

Crassostrea virginica that had been transplanted to Chincoteague Bay, Maryland, from Apalachicola Bay, Florida, were reported by Burton (1963) to be parasitized by *Tetragonocephalum* (= *Tylocephalum*) sp. The worms measured 105–231 μm, the largest being 210 by 231 μm. They occurred in thick cysts, apparently formed by the oyster, primarily in the connective tissue surrounding the large intestine, but some were in the gill tissue and other sites.

A more detailed study of these metacestodes and associated pathology in the American oyster from Pearl Harbor was initiated by Cheng (1966). He found there was no appreciable host response while the parasites (procercoids?) were passing through the molluscan gut wall, although there might be mechanical trauma to some of the tissues. The metacestode continued to penetrate until it reached the underlying connective tissue and Leydig cells. Once it reached the

> . . . stratum of connective tissue at the base of the lining epithelium, a conspicuous layer of connective tissue begins to form around it . . . and eventually a complete encapsulating wall of connective tissue is formed . . .
>
> (Cheng, 1966).

Leucocytes may be found intermingled among the fibres. Metacestodes encapsulated by connective tissue fibres also occurred between the digestive diverticula in the digestive gland. Other metacestodes occurred in the Leydig tissue zone between the alimentary tract and the digestive gland. Here, apparently healthy metacestodes were only slightly encapsulated, whereas those in the process of being resorbed were in heavy capsules of connective tissue fibres infiltrated with numerous leucocytes. Cheng (1966) considered that this supported the hypothesis that encapsulation occurs only with the presence of connective tissues and myofibres until ultimately the parasite is resorbed. Sparks (1963) did not believe that encapsulation necessarily meant death of the parasite as did Cheng (1966), although Cheng states that Sparks failed to recognize the presence of this phenomenon. Cheng (1966) concluded that the American oyster was not a fully compatible host for this parasite.

Metacestodes of *Tetragonocephalum* (= *Tylocephalum*) sp. also were reported from the clam *Tapes semidecussata* by Cheng and Rifkin (1968). They found most metacestodes encapsulated in the digestive gland rather than the gut periphery, and assumed, therefore, that the infective stage had penetrated from the exterior rather than from the gut lumen. This assumption requires more supportive data, however, since it would be the first known record of any cestode normally infecting a host without being ingested (Freeman, 1973, 1982b). In 20 clams studied systematically by Cheng and Rifkin (1968), 56% of the metacestodes were undergoing resorption, and even healthy metacestodes were encapsulated. Except for the presence of more leucocytes, the process was essentially as described by Cheng (1966). Resorption was associated with a conspicuous massing of leucocytes, ultimately resulting in a dense layer of packed leucocytes. Many so-called brown cells were also present, and gradual parasite disintegration followed. Such resorption was apparently more common and marked in this clam than reported by Cheng (1966) for the American oyster. Unfortunately, it was not possible to tell from such a study whether the apparently healthy metacestodes that were present were recently acquired, or resistant for some other reason. Presumably encapsulation and resorption was not fast enough, even in this clam, to eliminate completely all living metacestodes.

A detailed histochemical study of the cysts encapsulating metacestodes of *Tetragonocephalum* (= *Tylocephalum*) sp. in American oysters was reported by Rifkin and Cheng (1968). As before, they found some metacestodes that were not encapsulated and others that were interpreted as representing a progressive series of steps in capsule formation (Fig. 3 A-D). According to these authors, encapsulation begins by a metacestode mechanically pushing and compressing the surrounding Leydig cells (Fig. 3 A). A thickening of the intercellular substance between the compressed Leydig cells follows (Fig. 3 B). Further compression is associated with the trapping of some leucocytes into an increasingly dense area surrounding the parasite (Fig. 3 C).

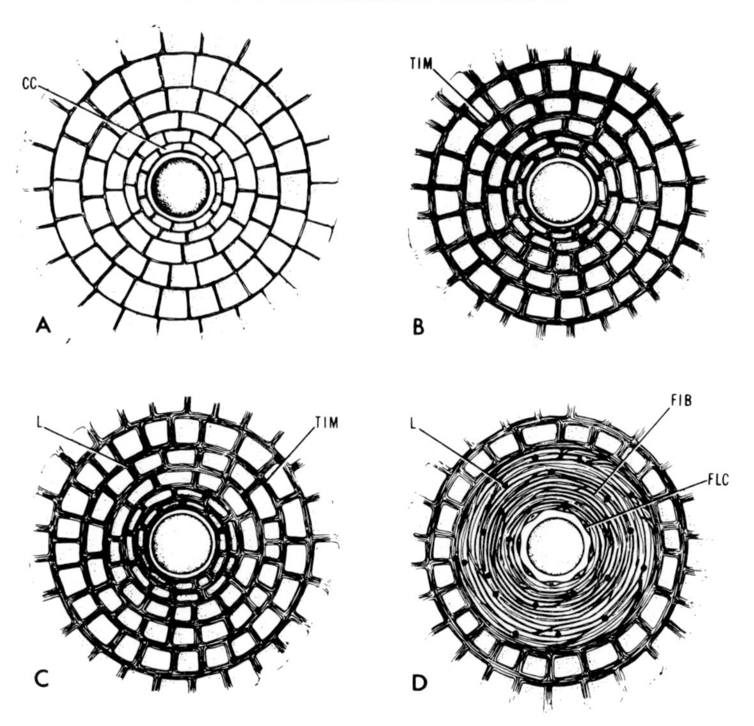

Fig. 3 A–D. "Diagrammatic drawings showing steps involved during the formation of fibrous capsule surrounding *Tylocephalum* metacestode in *Crassostrea virginica*: A, appearance of compressed Leydig cells around metacestode; B, appearance of thickened intercellular material between Leydig cells; C, infiltration of leucocytes into the area; D, deposition of more or less concentric lamellae of fibres, originating as thickened intercellular material, around metacestode and the appearance of innermost layer of fibroblastlike cells" (from Rifkin and Cheng (1968), *Journal of Invertebrate Pathology*; reproduced by permission. CC, compressed host cells; FIB, capsular fibres; FLC, fibroblastlike cells; L, leucocyte nucleus; TIM, thickened intercellular material between Leydig cells).

Shortly thereafter fibroblast-like cells line the innermost part of the cyst wall (Fig. 3 D). Now the parasite is completely immobilized. Fibres begin to appear as concentric lamellae, originating from the thickened intercellular substance between the immediately surrounding, as well as more peripheral, Leydig cells (Fig. 3 D; Rifkin and Cheng, 1968). The fibres were considered to be reticular rather than collagenous. The fibrous capsules surrounding metacestodes immediately adjacent to the gut wall were conspicuously thicker than those around parasites in the region of the digestive gland. It was not determined

> ... what stimulates the intercellular material between surrounding Leydig cells to thicken and become fibrous or what stimulates the migration of leukocytes into the area of the fibrous capsule. (Rifkin and Cheng, 1968).

Electron microscopy has also been used to examine constituents of cysts encapsulating *Tetragonocephalum* (= *Tylocephalum*) sp. in the American oyster (Rifkin *et al.*, 1969). Three cell types (leucocytes, fibroblast-like cells and brown cells) and two types of extracellular fibres were observed. The exact origin of the fibres was not established, but they appeared to form extracellularly, primarily in association with either fibroblast-like cells or leucocytes. The fibres showed no periodicity suggesting that they were non-collagenous. On the basis of structural similarity, these authors considered that either the fibroblast-like cells and leucocytes differentiated from a common precursor cell or, more likely, that the fibroblast-like cells differentiated from leucocytes. Leucocytes were common, intermingled with the fibres throughout the cyst, whereas the fibroblast-like cells were common in the innermost part of the cyst. The so-called brown cells of various authors were confined to the peripheral zone of the cyst. Rifkin *et al.* (1970) found that similar metacestodes had microvilli with vesicular tips. Earlier Rifkin *et al.* (1969) suggested

> ... that the long filamentous microvilli projecting from the cestode's tegument are capable of undulatory movements and during this process many of the fibers of host origin become attached to the unit membrane surrounding each villus.

Whether such undulatory movement, and possibly attachment, was a nutrient gathering mechanism by the parasite, or some defensive mechanism to prevent deleterious macromolecules from reaching the cestode surface, was not established.

The excellent studies by Cheng and his colleagues concerning the process of encapsulation and death of the metacestodes of *Tetragonocephalum* (= *Tylocephalum*) sp. in bivalves are among the most detailed for any metacestode–invertebrate host relationship. It is unfortunate that they could not be based on experimentally infected bivalves. To date, the exact identity of the parasite remains unknown, and the time scale of the processes they described in such detail is uncertain, but these studies merit verification. For a detailed review of five types of encapsulation processes see Cheng and Rifkin (1970), or for a synopsis see Malek and Cheng (1974).

Anantaraman and Krishnaswamy (1959) reported metacestodes of another lecanicephalid, presumably *Polycephalus* sp. since they had a crown of eight unarmed tentacles, from the ganglia of the ventral nerve of the malacostracan *Squilla holochista*. Presumably, some degree of pathology would be associated with infection of such a vital area, but none was mentioned.

A small, solid, pyriform "larva" with an apical sucker was originally described (Kruse, 1959) from the internal lining of the mid-intestine of *Penaeus* spp. According to Kruse (1959), Hutton *et al.* (1959) assigned what appeared to be the same organism "... to the Lecanicephala". Villella *et al.* (1970) considered that this organism was possibly *Polycephalus* sp. This metacestode will be considered more fully in the section covering Trypanorhyncha.

IV. TETRAPHYLLIDEA Carus, 1863, DIPHYLLIDEA Beneden *in* Carus, 1863, and TRYPANORHYNCHA Diesing, 1863

The adults of these orders, as with the Lecanicephalidea and one other minor order, mature in the gut lumen of elasmobranchs. Most adults are identified to the generic level by characteristic attributes of the adult holdfast. Some metacestodes, particularly trypanorhynchs, may be placed in a family or genus while in the penultimate host based on morphology of the tentacles. However, many tetraphyllids undergo final holdfast differentiation (i.e. to the plenametacestode of Freeman, 1981) in the gut lumen of the final host. Thus, identification of most metacestodes from naturally infected intermediate hosts, without confirming experimental feedings, is at best a guess (Euzet, 1959; Hamilton and Byram, 1974). Apparently, a few tetraphyllid metacestodes may be classified at the level of order in naturally infected first hosts (e.g. Wundsch, 1912; Dollfus, 1964b), but experimental feedings are necessary to identify most such metacestodes (e.g. see Williams, 1966, 1968, 1969; Mudry and Dailey, 1971). For a detailed survey of the literature covering the identification and distribution of tetraphyllidean and trypanorhynchan metacestodes and related forms in marine plankton and invertebrates including ctenophores, coelenterates, chaetognaths, lamellibranchs, gastropods, cephalopods, annelids, decapods and other crustacea, see Dollfus (1923b, 1924, 1929, 1931, 1936, 1964b, 1967, 1974, 1976).

Three- and probably four-host cycles, with or without paratenic or transport hosts, must be common in many tetraphyllid and trypanorhynch genera, although only parts of a few life-cycles have been verified experimentally (e.g. Ruszkowski, 1934; Riser, 1951, 1956a,b; Reichenbach-Klinke, 1956; Euzet, 1959; Schmidt, 1970; Mudry and Dailey, 1971; MacKenzie, 1975; Overstreet, 1979). Most published reports concerning metacestodes of these two orders consist of host records and metacestode morphology; data on pathogenicity or pathology are included as incidental findings only occasionally.

As long ago as 1912, Wundsch described two types of plerocercoids (plerocercoïdes) from the haemocoel of naturally infected copepods (*Calanus finnmarchicus*) collected from the North Sea. Presumably both were tetraphyllids, one type possibly belonging to the Phyllobothriidae and the other to the Oncobothriidae. The plerocercoids were approximately 0.10 mm and 0.05 mm long, respectively with four "suckers" on each holdfast and with or without an apical sucker. Mixed infections of 2000 and 3000 individuals filled the thorax of some copepods. Even at these levels of parasitism the copepods exhibited undiminished activity. Grossly, however, the gonads appeared suppressed, presumably due to the massive infections, but no deleterious effects were evident microscopically (Wundsch, 1912).

Marshall *et al.* (1934) also found marine copepods, *Calanus finnmarchicus* from off Scotland, infected with several hundred tetraphyllid and "cyclophyllid" plerocercoids in the haemocoel. Such copepods were a deep red colour due to development of red pigment in the chromatophores associated with the fat bodies. Again, the ovaries remained undeveloped in such copepods. More recently Dollfus (1964b) recorded large numbers of tetraphyllidean metacestodes, similar to those with apical suckers described by Wundsch (1912), from the haemocoel of *Eucalanus pseudattenuatus*. Most were in oval cysts that ruptured easily. Infection levels of about 300 such plerocercoids in a copepod were associated with parasitic castration and ovarian degeneration.

Euzet (1959), in an extensive study, showed that tetraphyllid eggs lack an operculum, usually have delayed development of the oncosphere, and do not develop a coracidium. Kay (1942), Euzet (1955), Riser (1956b), Williams (1969) and Mudry and Dailey (1971) found that the eggs of some species are fully larvated while within the uterus. The larvated eggs may be similar morphologically to those of the freshwater Proteocephalidea which are known to infect only copepods (Freze, 1965). The resulting metacestodes are quite similar to the "four-suckered" tetraphyllid plerocercoids described above (cf. Wundsch, 1912). Euzet (1959) fed larvated eggs from various tetraphyllids to shrimps, crabs, amphipods, gastropods, lamellibranchs and copepods. Eggs hatched only in some copepods, where those larvae, e.g. *Phyllobothrium* sp., that penetrated the gut wall into the haemocoel metamorphosed and initiated growth to the procercoid. All other successful experimental feedings of tetraphyllid eggs also have been to marine copepods (Reichenbach-Klinke, 1956; Riser, 1956b; Mudry and Dailey, 1971). However, none of these metacestodes developed beyond a procercoid, with the holdfast consisting of a single apical sucker, and none resulted in obvious pathology.

Brown and Threlfall (1968) state, "The first intermediate hosts for *Phyllobothrium* sp. have been reported to be crustaceans, particularly euphausids, on which it is postulated that the squid feed . . ." There is no experimental evidence, however, that euphausids, e.g. shrimp, or any malacostracans can function as a first host for any tetraphyllidean. If euphausids are involved with the life-cycle of this *Phyllobothrium* sp., they could be second hosts that have eaten infected copepods. This is plausible since penaeid shrimp apparently may become infected if they eat copepods infected with trypanorhynchan procercoids (Overstreet, 1979, 1982).

Results to date suggest that most, if not all, first hosts of tetraphyllids, of the four families recognized by Schmidt (1970), may be copepods or possibly close relatives. Most likely, the resulting procercoids require further development in the gut lumen or tissues of at least a second host before transfer to the gut of the final elasmobranch host. Similarly, it seems unlikely that more advanced metacestode forms, such as the plerocercoids described by Wundsch (1912), can get to the elasmobranch host directly from the copepod without a second, obligate or transport host. This is unlike those freshwater proteocephalideans that have two-host cycles (*vide infra*). However, the latter, as with the tetraphyllideans, usually have an extended period of metacestode growth to the plenametacestode in the gut lumen of the final host before adult growth ensues (Freeman, 1964, 1981).

A variety of tetraphyllid metacestodes have been recorded from both parenteral and enteral sites in diverse marine invertebrates other than copepods (Dollfus, 1923b–1976; Fuhrmann, 1931; Wardle and McLeod, 1952; Euzet, 1959; Joyeux and Baer, 1961). Whether any such invertebrates are first hosts in the cycle is unknown, although this seems unlikely based on present information. Many such metacestodes look somewhat like larger versions of the metacestodes reported by Wundsch (1912), except that the bothridia may be multiloculate. They are typical plerocercoids, usually recorded under the rubric *Scolex polymorphus* or *Scolex pleuronectis* etc., since the genus, let alone the species, to which they belong is uncertain. For an excellent review of such forms from marine invertebrates up to that time see Dollfus (1936). These plerocercoids frequently occur in the gut lumen and more rarely in associated tissues. Although very common, it is remarkable that such plerocercoids seldom cause overt pathology in the invertebrate hosts. Fisher (1946) reported, however, that similar plerocercoids occurred in "hernia-like swellings" in the wall of the siphon of echiuroid annelids.

Plerocercoids, greatly resembling those labelled as *Plerocercoïdes aequoreus* Wundsch by Dollfus (1936) and *Scolex pleuronectis* by Wardle and McLeod (1952), were found encapsulated in the muscular wall of the stomach or the oesophagus of *Loligo opalescens* off the west coast of North America (Fields, 1965). It should be noted that *P. aequoreus* from a copepod is minute compared to *S. pleuronectis* reported from larger invertebrates, although they are similar in morphology. N. W. Riser (personal communication cited in Fields, 1965) considered that the specimens from *L. opalescens* were probably *Phyllobothrium* sp.

Brown and Threlfall (1968) found four species of tetraphyllid metacestodes in just under 50% of a series of short-finned squid, *Illex illecebrosus illecebrosus*, that they examined off Newfoundland, Canada. Plerocercoids of *Phyllobothrium* sp. were most common, apparently being confined to the lumen of the caecum under normal circumstances. They apparently were not pathogenic. Plerocercoids of *Dinobothrium plicitum*, encysted in the walls of the intestine and caecum, were next most common. Plerocercoids occurred within a thin non-cellular layer surrounded by a mass of hypertrophied cells. Since there was no correlation between size of host, based on mantle length, and the intensity of infection, this suggested that harmful effects were negligible. Brown and Threlfall (1968) agree with other authors, whom they cite, that cestodes rarely cause pathology in cephalopods.

Sparks and Chew (1966) found numerous littleneck clams (*Venerupis staminea*) lying exposed on gravel beds on Bird Island, Humboldt Bay, California. Since these clams typically spend their entire post-larval lives buried within the gravel, they suspected that the exposed clams may have some disease. Plerocercoids identified as *Echeneibothrium* sp. were found encysted within the foot (Fig. 4). The cysts had thick walls, apparently consisting of fine collagenous fibres and numerous leucocytes. Occasionally

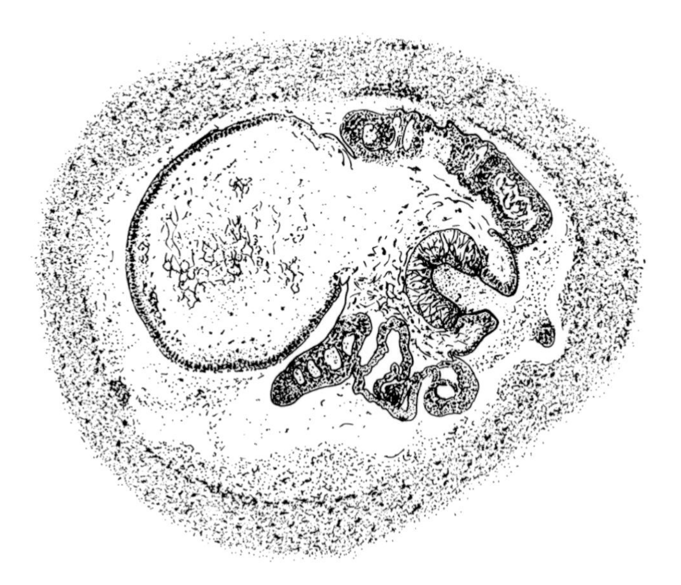

Fig. 4. Drawing of single metacestode of *Echeneibothrium* sp. encysted in the body of the littleneck clam, *Venerupis staminea.* Note the heavy cyst wall consisting of a ". . . compact network of fine collagenous fibres and numerous leucocytes." Amorphous material and leucocytes occur within the cyst (after Sparks and Chew (1966), *Journal of Invertebrate Pathology*; by permission)

there were also clumps of leucocytes within the cysts. As many as 35 metacestodes were counted per cross section of the foot (Fig. 5). The authors concluded that the heavy infections with metacestodes produced disease which caused this aberrant behaviour.

Two species of encysted plerocercoids, identified as *Echeneibothrium* spp., were found in the clam *Protothaca staminea* by Warner and Katkansky (1969a). Each clam contained numerous cysts of two sizes. Larger ones were in the viscera, while smaller ones were more peripheral and in the mantle. Some of the cysts were yellowish, and collagen fibres were demonstrated in the cyst wall when stained with Mallory's aniline blue collagen stain modified for oyster tissue. No gross pathology was evident. Subsequently, Katkansky and Warner (1969), Katkansky *et al.* (1969) and Warner and Katkansky (1969b) found metacestodes of *Echeneibothrium* spp., or a closely related form, encysted in the tissues or free in the gut lumen of four other marine clam species from California. Even with as many as 270 cysts in the mantle and foot, none of the clams showed obvious pathology other than

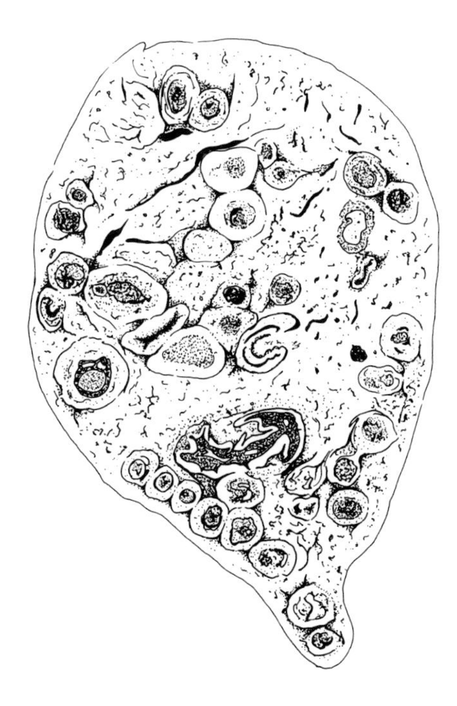

Fig. 5. Drawing of cross section of entire littleneck clam, *Venerupis staminea.* Note numerous, closely packed host cysts with metacestodes of *Echeneibothrium* sp., throughout all body tissues (after Sparks and Chew (1966), *Journal of Invertebrate Pathology*; by permission).

relatively thin-walled cysts around most metacestodes. Again, collagenous fibres were detected in some cyst walls when stained as described above. Obviously, based on findings to date, tetraphyllid metacestodes, even when present in large numbers, are usually only mildly pathogenic when in the tissues, and even less so when in the gut lumen of various invertebrates.

Echinobothrium affine Diesing matures in batoid elasmobranchs. It belongs to the small order Diphyllidea Beneden, which is closely related to the Tetraphyllidea. The holdfast, with spines, hooks and two bothridia, is most characteristic. Dollfus (1964a) reported metacestodes of this species from the haemocoel of a gravid female green crab, *Carcinus maenus*, off the coast of France. The metacestodes apparently reproduced, asexually, and appeared as a bushy mass of clumped filaments near the hepatopancreas. The lobes of the latter organ were irregular and somewhat bloated, but much reduced in size, and the organ occupied scarcely a quarter of its normal space.

The Trypanorhyncha (=Tetrarhynchidea) is the other major order of marine cestodes with adults maturing in the gut of elasmobranchs. The adults and late metacestodes, called plerocerci, usually can be recognized to the level of genus, and frequently species, based on the characteristic holdfast with four, spiny, eversible tentacles (proboscides) and two or four bothridia. Only parts of the earlier aspects of the life-cycles of species in a few trypanorhynchan genera have been established experimentally, although numerous more advanced metacestodes are known from various larger invertebrates, teleost fishes, reptiles and marine mammals (Dollfus, 1923a,b, 1929, 1936, 1942, 1946, 1958, 1969, 1974, 1976; Wardle and McLeod, 1952).

Unlike tetraphyllids, all of which lack operculate eggs and coracidia, some trypanorhynchs have both, whereas eggs of other species are non-operculate and have typical oncospheres without coracidia (Ruszkowski, 1934; Riser, 1951, 1956a; Schmidt, 1970; Mudry and Dailey, 1971; MacKenzie, 1975; Overstreet, 1979, 1982). Experimental feeding of either larvated non-operculate eggs or coracidia of trypanorhynchs succeeded only in marine copepods. The degree of metacestode development, following metamorphosis, varied from an undifferentiated procercoid with no obvious holdfast and an incipient cercomer (*Grillotia* sp., see Ruszkowski, 1934), to more advanced metacestodes. A second type had a prominent sucker at one end, and no cercomer at the other, where the oncospheral hooks were scattered (*Lacistorhynchus* sp., see Riser, 1956a; Mudry and Dailey, 1971). Morphologically this metacestode is reminiscent of those of *Tetragonocephalum* (= *Tylocephalum*) sp. and certain tetraphyllideans discussed earlier in this chapter. A third type of metacestode was a unique elongate form with a capsule-like structure at one end (*Parachristianella* sp., see Mudry and Dailey, 1971). Within the capsule developed a body that contained an incipient adult-type holdfast with four spiny tentacles and bothridia. This led Mudry and Dailey (1971) to suggest that this species may require only one intermediate host. Overstreet (1978), in discussing trypanorhynchan life-cycles, referred to the first hosts as small crustaceans, suggesting that crustaceans other than copepods, e.g. shrimp, perhaps may be first hosts. To date only copepods have been infected with trypanorhynchan oncospheres under experimental conditions.

Overstreet (1979, 1982) reported that it was possible to infect harpacticoid copepods with oncospheres of *Prochristianella hispida*. More advanced metacestodes of this species have been described from the hepatopancreas of penaeid shrimp. The same species of copepod could also be infected with coracidia of *Poecilancistrium caryophyllum*. Of significance is the observation that up to 20 procercoids of *P. hispida* were tolerated by an infected copepod, whereas the copepod usually died quickly when only five coracidia of *P. caryophyllum* entered its haemocoel (Overstreet, 1979). It was noted that the oncosphere of *P. caryophyllum* is larger and tends to locate more anteriorly in the copepod than does *P. hispida*.

As pointed out above under Lecanicephalidea, Kruse (1959) described a procercoid-like metacestode with a well developed apical sucker at one end, and at the other end a small osmoregulatory bladder emptying to the outside through a prominent pore. These metacestodes occurred in the internal lining of the mid-intestine of 16.4% of the brown shrimp, *Penaeus aztecus*, and the pink shrimp, *P. duorarum*, examined. Presumably these could be either tetraphyllid or trypanorhynch procercoids as described by Mudry and Dailey (1971). Kruse (1959) did not state whether these procercoids were encysted or not so it is not known if they were migratory forms. Kruse (1959) also described fully the differentiated plerocerci of *Prochristianella penaei*, which occurred in the "Digestive gland and tissues surrounding the digestive gland and stomach [with] blastocysts frequently penetrating wall of digestive gland." There was an average of 8.1 plerocerci/shrimp in 96.3% of 301 shrimps (*Penaeus aztecus*, *P. duorarum*, and the white shrimp *P. setiferus*) examined. Earlier, Dollfus (1946) had reported the plerocercus of *Prochristianella trygonicola* from the digestive gland of the anomuran decapod *Upogebia stellata* from the coast of France, which suggests that other crustaceans possibly may be potential second hosts in the life-cycle of *Prochristianella* spp. Notwithstanding that in these various crustacea the plerocerci typically affected the digestive gland by rupturing its wall, no other deleterious effects on such infected hosts were reported.

A more detailed histopathological study of the host response of white shrimp, *Penaeus setiferus*, to natural infections with *Prochristianella penaei* was reported by Sparks and Fontaine (1973). They found metacestodes located

> . . . between the acini of the outer portion of the hepatopancreas, and in the hemocoel immediately peripheral to the hepatopancreas and separated from

that organ by the connective tissue sheath that surrounds the hepatopancreas. (Fig. 6).

The range of response in the hepatopancreas (=digestive gland) varied from a thin cyst to full encapsulation, which ultimately resulted in reduction of the metacestode to a melanized necrotic mass. The mature cyst included numerous haemocytes, fibroblasts, and small fibres interpreted as collagen, with haemocytes in the innermost layers being necrotic and forming a thick brown inner nodule. Tubules near metacestodes in the hepatopancreas were smaller and more basophilic than normal, and frequently were destroyed and incorporated as part of the cyst wall. A similar, but much less intense, encapsulating reaction occurred around metacestodes in the haemocoel. The cyst wall was thinner, although it incorporated all the components of the reaction in the hepatopancreas. The part of the cyst closest to the hepatopancreas showed intense melanization, but in none of these cysts was the metacestode destroyed, as occurred within the hepatopancreas (Fig. 6). Sparks and Fontaine (1973) concluded:

> . . . that the various stages of host response are clearly time related; thus the worm entering the hepatopancreas is quickly encapsulated, the cyst is progressively thickened until the plerocercoid is destroyed and eventually resorbed.

These authors also suggested that brown shrimp may respond to infections differently from white shrimp, since infections in the former did not decrease in intensity with age as presumably would occur in white shrimp. Overstreet (1978) points out, however, that an earlier study (Overstreet, 1973) indicated that there was no decrease in intensity of infection during a 2–3-month period that shrimp were reared in Louisiana ponds. Although previous workers had concluded that the hepatopancreas was the normal location for these trypanorhynchs, the observations of Sparks and Fontaine (1973) suggest that the haemocoel is more favourable than the hepatopancreas for metacestode survival until it can be eaten by the final host.

Trypanorhynchan infections in *Penaeus duorarum* and *P. brasiliensis* in Biscayne Bay, Florida, were studied by Feigenbaum and Carnuccio (1976). They found that *Prochristianella penaei* and *Parachristianella monomegacantha* were most prevalent, with *Parachristianella heteromegacanthus* [sic] much less common and *Renibulbus penaeus* least common. They frequently found heavily encapsulated, moribund cestodes in the hepatopancreas of both host species. They accept the conclusions of Sparks and Fontaine (1973) that this reaction is progressive and that the metacestodes are eventually destroyed and resorbed. Feigenbaum and Carnuccio (1976) pointed out that numbers of metacestodes in both species of shrimp depend on the numbers of metacestodes in the water, the diet of the hosts, and the

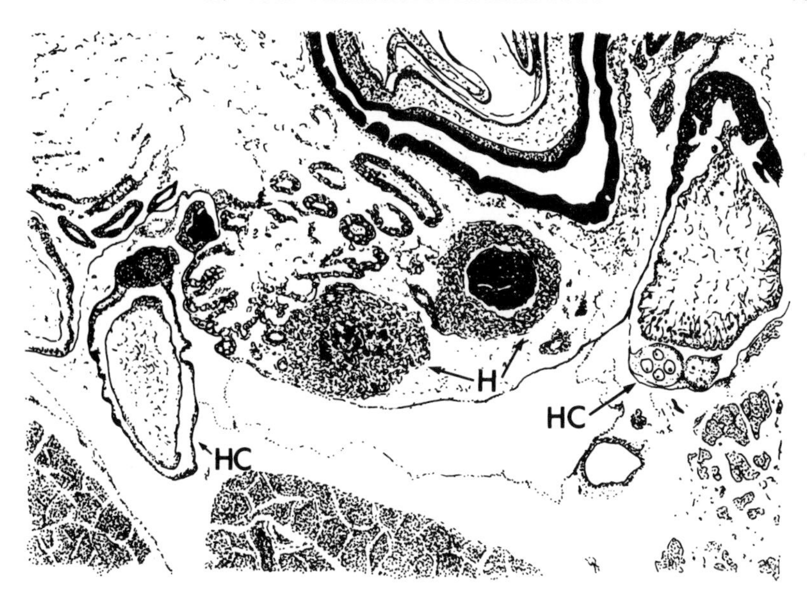

Fig. 6. Drawing that shows, "The difference in intensity of host response to the plerocercoid larvae of *Prochristianella penaei* in the hepatopancreas and the hemocoel of *Panaeus setiferus.*" (after Sparks and Fontaine (1973), *Journal of Invertebrate Pathology*; by permission). H, thick-walled host cyst within hepatopancreas; HC, thin-walled host cyst within haemocoel, with thicker end near hepatopancreas.

immunoreactions of the hosts to the metacestodes. They maintained that their method of sampling of shrimp minimized the first variable, but that the relative importance of the remaining two could not be assessed from their data. Since there was little correlation between host size and intensity of infection in their data, they suggested that possibly defence reactions by the shrimp largely accounted for the different infection intensities in the shrimp examined.

In all probability, other three-host cycles utilizing two invertebrate intermediate hosts will be discovered, and undoubtedly other three-host cycles utilizing a teleost fish as a second intermediate host will be established. Finally, assuming that all trypanorhynchs have a copepod, or equivalent, first host, then there likely are four-host cycles as well, e.g. *Callotetrarhynchus* sp., which must have two teleost fishes as second and third intermediate hosts before reaching adulthood in the final elasmobranch host (Nakajima and Egusa, 1969a,b,c, 1972).

There obviously is much more that needs to be known about the life-cycles of trypanorhynch cestodes before their pathogenicity to invertebrate hosts is fully understood. It may be that cycles with two invertebrate hosts, as well as some with two vertebrate intermediate hosts, may be more common than previously realized. Overall, however, the little information available suggests that in natural infections, trypanorhynch metacestodes appear to be tolerated, although occasionally they are overcome by their invertebrate hosts.

V. PROTEOCEPHALIDEA Mola, 1928

This order, also cited as Proteocephala Wardle and McLeod, 1952 and Ichthyotaeniidea Joyeux and Baer, 1961, includes a large group of cestodes that mature in the gut lumina of freshwater fishes, amphibians and reptiles. Freze (1965) did a comprehensive review of this group, recognizing it as the suborder Proteocephalata Spasskii, 1957 of the order Tetraphyllidea. He recognized two superfamilies, namely, Proteocephaloidea Southwell which includes most of the better known species from fishes, amphibians and reptiles, and the Monticellioidea Freze which includes a group of poorly known species from tropical and subtropical silurid fishes. The life-cycles, in whole or in part, of a relatively diverse group of proteocephaloids are now known, whereas not a single monticellioid life-cycle has been established (Freze, 1965). In all experimentally verified life-cycles the first hosts have been copepods (Freze, 1965). Proteocephaloid eggs have failed to infect cladocerans, rotifers, ostracods, conchostracans, anostracans, amphipods, aquatic oligochaetes, acarina, planaria, chironomid and chaeoborid larvae, and a mayfly nymph (Wagner, 1954; Freeman, 1964; Fischer, 1968, 1972). These feeding experiments support the conclusion that proteocephalid oncospheres, even if they succeed in entering the haemocoel/coelom of invertebrates other than copepods, cannot metamorphose and grow to plerocercoids.

Gruber (1878) was the first to discover what he believed to be metacestodes of *Proteocephalus torulosus* (then placed in the genus *Taenia*) in naturally infected *Cyclops brevicaudatus.* He found large metacestodes above the intestine, nearly filling the space between the eye and the abdomen. He noted that, in spite of having a significant part of the haemocoel filled by the parasite, the copepod did not appear to suffer significantly as it moved vigorously around the aquarium. Gruber did observe, however, that in infected females the orange-red fat droplets, normally filling the whole animal, had disappeared and that the ovaries had atrophied. No infected male copepods were found, and it was Gruber's judgement that such small copepods probably would be killed before the metacestodes reached full size.

Meggitt (1914) first established experimentally that a copepod, *Cyclops varius*, can serve as the first host of a proteocephalid, *Proteocephalus*

(=*Ichthyotaenia*) *filicollis*. He concluded that the presence of the metacestode led to starvation and death of the copepod. The orange globules in the copepod disappeared, and the ovary degenerated and dwindled to half its usual size. The copepod's swimming activity was increasingly affected. Ultimately the copepod died. Some 50 years later the present writer (Freeman, unpublished) kept copepods after exposure to oncospheres of various species of proteocephalids for as long as three months with results similar to Meggitt's. Particularly notable was the disappearance of the orange-red and, occasionally, bluish globules in the haemocoel. They disappeared in both infected and uninfected copepods, suggesting this may have been due to the semistarvation conditions under which the copepods were kept. This was particularly true with *Cyclops bicuspidatus thomasi* which preyed on other copepods present, including its own kind. The number of *C. bicuspidatus thomasi* ultimately remaining appeared to be proportional to the volume of the container in which they were kept. Individually isolated copepods survived in as little as 10 ml of water for moderately long periods of time unless too heavily infected.

In a detailed study of the life-cycle of *Proteocephalus torulosus* (Batsch) (=*Ichthyotaenia torulosa*), based on natural infections in a fish pond, Wagner (1917) also reported the effects of the metacestodes on *Diaptomus castor*. Most copepods had two or three plerocercoids both dorsal and ventral to the intestine and near the ovaries. The plerocercoids invaded the ovaries, apparently causing the gonads to atrophy. Some developing plerocercoids entered the head pushing the eye downward and to one side. Other plerocercoids attempted to penetrate into the abdomen. Plerocercoids crept backward and forward among the organs in the haemocoel, but were most common near the intestine and gonads. Female copepods were more heavily infected than the smaller males. Heavily infected copepods with up to 10 metacestodes in various stages of development usually occurred at the bottom of the aquarium and died after several days, whereas those with one to four plerocercoids swam around in a lively manner and survived longer.

By 1929 it had been demonstrated experimentally that at least six species of *Proteocephalus* and two species of *Corallobothrium*, that mature in fish, utilize seven species of cyclopids and one species of calanoid as first intermediate hosts (Hunter, 1929). There was general agreement that copepods with heavy infections become lethargic and die, whereas one to as many as four metacestodes in larger copepods are tolerated much better. Most workers (e.g. Essex, 1928) pointed out, however, that infections with more than one proteocephalid metacestode are rare in naturally infected copepods from a large lake or river. Essex (1928) assumed that single infections are normal and concluded, therefore, that in nature the effect of such parasites on copepods is slight.

Herde (1938) and Thomas (1941) found that *Ophiotaenia perspicua* matures in snakes, and utilizes *Cyclops obsoletus*, *C. prasinus*, *C. varicans* and *C. viridis* as first hosts. Herde observed that a small copepod with a single metacestode might be more visibly affected than a larger copepod with several metacestodes. According to Herde (1938), affected copepods were "... definitely less active, less responsive to stimuli, and therefore more easily caught with a pipette". He also made the observation, familiar to many who have infected copepods with proteocephalid oncospheres, that during the first week or so after a feeding experiment was begun, nearly every copepod examined was infected. Yet, after another week or more, the few infected copepods that remained were large and harboured few metacestodes. Abundant disintegrating remains presumably represented those that had been heavily infected. Thomas (1941) placed a ripe ploglottid in a dish with approximately 100 copepods. The copepods ripped open the proglottid and devoured the eggs avidly. Within a week, most of these copepods had died from too heavy infections. In comparison, when fewer eggs were supplied, about a 65% infection resulted, with one to eight metacestodes per copepod (Thomas, 1941).

It has been stated that oncospheres claw their way through the copepod gut wall (Wardle and McLeod, 1952). If this were the case, copepods could possibly be severely injured, if not killed, by such action. Thomas (1941) reported, however, that oncospheres used their hooks to work their way through the intestinal wall by pushing aside the cells of the intestinal wall and then, using additional amoeboid movement, they entered the haemocoel. This process required 20–30 min. Freeman (1964) described in more detail the oncosphere of *P. parallacticus* entering the haemocoel of *Cyclops vernalis*. He concluded that entry was by cell displacement rather than cell rupture. Thus, such penetration would be expected to minimize trauma.

As a rule, most species of cyclopid copepods are easy to infect with most proteocephalids, but growth and differentiation of the metacestodes usually ensues free in the haemocoel of only a few species from a particular habitat (Freeman, 1964; Fischer, 1968; Befus and Freeman, 1973). Obviously, there is some degree of host restriction. Oncospheres may become encysted in the gut wall (Freeman, 1964; Fischer, 1967; Befus, 1972; Befus and Freeman, 1973) or elsewhere (Fig. 7). Presumably, encystment is a defence mechanism since even species of metacestodes that frequently complete their development in a particular species of copepod can be found overcome (Freeman, 1964; Fischer, 1968). The cyst wall appears to be produced by migratory cells (Freeman, 1964; Fischer, 1967), but the exact mechanism by which this occurs remains unknown. As a rule, an encysted oncosphere or metacestode does not survive. According to Fischer (1967) metacestode degeneration and resorption occurred at two separate times. Initially it might occur before metamorphosis began. Metacestode de-

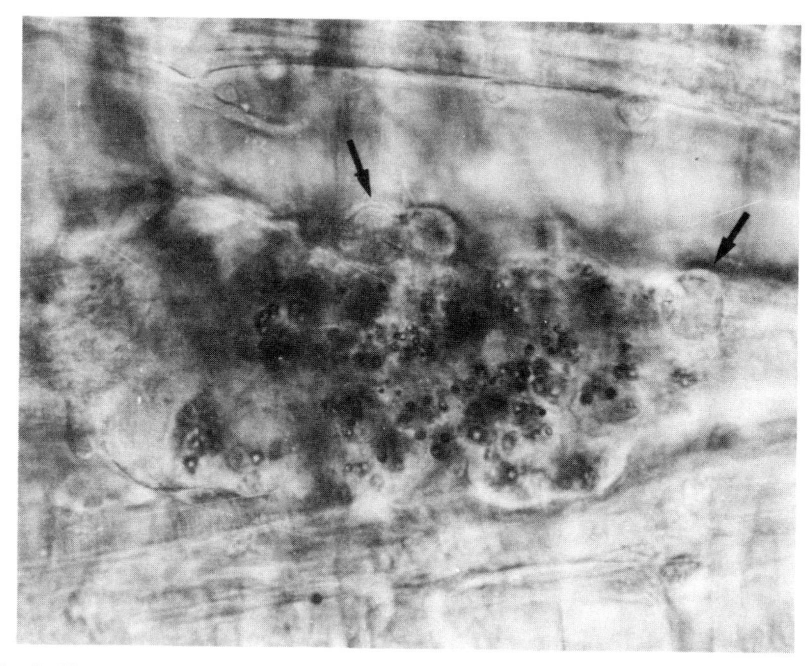

Fig. 7. Photomicrograph of *Corallotaenia minutia* (arrows) embedded in gut wall of a copepod; such encysted forms die and degenerate (from Befus and Freeman (1973), *Canadian Journal of Zoology*; reproduced by permission).

generation and resorption also might occur after metamorphosis, when an abnormal cavity developed in the parenchyma of the metacestode. Such metacestodes usually ceased normal growth, and finally died and were resorbed. Ultimately, only one or more hooks and a few fibres remained. Befus (1972) observed that oncospheres of *Corallotaenia minutia* frequently became encysted in the gut walls of *Cyclops bicuspidatus*, *Mesocyclops edax*, and *Tropocyclops prasinus* where they died and degenerated (Fig. 7). This is unlike *Corallobothrium parafimbriatum* which was never seen encysted in the copepod gut wall.

Almost every worker who has exposed copepods to proteocephalid oncospheres quickly learns that there is a maximum number of metacestodes that can be tolerated. The number can vary from only one (Herde, 1938) to many. Fischer (1968) reported that one *Cyclops bicuspidatus thomasi* had at least 12 dead oncospheres of *Proteocephalus fluviatilis*, 30 undifferentiated oncospheres, and seven developing metacestodes. It is unlikely that such an infected copepod would survive long enough for metacestodes to become fully differentiated plerocercoids, since Fischer (1968) also found that copepods with 15 living parasites survived less than a week.

What actually kills the copepod infected with proteocephalids has not been established. It does not appear to be sheer mechanical damage. There may be atrophy of the gonads as frequently reported. Also, when the metacestodes are unsuccessful they become engulfed in host tissue, but by and large such tissue reaction is not very extensive. Conceivably, there also could be toxic reactions. A definitive study of the pathology of proteocephalid metacestodes in copepods remains to be done.

VI. PSEUDOPHYLLIDEA Carus, 1863

Janicki and Rosen (1917), in their classic study, first established that pseudophyllids may have a three-host life-cycle, including a copepod, a fish, and a mammal. It may be that a second fish host is ecologically obligatory as a transport host in certain lakes. On the other hand, it has been shown that various other pseudophyllids minimally require only the copepod and the final host, e.g. *Bothriocephalus* spp. (see Thomas, 1937; Jarecka, 1959, 1964) and *Cephalochlamys namaquensis* (see Thurston, 1967) to complete their life-cycles. Here again a vertebrate host between the copepod and the final host may not be essential physiologically, but may be required ecologically. Thus far, however, no pseudophyllid life-cycle has been described that utilizes a *second invertebrate* intermediate or transport host. This section will therefore be concerned with the pathology associated with the metacestode, usually a procercoid, infecting the copepod host.

Much work has been done on the life-cycles of certain pseudophyllids because they might infect man (e.g. *Diphyllobothrium* spp. and *Spirometra* spp.) or because the presence of plerocercoids in flesh may detract from the value of commercially important freshwater fish (e.g. *Triaenophorus* spp.). Humes (1950) and Guttowa (1961) summarized the data on the copepods which have been shown experimentally to serve as hosts for plerocercoids of *Diphyllobothrium latum*. Neither author indicated that copepods normally die as a result of being infected with such procercoids.

The single most detailed analysis of the relationship between *D. latum* and various copepods was made by Guttowa (1961). In part she built on the work and concepts that began with Michajlow (1932). Factors such as the temperature at which the oncosphere develops, the time the coracidium has been out of the egg, the species and developmental stage of the copepod that ingests the oncosphere, and the number of larvae ingested were all considered to have a bearing on the cestode–copepod relationship. Guttowa (1961) followed Michajlow (1932) and Kisielewska (1959) in considering that pseudophyllid oncospheres face two barriers in order to infect copepods. The "first selective barrier" includes factors in the gut that can destroy oncospheres before they can penetrate the copepod gut wall, and the "second selective barrier" includes factors that can destroy larvae after they enter the copepod haemocoel. Thus, she considered copepods as being:

(1) "obligatory hosts" if they readily ingest coracidia and the oncospheres penetrate into the haemocoel where they develop normally into procercoids, whether there is a small (1–3 cestodes) or large (10–15 or more cestodes) population per copepod;

(2) "auxiliary hosts" if they ingest many coracidia, most of which are killed in the gut lumen; yet some oncospheres may penetrate the gut wall and develop to normal procercoids;

(3) "accidental hosts" if they ingest coracidia, and some oncospheres may even hatch and survive and enter the haemocoel, but none grow to normal procercoids; and

(4) all the remaining species of copepods, which usually constitute the majority in any water body, which are resistant to infection, i.e. the oncospheres do not enter the haemocoel even if coracidia are eaten (Guttowa, 1961).

Thus, for copepods in categories (3) and (4), coracidia function solely as food, a point usually overlooked. Guttowa (1961) concluded that in *Eudiaptomus coeruleus* v. *vulgaris*, an obligatory host, even heavy infections of 10–15 metacestodes did not adversely influence the vital functions of the copepods, since they survived an average of 20 days to a maximum 6 weeks. Furthermore, in such infections all metacestodes differentiated into procercoids, although they were somewhat smaller in size. This corroborated the observations of Rosen (Janicki and Rosen, 1917) who maintained that of the oncospheres that penetrated into the haemocoel of either *Cyclops strenuus* or *Diaptomus gracilis* only two, at the most, developed fully. Guttowa (1961) found both *D. gracilis* and some *C. strenuus* to be auxiliary hosts. Mueller (1966), based on his studies concerned with the mass culture of copepods (primarily *Cyclops vernalis* infected with *Spirometra mansonoides*), came to much the same conclusion that

> . . . resistant copepods have two lines of defense against coracidia, the one in the gut, the other in the humoral composition of the blood. Only those copepods lacking both these mechanisms serve as adequate hosts for development of the procercoid.

Guttowa (1967) found *Eudiaptomus gracilis* naturally infected with plerocercoids of *Proteocephalus* (possibly *P. percae*) and easily superimposed an infection of *D. latum* in some of these copepods. The *D. latum* grew normally for the first 8 to 12 days at which time their development became inhibited and increased mortality of co-infected hosts became evident. Obviously, *E. gracilis*, which is an auxiliary host for *D. latum*, was not protected by previous infection with *Proteocephalus* sp., and, furthermore, this indicated that the two parasites together were more pathogenic than either infection alone.

It is noteworthy that certain workers have concluded that pseudophyllid procercoids, frequently even in heavy infections, caused little mortality among their copepod hosts (Michajlow, 1953; Watson and Price, 1960; Guttowa, 1961; Kuperman, 1973). Yet others (Li, 1929; Thomas, 1937; Miller, 1943; Mueller, 1959; Thurston, 1967; and, according to Kuperman (1973), Ammann, 1955; Morozova, 1955; and Dubinina, 1965) observed that heavy infections, with various species of pseudophyllids, killed copepod hosts or in other ways were deleterious to their survival. The severity of the reaction frequently varied from one copepod to the next, however, since other infected individuals of the same species, as Thomas (1937) stated, "... carried large numbers without difficulty". Mueller (1959) concluded from mass culture of copepods infected with *Spirometra mansonoides* that infected copepods weakened, became sluggish, and many sank to the bottom where they were subjected to stagnation. He concluded that undue fouling is objectionable, yet starved copepods produced only miniature procercoids or succumbed shortly after infection. Mueller (1959) found that after 2 or 3 weeks, a culture that originally showed 80–90% infection might have only 30% of the copepods infected. He concluded this may have been due to over-infection, or bacterial attack, or possibly due to rapid growth and maturity of uninfected nauplii which would change the observed percentage infected. That bacterial attack may be an important factor was suggested by the fact that adding tetracycline to the culture water resulted in a much higher yield (70–80%) of infected copepods.

Rosen (Janicki and Rosen, 1917) pointed out that as soon as the procercoid in *Cyclops strenuus* had lost its cercomer, it carried out rapid movements which appeared to impede the normal swimming movements of the copepod. It settled to the bottom and crept about slowly. Miller (1943) made similar observations on copepods infected with *Triaenophorus* spp. Such evidence suggests that infected copepods may be weakened although not directly killed. Conceivably, such weakened copepods may be more susceptible than non-infected copepods to other adverse effects, e.g. reduced oxygen content of the water.

Klekowski and Guttowa (1968) and Guttowa (1971) followed the oxygen consumption of *Eudiaptomus gracilis* infected with *D. latum* and found that 6 to 10 days after initial infection there was a notable drop in oxygen consumption. They stated that,

> During this period of organogenesis, the parasite is subjected to increased material changes and particularly to increased demand for building material such as amino acids as well as for high-energy compounds such as glycogen

. . . [which] can disturb the physiological equilibrium of the host.
(Klekowski and Guttowa, 1968).

Later it was shown that infection of various copepods by *Triaenophorus nodulosus* resulted in an increase in proline and methionine in the haemolymph (Guttowa, 1971). The concentrations of these amino acids in procercoids of *T. nodulosus* was most similar to those in *E. gracilis* and *Cyclops vicinus*. This suggested that *T. nodulosus* may use not only the host's food sources, but its metabolic products as well (Guttowa, 1971).

There are other effects that indicate that pseudophyllid procercoids may seriously affect copepods even if they are not killed by such infections. Clarke (1954) observed that a fourth stage nauplius of *Cyclops serrulatus*, infected with a mature but miniature procercoid of *Schistocephalus solidus*, was still a nauplius three weeks later. His experience indicated that within that time, the copepod normally should have moulted several times and grown to an adult. This led Clarke to follow carefully the growth changes of recently infected and uninfected fourth stage nauplii. Within 15 days, the uninfected nauplius had completed all moults to become an adult, whereas the infected copepod at that time was a smaller second-stage copepodid. Shortly after this observation the procercoid "disappeared", yet it clearly had not died and been slowly resorbed. The now uninfected copepodid then completed the necessary moults within the next 15 days or so, and became a fully differentiated, albeit slightly smaller than normal, adult. Subsequently Clarke discovered that other partly and fully developed procercoids "disappeared" from copepods without undergoing slow degeneration.

Mueller (1959) showed subsequently that intact procercoids of *Spirometra mansonoides* rapidly escape from copepods killed by heat (37°C) or iodine solution; this technique was used to recover the large numbers of sterile procercoids needed for culture experiments. Possibly such escape of procercoids may occur from naturally infected copepods as well. Undoubtedly, such free procercoids, or weak and crawling copepods with procercoids, are ingested by bottom-dwelling invertebrates. Whether they can transfer to such potential "reservoir hosts" as occurs with certain hymenolepids (Ryšavý, 1961) remains to be determined.

Mueller (1966) observed that infected *Cyclops vernalis* never seemed to develop egg sacs and might not moult to the adult stage. He also observed that 70% of a fresh stock of copepods from the field might become infected. Yet, when regenerated populations of the same copepod culture were exposed to infection as often as four successive times, only 10% or less of the copepods now became infected. He suggested, therefore, that since restriction of copepod reproduction probably accounted for this change of susceptibility to infection, it follows that susceptibility may be a genetic trait. Conceivably, something of this nature could account for the observation of Vogel (1929) that *Cyclops strenuus* from one natural source occasionally could be infected with *D. latum*, whereas those from another source were completely resistant.

Overall, it is evident that pseudophyllid procercoids probably are, at most, mildly pathogenic to copepod obligatory hosts since natural multiple infections are scarce. In other categories of hosts, the oncospheres are usually killed before they can do much damage.

VII. CYCLOPHYLLIDEA Beneden *in* Braun, 1900

This single cumbersome order includes all the cestodes that have their complete life-cycles in terrestrial hosts, although a number of species have returned, presumably secondarily, to utilizing a variety of freshwater invertebrates as first hosts. Except for the family Taeniidae, and some species in the Dilepididae (*sensu lato*), and, exceptionally, *Hymenolepis nana*, the remaining cyclophyllideans have invertebrates as first hosts. Most life-cycles have two hosts. A few species may have a second fish host between the first invertebrate and the final vertebrate host. A few hymenolepids and dilepidids may have molluscan so-called "reservoir hosts" between the first and final hosts.

Hall (1929) indicated that the first invertebrate hosts known for Cyclophyllidea (*sensu lato*) included: among insects the Coleoptera, Mallophaga, Siphonaptera, Lepidoptera, Dermaptera, Diptera and Odonata; and among other arthropods the Myriapoda, Ostracoda, Copepoda, Amphipoda and Decapoda. Since then other insect first hosts added to the list include: the Collembola, Hymenoptera, Dictyoptera, Orthoptera, Mecoptera, Corrodentia; and among other arthropods the Branchiopoda, as well as the oribatoid and tyroglyphoid mites. Other invertebrates now known to be first hosts include both aquatic and terrestrial molluscs and annelids.

In the cyclophyllidean life-cycle, as with all eucestodes, the egg must be ingested by a suitable host, and in the gut the oncosphere must become activated and escape its protective envelopes. The oncosphere must now withstand the digestive mechanisms in the host's gut and it must reach and breach the gut wall. Normally only after the oncosphere has entered its proper growth site do metamorphosis and growth of the metacestode ensue. It is evident to most students of cyclophyllidean life-cycles that moderate infections in the proper invertebrate hosts rarely elicit obvious damage to either host or parasite.

That invertebrate hosts can mount protective reactions at various points of the infection sequence becomes particularly evident when an oncosphere enters an unsuitable host. Thus, there appears to be the same problem of recognizing self from non-self among invertebrate hosts as occurs in vertebrate hosts before the host mounts a defence against the parasite. The mere observation that a parasite can develop without discernible encapsulation by the host is insufficient evidence, however, for accepting that the parasite was not recognized as foreign by the host (Nappi, 1975; Damian, 1979). Apparently whether an invertebrate host will react and how it will react against a parasite is more complicated than merely recognizing non-self from self, and then encapsulating it. Damian (1979) also pointed out that invertebrate immune systems are inducible but apparently ". . . the qualities of specificity and memory are much less well-developed than in the vertebrates." Obviously much the same holds here as described by Guttowa (1961) and Mueller (1966, *vide supra*) under the Pseudophyllidea.

When Salt (1963) published the first comprehensive review of the defence reactions among insects to metazoan parasites, the only report that concerned a metacestode was by Chen (1934). This article considered the development of the dilepidid, *Dipylidium caninum*, in massive infections in the cat flea *Ctenocephalides felis.* Since only the larval flea has mouth parts adapted for chewing, infection must take place during this stage. The metacestode does not transfer to the final dog or cat host, however, until it ingests the infected adult flea. The oncosphere must, therefore, remain alive and metamorphose to a fully differentiated metacestode while the flea undergoes its own growth and metamorphosis through the larval, pupal and adult stages. That Chen (1934) utilized massive infections was evident, since there were as many as 97 and 56 oncospheres found in the larval and pupal fleas and 55 developing metacestodes in the adult fleas. Numerous flea larvae died the day of infection, supposedly due to damage to the gut wall. Dead and dying flea larvae were grossly recognizable by their red colour due to the amount of vertebrate blood from the ingested food that had entered the haemocoel via the damaged gut wall. Chen also found what he called "erratic parasites" in the ganglia and hypodermis. Since they were observed only in dead larvae, he assumed they were contributing factors associated with host death. Most deaths of larval fleas occurred during the first day of infection. A second peak of flea mortality occurred just before pupation when approximately 20% of the infected fleas died. Of the infected fleas that emerged as adults, only another 5% died subsequently, although the haemocoel now was packed with metacestodes in early stages of development. Apparently no fully developed acanthacetabuloplerocercoids, usually erroneously called cysticercoids (Freeman, 1973), were present in any of the fleas examined by Chen (1934).

Marshall (1967) followed the growth of four species of *Hymenolepis* and

D. caninum in the cat flea. In contrast to the massive infections with *D. caninum* studied by Chen (1934), Marshall intentionally based his conclusions regarding metacestode growth patterns on fleas infected with five or fewer metacestodes of *D. caninum* each. Marshall found that, although there was an increase in the size of *D. caninum* in the larval and prepupal flea, differentiation of the holdfast did not begin until late in the pupal flea, and consequently most metacestode differentiation and growth occurred in the adult flea. There was a noticeable reduction in size and retardation in the rate of development of *D. caninum* in fleas with more than 20 metacestodes. One to 20 metacestodes per flea was considered to constitute light to moderate infections by Marshall (1967). The data from Marshall and Chen are therefore not fully comparable, because unlike Chen, Marshall purposely excluded the more heavily infected fleas from his calculations. Marshall concluded that host mortality coincided with periods of rapid parasite growth. He could not confirm Chen's (1934) observations that flea larvae could be killed by extensive damage to the gut wall to the extent that the fleas' intestinal contents could enter the haemocoel.

Chen (1934) recognized three types of leucocytes in the haemolymph of fleas, namely, pre-amoebocytes, amoebocytes and macrocytes. They occurred in all stages of the flea life-cycle, when infected with *D. caninum*, but macrocytes were not evident in uninfected pupae. The pre-amoebocytes and amoebocytes were most common. Both the amoebocytes and macrocytes apparently were phagocytic. The primary reaction against the oncospheres and developing metacestodes was the attraction to them of the pre-amoebocytes and amoebocytes. The amoebocytes proliferated and attached both to the parasite and to each other. Then, usually by the sixth day of infection in the flea larvae, typically fusiform macrocytes formed a somewhat interlaced capsule several layers thick around the parasite. Chen believed that the macrocytes were derived directly from the amoebocytes, and that the encapsulating cyst never became a definite syncytium. Once encysted, the parasite was invariably killed and digested, with about 30% of the parasites in flea larvae being killed this way. The encapsulated parasite became a yellowish mass which then turned brown and finally dark brown. This is the process of melanization commonly seen in various invertebrates (Salt, 1963, 1970; Tripp, 1969; Nappi, 1975). Apparently in the flea pupal stage, the amoebocytes were involved in dissolving and removing larval tissues during metamorphosis to the adult flea. The cellular attack on metacestodes was suspended during the pupal stage, but by the sixth day of flea adulthood, attenuated leucocytes again began encircling the parasites. Unlike in the larval flea this capsule was a single, thin-webbed, inconspicuous layer consisting primarily of macrocytes with some amoebocytes. Approximately 5% of the metacestodes in adult fleas were killed this way.

Chen (1934) believed that the pigment resulting from the death of these parasites was expelled from the flea haemocoel principally into the gut lumen.

The life-cycles of various hymenolepids have been extensively studied, particularly since the 1950s. They typically utilize as first hosts a variety of terrestrial or aquatic arthropods, but some develop in oligochaetes as well. Molluscs also serve as reservoir hosts (Ryšavý, 1961; Kotel'nikov, 1963; Valkounová, 1973). A few species, especially *Hymenolepis diminuta* (Rudolphi, 1819), *H. nana* (Von Siebold, 1852), *H. microstoma* (Dujardin, 1845), and *H. citelli* (McLeod, 1933) are now well established in laboratories around the world. They all develop well in various arthropods readily maintained in the laboratory, especially beetles, but also fleas and other insects (e.g. see Schiller, 1959; Marshall, 1967; Heyneman and Voge, 1971; Arai, 1980; Schom *et al.*, 1981). The life-cycles of a variety of species in other cyclophyllidean families have also been investigated in the last 20 years or so, but none has received as much attention as the hymenolepids. Consequently the remainder of this chapter will be restricted to various more recent studies on the potentially pathogenic relationships between hymenolepids and various terrestrial arthropods.

The "normal" development of *Hymenolepis nana* and *H. diminuta* in the adult confused flour beetle (*Tribolium confusum*) described by Voge and Heyneman (1957) can serve as a baseline. The beetles were kept at 30°C and, on the average, 10 to 15 (maximum 55) cysticercoids of *H. diminuta* and 35 to 45 (maximum 75) cysticercoids of *H. nana* per beetle constituted a heavy infection. *H. nana* could complete its development to an infective cysticercoid in an average 5.5 days, whereas *H. diminuta* required approximately 8 days. Crowding prolonged the rate of development of *H. nana*, but not that of *H. diminuta*. When *H. diminuta* were crowded, however, the sheaths surrounding the holdfasts were thinner and the tail, i.e. cercomer, smaller. In neither species were holdfasts reduced in size by crowding. This study made no reference to any type of host reaction against the oncospheres or metacestodes in the beetle gut lumen, gut wall, or in the haemocoel.

Earlier work suggested that the adult beetle *Tenebrio molitor* was a good host for *H. diminuta*, whereas the larval stage of this beetle was not. Voge and Graiwer (1964) set out using *H. diminuta*, to determine if this was because: (1) oncospheres failed to hatch in the larval gut, or (2) the gut structure prevented penetration by the oncospheres, or (3) the metacestodes could not undergo sustained growth in the haemocoel of the beetle larva. These authors discovered that oncospheres hatched well in the larval gut and that a few oncospheres penetrated the gut wall, and once in the haemocoel, grew to viable cysticercoids. They found that if oncospheres were hatched *in vitro* and injected into the larval haemocoel, they also developed very well except when too crowded. Histological study indicated that the larval gut has a larger lumen than does the adult and that the intestinal emptying time was $2\frac{1}{2}$ times faster in the larva than in the adult. They also showed that the gut wall of the larva (1) had a thick peritrophic membrane, (2) overall was thicker than that of the adult, and (3) lacked the thin-walled diverticula so characteristic of the adult midgut. Hatched oncospheres were found in the larval gut, but none were encountered in the relatively rare process of penetration. They concluded that the rapid emptying time and the thicker peritrophic membrane and gut wall in the larva made it difficult for the oncospheres to reach the larval haemocoel, and therefore larval *T. molitor* is an inefficient host of *H. diminuta*. Lethbridge (1971) subsequently showed that it was the thickness of the gut wall *per se* and not the thickness of the peritrophic membrane or the rapid emptying time of the gut that made larval *T. molitor* a poor host. He also showed that in the adult beetle, most oncospheres enter the haemocoel via the thin-walled diverticula.

Cavier and Léger (1965) initiated a series of papers by Léger that reported on cyclophyllid metacestode–invertebrate host relationships. Cavier and Léger (1965) compared what occurred in beetle and gryllid hosts (*T. molitor* and *Gryllus domesticus*), which normally become infected by eating eggs of *H. nana* var. *fraterna*, with hosts such as cockroaches (*Periplaneta americana* and *Leucophaea maderae*) which are not orally infectable with this cestode. The latter hosts they infected by injecting oncospheres, hatched *in vitro*, through the tegument into the haemocoel. As controls, *T. molitor* and *G. domesticus* also were injected similarly. No reaction phenomena were observed in the latter hosts whether eggs were eaten or hatched oncospheres were injected. Distinctive reactions were observed in the cockroaches, however, where metacestodes developed more slowly and some exhibited various anomalies. Nevertheless, when some cysticercoids from *L. maderae* were fed to mice they developed to adult cestodes. When infected cockroaches were dissected there was an abundance of black granulations visible to the naked eye. Microscopically these granulations consisted of black pigment surrounded by a thin non-pigmented cellular capsule. Early in the infection the capsules contained empty shells or whole eggs, but by 7 days cysticercoids with blackish deposits were evident in some capsules from *P. americana*. By 14 days groups of metacestodes were united by their holdfasts in common capsules which had eggs and shell debris surrounded by blackish granulations in the centre. The capsule itself consisted of several concentric layers of cells, much as earlier described by Chen (1934). Such capsules were found attached to the internal organs and the wall of the haemocoel. Nevertheless, *L. maderae* itself apparently was not adversely affected since it exhibited normal behaviour 14 days after injection. The authors pointed out, however, that it was not determined why the cockroaches were not infectable orally.

In 1970 Ubelaker *et al.*, and later Collin, reported ultrastructural studies on different hymenolepids growing in *T. confusum*. It was shown that 3 days after *Hymenolepis citelli* eggs had been eaten by the beetle, a layer of cells, presumably haemocytes, three to four cells thick, surrounded the developing metacestodes (Collin, 1970). Microvilli from the metacestode tegument were seen joining to the haemocytes. Two days later, fewer host cells were attached to the metacestodes, and frequently these cells were cytolysed. Eight days after eggs of *H. diminuta* were eaten by *T. confusum*, Ubelaker *et al.* (1970) studied the ultrastructure of the surface of the canal into which the holdfast of the differentiating cysticercoid was withdrawn. There they found numerous branched microvilli with terminal vesicles on the surface of the metacestodes. They also found degenerating amoebocytes in close association with the surface of these metacestodes. They suggested that these vesicles probably were secretory in nature and that their contents, when released, might be responsible for lysis of the haemocytes. The exact nature and function of these vesicles was not determined. The active involvement of such metacestode microvilli is similar to that reported earlier by Rifkin *et al.* (1969, *vide supra*) from the metacestode of the lecanicephalid *Tetragonocephalum* (= *Tylocephalum*) sp. which may attach to the encircling host fibres. Both Rifkin *et al.* (1969) and Ubelaker *et al.* (1970) suggested that this may constitute a nutrient source for the metacestode, or possibly some type of defensive mechanism against the host.

Among other things, Léger and Cavier (1970) further investigated why eggs of *H. nana* fed to *P. americana*, *L. maderae* and *Blabera fusca* failed to establish and grow. They showed that eggs fed to these cockroaches did hatch within the midgut. Furthermore, when such oncospheres, or those hatched *in vitro*, were injected directly into the cockroach haemocoel they grew into cysticercoids. They then compared the gut wall, particularly the peritrophic membrane, of these cockroaches with those of various orthopterans, *Gryllus domesticus*, *G. bimaculatus*, *Locusta migratoria* and *Schistocerca gregaria*, all of which were orally infectable and which produced cysticercoids comparable to those in *T. molitor*. They considered that the gut wall, and particularly the relative thickness of the peritrophic membrane in cockroaches, prevented *H. nana* from entering the gut. This was similar to what Voge and Graiwer (1964, *vide supra*) indicated may account for the resistance of larvae of *T. molitor* to infection with *H. diminuta*. This is of considerable interest because more recently Brooks (1975) discussed the relationships that occur between certain insects and their microbial, mutualistic symbiotes. She pointed out that such insects, e.g. cockroaches and flies, are inherently non-susceptible to most pathogenic micro-organisms, whereas other insects that apparently are susceptible to various infectious agents, including protozoa and helminths, lack such symbiotes. She also indicated that although insects are not considered to be immuno-competent, some can produce a lysozyme-like substance which may be involved with immunological phenomena in insects. Whether such a substance is involved in those insects naturally more resistant to infections with cestodes remains to be determined. The question remains, nevertheless, is this a problem of the insect recognizing self from non-self? Furthermore, if this be true, then how do oncospheres and metacestodes avoid being recognized as non-self when in proper hosts?

The host cellular responses of adult flour beetles, *T. confusum*, to *H. diminuta*, *H. microstoma* and *H. citelli* following single oral infections, re-infections and cross-infections were studied by Heyneman and Voge (1971). The beetles responded with cellular responses in the haemocoel from 3–5 days after infection against all three species of cestodes. Against *H. citelli*, the response occurred as a layer of host cells one to three cells thick. Against *H. diminuta* and *H. microstoma*, it varied from no evident host cells to a few attached cells or to attached, scattered clusters several cells deep. By 5 days the capsule-like matrix had disappeared from around *H. citelli*, and few, if any, host cells were evident around *H. diminuta* and *H. microstoma*. Second or third re-infections with *H. diminuta*, following initial infection with the same species, showed no reduction in numbers of metacestodes that established and developed; nor was metacestode development delayed. The results were the same for cross-infections between *H. diminuta* and *H. microstoma*. Heyneman and Voge (1971) concluded that the coating response with host cells by *T. confusum* 3–5 days after infection is ineffective protection against these three species of cestodes.

Pesson and Léger (1975, 1977) examined further what happens to oncospheres of *H. nana* var. *fraterna* when they are eaten by the cockroach, *L. maderae*, a normally refractory host. Pesson and Léger (1975) concluded that the peritrophic membrane was less important as a factor limiting infection than they originally believed (Léger and Cavier, 1970). They considered, however, that the gut wall was both a mechanical and a physiological barrier. Within 24 h of initial infection a few oncospheres managed to penetrate into the gut wall. Some had already died, but a few living oncospheres had reached the connective tissue and muscle layers. Most living oncospheres had elicited no obvious cellular reaction (Pesson and Léger, 1975), but other oncospheres were encircled with a double layer of cells with large nuclei. This they interpreted as being an epithelial type of reaction. In succeeding days, a few oncospheres managed to penetrate into the haemocoel where they became engulfed in layers of haemocytes (Fig. 8). By 6 days the haemocytic reaction diminished and slight melanization was evident (Fig. 9). By 12 days no melanized formations were evident

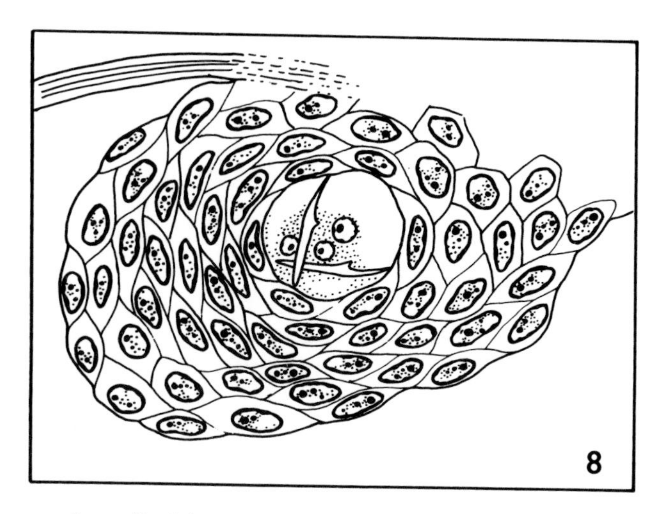

Fig. 8. A somewhat stylized drawing of haemocytic reaction on the haemocoel side of the gut wall of *Leucophaea maderae* 48 h after it ingested eggs of *Hymenolepis nana* var. *fraterna* (Figs 8 and 9 from Pesson and Léger, 1975).

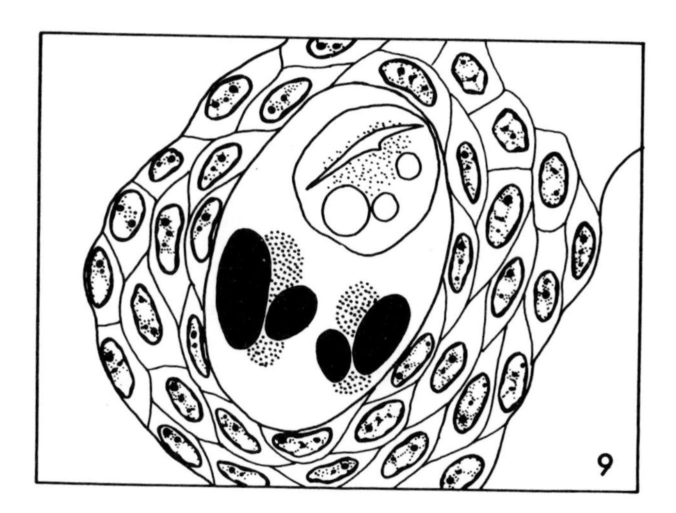

Fig. 9. A somewhat stylized drawing, same as Fig. 8 only 6 days after oncospheres were ingested, showing melanization of oncosphere. (Figs 8 and 9 from B. Pesson and N. Léger (1975). *Annales de Parasitologie humaine et comparée*; by permission.)

either in histological sections or in the haemocoel of dissected cockroaches. They considered the cellular reaction within the gut wall to be facultative or temporary, whereas they considered the haemocytic reaction followed by melanization in the haemocoel to be an immunological barrier (Pesson and Léger, 1975). Apparently no oncospheres that penetrated through the gut wall survived and grew to cysticercoids. This is unlike some artificially hatched oncospheres that grew to cysticercoids following injection through the body wall directly into the haemocoel. Pesson and Léger (1975) suggested, therefore, that it is the oncospheral passage through the gut wall of the cockroach that apparently initiates a process that first inhibits growth of the parasite and subsequently leads to its disintegration.

Subsequently, Pesson and Léger (1977) observed that indeed within 24 h after ingesting eggs, when some oncospheres were near the outer epithelium but still in the gut wall of the cockroach, two types of cellular reactions were evident. One was the cellular encirclement of the oncosphere within the gut wall proper, and the other was a massing of haemocytes near the oncosphere, but on the haemocoel side of the gut wall. This suggested that the oncospheres in the gut wall were recognized by haemocytes from a distance. Pesson and Léger (1977) attempted to defeat these cellular reactions in two ways, firstly by repeated feedings of numerous eggs, and secondly by exposing the cockroaches before infection to irradiation of 10 000 or 20 000 Roentgens from ^{60}Co. One of 10 cockroaches following repeated feedings with eggs had a few fully differentiated cysticercoids in the haemocoel. To these authors this suggested that the specific and non-specific defensive reactions were being overwhelmed by the numbers of penetrating oncospheres. The irradiation apparently did not significantly alter the reaction within the gut wall since only once in 25 attempts did a few metacestodes develop in the haemocoel. The irradiation drastically suppressed the haemocytic reaction in the haemocoel, however, since oncospheres that were injected directly into the haemocoel following irradiation now developed into cysticercoids much as in natural hosts. Pesson and Léger (1977) contended, therefore, that the cellular reaction within the gut wall and that within the haemocoel are fundamentally different.

Pesson and Léger (1978) and Pesson *et al.* (1978) then studied the ultrastructure of the metacestodes of *H. nana* var. *fraterna* in adults of both *L. maderae* and *T. molitor*, a non-susceptible and susceptible host, respectively. Oncospheres were injected into the haemocoels of both species of hosts, and some beetles were also fed gravid cestode segments (Pesson *et al.*, 1978). Other cockroaches were either injected 4 days before infection with a water-soluble antigen prepared from adult cestodes, or they were exposed to 10 000 R from ^{60}Co 5 h before infection (Pesson and Léger, 1978). Infection again was by injection of activated oncospheres into the haemocoel. The beetles were killed 5 days and 12 days after eating eggs or following injection with oncospheres. Metacestodes 5 days after oral infection had abundant microvilli on their teguments, whereas after 12 days the microvilli were slender, sparse and irregularly scattered. Similarly, in beetles 5 days after receiving injected oncospheres the metacestodes had abundant microvilli as well as a few attached haemocytes, some undergoing lysis. After 12 days, however, the microvilli were still abundant but now were more slender with a few scattered haemocytes attached. The metacestodes from cockroaches injected with oncospheres and killed 12 days and 30 days later were markedly different. After 12 days, hosts with massive infections had a few free metacestodes with abraded teguments. Within some reaction-capsules the metacestodes had scattered villi joined to haemocytes. Even after 30 days, the encapsulated metacestodes had a tegument with haemocytes in close contact. By now most free metacestodes had partially or completely lost their tegument, but where the tegument was intact it was either smooth or had very small microvilli. Even with these lesions some cysticercoids free of cellular reaction were infective to the final host (Pesson *et al.*, 1978).

Cockroaches first injected with antigen from whole adult *H. nana* and challenged 4 days later with an intrahaemocoel injection of oncospheres, survived only a maximum 7 days (Pesson and Léger, 1978). In contrast, control cockroaches injected only with antigen survived to the end of the experiment. The reaction to the injected oncospheres and developing metacestodes after 6 days or 7 days, i.e. just before death of the hosts, consisted of a few sites rich in haemocytes but weakly melanized. There also were free-living metacestodes with thin teguments bearing few microvilli. Such microvilli were short, stocky and slightly ramified. Occasionally, numerous small vesicles occurred near these microvilli. The cercomers on free cysticercoids were enlarged and hollow with only a peripheral parenchyma consisting of a few cells. This was similar to that seen in cysticercoids encapsulated in a haemocytic reaction. Oncospheres injected into irradiated cockroaches developed much as in natural hosts. Again the tegument after 7 days and 12 days was rich in microvilli (Pesson and Léger, 1978). These authors concluded that under conditions with weak haemocytic reactions there is continued conflict between the host and metacestode. This is evident as maximal production of microvilli on the metacestodes. They also presented other arguments which they believed supported the conclusion that the haemocytic reaction by itself does not constitute a natural barrier of resistance to *H. nana* in *L. maderae*. Pesson and Léger (1978) continued to maintain that other factors in the region of the intestinal wall, and perhaps other diffusible substances, are what inhibit development of these metacestodes.

Lackie (1976) reported an extensive study of the haemocytic reactions of various adult or larval beetles or adult cockroaches, locusts and gryllids following intrahaemocoel injection with activated oncospheres of *H. diminuta. H. diminuta* established and grew with no appreciable reaction in larval *T. molitor* and in the adult locust (*Schistocerca gregaria*), but only rarely in the adult cockroach (*P. americana*). *H. diminuta* failed to develop in other cockroaches (*L. maderae, Henschoutedenia flexivitta, Blatta orientalis* and *Pycnocelus surinamensis*). Lackie (1976) also transplanted cysticercoids or tissues from several insect species into the haemocoels of other insect species. Such transplants were tolerated without haemocytic reactions only between insect species in which *H. diminuta* normally survived. Supposedly inert latex beads injected into *S. gregaria* were engulfed by numerous haemocytes within 30 min, whereas developing metacestodes injected into the same hosts had only a few attached haemocytes from 30 min to 8 h later. With electron microscopy it was shown that metacestodes grown *in vitro* had microvillar coats similar to those grown *in vivo*. Lackie (1976) suggested, therefore, that the oncospheres and metacestodes of *H. diminuta* either have surfaces inherently compatible with surfaces of host tissues, or somehow can inhibit haemocyte capsule formation in those hosts in which normal growth occurs. Other experiments suggested that haemocyte lysis was not a major protective mechanism. Lackie (1976) concluded that the experiments with tissue transplants strongly suggest that the parasite

> . . . either possesses a surface which is inherently compatible with its host's tissue surfaces, or is able to inhibit capsule formation by haemocytes of certain insect species.

This would refute, therefore, the assertions by Collin (1970) and Ubelaker *et al.* (1970) that such microvillar lysis is a protective measure. Rather, Lackie (1976) suggested such lysis is an incidental happening associated with digestive or excretory processes occurring at the metacestode surface. There is an element of concurrence between the conclusions of Lackie (1976) and of Pesson and Léger (1978).

A subsequent study by Lackie (1981) indicated that serum of a refractory host, *P. americana*, strongly agglutinated oncospheres of *H. diminuta*, whereas the serum of a suitable host, *S. gregaria*, did not. The function of oncospheral agglutination by insects is unknown, but Lackie suggested that it may be of some relevance since both the haemocytes and serum react against *H. diminuta* in *P. americana*, but neither reaction occurs in *S. gregaria*. Neither study by Lackie (1976, 1981) was directly concerned with the pathology or morbidity of the hosts involved, but they are briefly discussed here because they shed light on the immune processes which ultimately may be shown associated with a disease process.

The beetle *T. confusum* readily ingests eggs of *H. diminuta*, and "normal" development to cysticercoids apparently ensues in the beetle haemocoel (Voge and Heyneman, 1957; Heyneman and Voge, 1971). Although there is a limited haemocytic response early in the infection, it is considered to be an ineffective protection (Heyneman and Voge, 1971). The level of pathogenicity, if any, up to this point was not established. Keymer (1980) reported experiments designed specifically to measure the effects of oral infections with *H. diminuta* on the survival and fecundity of *T. confusum*. One procedure was systematically to expose similar groups of beetles to as many as six separate feedings of eggs at uniform intervals and under standard conditions. The second procedure was to expose six groups of beetles to a known level of infection, and then feed them their regular food for 2 weeks. One group was then killed and the other five groups were starved. The groups were killed sequentially when approximately 5/6, 4/6, 3/6, 2/6 and 1/6 of the population survived. Resultant mortality and fecundity, the latter based on egg production per host with increasing parasite burden, were examined as dynamics of the host–parasite associations. Only the conclusions will be included here.

Keymer (1980) showed that there was a linear relationship between the number of exposures to eggs and the numbers of parasites per host. The size of the cysticercoids, based on total length, was non-linear although generally decreasing in length as the parasite burden per host rose. Similarly, the fecundity, i.e. number of eggs produced per infected female per day, declined, albeit not linearly with increasing worm burdens. Keymer (1980) concluded that reduction in host survival was directly proportional to the presence of and the number of parasites per beetle. However,

> . . . due to the effects of over-dispersion in the distribution of parasite numbers/host, this may not result in a linear relationship between *observed* host mortality and parasite burden. (Keymer, 1980).

She suggested that host survival is related to the numbers of parasites penetrating and damaging the gut wall. This is somewhat similar to the conclusion reached by Chen (1934, *vide supra*) concerning mortality associated with massive infections in the larval cat flea, *Ctenocephalides felis*, by oncospheres of *Dipylidium caninum*. Keymer (1980) reasoned that since the relationship between reduction in host fecundity and parasite burden is non-linear, then fecundity is not related to the numbers of penetrating oncospheres. She suggested, however, that host fecundity is affected by the increasing biomass of the parasites in the host. Overall, Keymer (1980) concluded: ". . . that cysticercoids of *H. diminuta* have deleterious effects on both the survival and fecundity of the intermediate host, *T. confusum*."

In their introduction, Schom *et al.* (1981) indicated that orally transmitted infections with *H. nana*, *H. diminuta*, *H. microstoma* and *H. citelli* all caused some mortality to the beetle host *T. confusum*. This occurred most frequently during the first 15 days of infection. Nevertheless the cause of mortality remains unknown, although it is accepted that *H. citelli* is the most pathogenic of the four species (Schom *et al.*, 1981). Beetle mortality, associated with infections by *Hymenolepis* spp., has been linked with (1) beetle starvation before infection, (2) level of parasite crowding, and (3) sex of the beetle. Schom *et al.* (1981) were interested primarily in the evolutionary implications associated with factors affecting host survival in the *T. confusum*/*H. citelli* relationship. Consequently their study generated data of interest here.

In all their experiments, groups of 25 to 30 adult beetles were starved either 1 day or 6 days and then allowed to feed *ad libitum* for 24 h on a few drops of eggs of *H. citelli* concentrated on filter paper (Schom *et al.*, 1981). The eggs were then removed and the beetles received their normal diet for the remaining 15 days of the experiments. Beetles were examined as they died during the next 14 days, and all survivors were killed on day 15 and examined. They found that beetle survival time tended to vary inversely with the number of metacestodes per beetle. Total mortality was higher following pre-starvation for 6 days (73–93%) than pre-starvation for 1 day (51%). Although some beetles died each day, the vast majority that died did so 8 days to 11 days post-infection. Generally beetles that survived for 15 days carried 14 or fewer metacestodes, whereas those that died earlier had 20 or more metacestodes. Total mortality between sexes was essentially the same, but overall, females survived an average 8.8 days and males an average 9.6 days. Furthermore, the cysticercoids appeared to develop faster in male than in female beetles.

Female beetles were more heavily infected than the males, however, which most likely accounted for the apparent differences in mortality between the sexes (Schom *et al.*, 1981). Why most mortality occurred between 8 days and 11 days, whether in males or females, was not established. An interesting feature pointed out by these authors was that *H. citelli* is known to elicit a stronger haemocytic response by *T. confusum* than either *H. nana* or *H. diminuta*. Schom *et al.* (1981) suggested, therefore, that beetle mortality may be related to the extent of the host's cellular response. This would appear to agree with the conclusion of Pesson and Léger (1978) that the haemocytic response does not of itself constitute a natural barrier of resistance. Conceivably, at least under certain circumstances, if not always, the haemocytic reaction may be more pathogenic than beneficial to the insect host.

VIII. CONCLUSION

This review indicates that relatively little is known concerning the broad pattern of pathogenesis associated with metacestodes developing in invertebrate hosts. This is particularly true of those species with marine life-cycles, e.g. of the orders Tetraphyllidea, Trypanorhyncha and Lecanicephalidea. The little that is known is inferred from natural infections, which may be based on possibly erroneous identifications. Experimental evidence is lacking since not even one complete life-cycle for any species has been confirmed experimentally in these three marine orders. This is true for the detailed studies done by Cheng and his colleagues on what is identified as *Tetragonocephalum* (= *Tylocephalum*) sp. from the American oyster. It is likely that this cestode(s) belongs in the order Lecanicephalidea, but morphologically and ecologically it might equally belong in the Tetraphyllidea. Regardless of what the species is, the pathogenesis associated with the metacestodes of this species is one of the few that has received detailed study. In part this occurred because the American oyster is an economically important animal.

In contrast, the complete life-cycles are known for several species of freshwater cestodes, e.g. of the orders Caryophyllidea, Proteocephalidea, Pseudophyllidea and aquatic Cyclophyllidea. A few experimental studies of caryophyllids in oligochaetes have provided valuable information. There also is some information on the effect of the metacestodes of proteocephalids and pseudophyllids reared experimentally in their copepod first hosts. Although the potential to do the necessary detailed experiments is available, very few studies have been done with the primary objective of determining how the metacestodes of these orders may be pathogenic in copepods.

Only with a limited number of cyclophyllidean cestodes, especially *Hymenolepis* spp. that infect terrestrial insects, have studies to determine the pathogenicity of these metacestodes been done under rigid laboratory-controlled conditions. From such studies, and from observations from other experimental or natural infections associated with life history studies, some general conclusions may be drawn.

(1) Massive infections with oncospheres, as can be done experimentally, may cause such serious damage to the host's gut wall as to kill it. Such heavy infections are unlikely, however, under most natural conditions.

(2) There is a limit to the number of metacestodes that can develop parenterally in invertebrates without seriously, or fatally, damaging the host. Damage to the host is related to the size, number, and location of the metacestodes. Even one large caryophyllid or spathebothriid may cause sufficient mechanical, or possibly other, damage to be fatal. Again, natural infections with single metacestodes are generally observed, however, and most such infections usually are not obviously pathogenic.

(3) Some marine cestodes, e.g. tetraphyllids and trypanorhynchs, may have at least two invertebrate hosts in their life-cycles. If metacestodes occur in the gut lumen of the second invertebrate host, e.g. *Scolex pleuronectis*, little damage to the host is evident, even in heavy natural infections. In parenteral sites lesions may be obvious.

(4) Some arthropod and mollusc first hosts are capable of mounting

strong haemocytic responses usually followed by melanization against parenteral metacestodes. There is some evidence suggesting that humoral reactions against metacestodes also may occur, particularly in arthropods. Thus far there is no evidence that one to several previous infections with oncospheres in an arthropod will reduce the number of oncospheres establishing from a subsequent challenge by the same species of cestode. Here there appears to be a dilemma in that, while a haemocytic response followed by melanization is strongest in "atypical" arthropod hosts and weakest in "normal" hosts, such a reaction may itself be pathogenic. There is some evidence suggesting that, as with vertebrates, there may be recognition of the metacestode as either non-self or self. When recognized as non-self the strong reaction may ensue, whereas when recognized as self the reaction, if any, is minimal and transient. Generally, however, the light parenteral infections which are normally seen in nature, are usually well tolerated. Is it possible, then, that both systems are ecologically mediated? Thus, the strong haemocytic and melanistic reaction would protect against an occasional accidental infection. On the other hand, the weak reaction also protects by not mounting a strong, potentially pathogenic reaction against a relatively common, but usually light, well-tolerated infection.

(5) It would appear then that, as indicated in the introduction, the interpretations of Bauer (1958), Read (1972) and Davies *et al.* (1980) relating to parasitism, pathogenicity and colonization, all may be partially correct as far as metacestode infections of invertebrate hosts are concerned. Thus, although even one metacestode may prove detrimental to a "normal" host (Bauer, 1958), this usually is not so. Then, in order not to attack the metacestode with a potentially harmful reaction to itself, the metacestode somehow becomes accepted as self. Therefore, the metacestode–invertebrate host relationship moves toward the metacestode-invertebratehost system (Read, 1972). This, then, could be the first step to potential colony formation (Davies *et al.*, 1980). Although the new relationship may now be nearly "one", it nevertheless is not a true colony, since most metacestodes do not reproduce. If the metacestodes did reproduce, however, the increase in numbers probably would ultimately kill the invertebrate host, destroying the colony.

IX. ACKNOWLEDGEMENTS

Sincere thanks to Ellen B. Freeman for invaluable editorial help during completion of the manuscript, to Miss Molly A. Schlosser for making the drawings for Figs 2, 4, 5 and 6, and to Mrs Amelia da Silva for cheerfully and accurately typing the several versions of this manuscript. Some of the work reported here was funded by the Natural Sciences and Engineering Research Council of Canada.

X. REFERENCES

Aduladze, K. I. (1964). Taeniata of animals and man and diseases caused by them. *In* "Essentials of Cestodology", (K. I. Skrjabin, ed.), Vol. IV. (Transl. M. Raveh and A. Storfer; A. Isseroff, ed., 1970. Clearinghouse, Springfield, VA).

Amin, O. M. (1977). Helminth parasites of some southwestern Lake Michigan fishes. *Proc. helminth. Soc. Wash.* **44**, 210–217.

Amin, O. M. (1978). On the crustacean hosts of larval acanthocephalan and cestode parasites in southwestern Lake Michigan. *J. Parasit.* **64**, 842–845.

Ammann, F. (1955). "Der Befall des Bodenseefische mit *Triaenophorus* unter besonderers Berücksichtigung des biologischen Cyclus", München. (*from* Kuperman, B. I., 1973. *Vide infra*).

Anantaraman, S. and Krishnaswamy, S. (1959). Une larve de Cestode dans le système nerveux de *Squilla holochista* Wood-Mason. *Annls Parasit. hum. comp.* **34**, 593–594.

Anderson, D. P. (1974). "Book 4: Fish immunology." *In* "Diseases of Fishes", (S. F. Snieszko and H. R. Axelrod, eds), T. F. H. Publications Inc., Nepture City, NJ.

Anderson, R. M. (1978). The regulation of host population growth by parasite species. *Parasitology*, **76**, 119–157.

Arai, H. P. (ed.) (1980). "Biology of the Tapeworm *Hymenolepis diminuta*", Academic Press, New York and London.

Baer, J. G. (1948). Contributions à l'étude des cestodes de sélaciens I-IV. *Bull. Soc. neuchâtel. Sci. nat.* **71**, 63–122.

Bauer, O. N. (1958). Relationships between host fishes and their parasites. *In* "Parasitology of Fishes", (V. A. Dogiel, G. K. Petrushevski and Yu. I. Polyanski, eds) (Transl. Z. Kabata, 1960, pp. 84–103; T. F. H. (Great Britain) Ltd, Surrey, England).

Befus, A. D. (1972). Aspects of the biology of *Corallobothrium parafimbriatum* sp. n. and *Corallotaenia minutia* (Cestoda: Proteocephaloidea) from the brown bullhead, *Ictalurus nebulosus.* M.Sc. Thesis, University of Toronto.

Befus, A. D. and Freeman, R. S. (1973). Life cycles of two corallobothriin cestodes (Proteocephaloidea) from Algonquin Park, Ontario. *Can. J. Zool.* **51**, 249–257.

Beklemishev, W. N. (1964). "Principles of Comparative Anatomy of Invertebrates", Vol. 1. Promorphology. (Transl., J. M. MacLennan; Z. Kabata, ed. 1969. University of Chicago Press, Chicago).

Brooks, M. A. (1975). Symbiosis and Attenuation. *In* "Pathobiology of Invertebrate Vectors of Disease", (L. A. Bulla, Jr and T. C. Cheng, eds). *Ann. N.Y. Acad. Sci.* **266**, 166–172.

Brown, E. L. and Threlfall, W. (1968). Helminth parasites of the Newfoundland short-finned squid, *Illex illecebrosus illecebrosus* (LeSueur) (Cephalopoda: Decapoda). *Can. J. Zool.* **46**, 1059–1070.

Burt, M. D. B. and Sandeman, I. M. (1969). Biology of *Bothrimonus* (=*Diplocotyle*) (Pseudophyllidea: Cestoda) Part I. History, description, synonomy and systematics. *J. Fish. Res. Bd Can.* **26**, 975–996.

Burton, R. W. (1963). A cestode microparasite of the oyster, *Crassostrea virginica*, from Florida. Proc. A. Shellfish Mortality Conf., Jan. 23–30, pp. 61–62. Bureau of Commercial Fisheries Biological Laboratory, Oxford, Md.

Cake, E. W. (1979). *Polypocephalus* sp. (Cestoda: Lecanicephalidae): A description of tentaculo-plerocercoids from bay scallops of the Northeastern Gulf of Mexico. *Proc. helminth. Soc. Wash.* **46**, 165–170.

Calentine, R. L. (1964). The life cycle of *Archigetes iowensis* (Cestoda: Caryophyllaeidae). *J. Parasit.* **50**, 454–458.

Calentine, R. L. (1965a). The biology and taxonomy of *Biacetabulum* (Cestoda: Caryophyllaeidae). *J. Parasit.* **51**, 243–248.

Calentine, R. L. (1965b). Larval development of four caryophyllaeid cestodes. *Proc. Iowa Acad. Sci.* **72**, 418–424.

Calentine, R. L. and DeLong, B. L. (1966). *Archigetes sieboldi* (Cestoda: Caryophyllaeidae) in North America. *J. Parasit.* **52**, 428–431.

Calentine, R. L., Christensen, B. M. and Christensen, L. A. (1970). Specificity of caryophyllaeid cestodes for their intermediate hosts. *J. Parasit.* **56**, 346–349.

Cavier, R. and Léger, N. (1965). A propos de l'évolution d'*Hymenolepis nana* var. *fraterna* chez des hôtes intermédiares inhabituels. *Annls Parasit. hum. comp.* **40**, 651–659.

Chen, H. T. (1934). Reactions of *Ctenocephalides felis* to *Dipylidium caninum*. *Z. ParasitKde.* **6**, 603–637.

Cheng, T. C. (1966). The coracidium of the cestode *Tylocephalum* and the migration and fate of this parasite in the American oyster *Crassostrea virginica*. *Trans. Am. microsc. Soc.* **85**, 246–255.

Cheng, T. C. (1967). Marine molluscs as hosts for symbiosis: With a review of known parasites of commercially important species. *In* "Advances in Marine Biology", (F. S. Russell, ed.), Vol. 5, pp. 69–70, 102–103, 254–261. Academic Press, London and New York.

Cheng, T. C. (1973). "General Parasitology", Academic Press, New York and London.

Cheng, T. C. and Rifkin, E. (1968). The occurrence and resorption of *Tylocephalum* metacestodes in the clam *Tapes semidecussata*. *J. Invert. Path.* **10**, 65–69.

Cheng, T. C. and Rifkin, E. (1970). Cellular reactions in marine molluscs in response to helminth parasitism. *In* "Diseases of Fishes and Shellfishes", (S. F. Snieszko, ed.), pp. 443–496. Spec. Publ. No. 5, Amer. Fish. Soc., Washington, D.C.

Clarke, A. S. (1954). Studies on the life cycle of the pseudophyllidean cestode *Schistocephalus solidus*. *Proc. zool. Soc. Lond.* **124**, 257–302.

Collin, W. K. (1970). Electron microscopy of postembryonic stages of the tapeworm *Hymenolepis citelli*. *J. Parasit.* **56**, 1159–1170.

Damian, R. T. (1979). Molecular mimicry in biological adaptation. *In* "Host–Parasite Interfaces", (B. B. Nickol, ed.), pp. 103–126. Academic Press, London and New York.

Davies, A. J. S., Hall, J. G., Targett, G. A. T. and Murray, M. (1980). The biological significance of the immune response with special reference to parasites and cancer. *J. Parasit.* **66**, 705–721.

Dollfus, R. Ph. (1923a). Le cestode des perles fines des Méléagrines de Nossi-Bé. *C. r. hebd. Séanc. Acad. Sci., Paris*, **176**, 1265–1267.

Dollfus, R. Ph. (1923b). Enumération des cestodes du plancton et des invertébrés marins. *Annls Parasit. hum. comp.* **1**, 276–300, 363–394.

Dollfus, R. Ph. (1924). Enumération des cestodes du plancton et des invertébrés marins. Note additionelle. *Annls Parasit. hum. comp.* **2**, 86–89.

Dollfus, R. Ph. (1929). Addendum à mon "Enumération des cestodes du plancton et des invertébrés marins". *Annls Parasit. hum. comp.* **7**, 325–347.

Dollfus, R. Ph. (1931). Nouvel addendum à mon "Enumération des cestodes du plancton et des invertébrés marins". *Annls Parasit. hum. comp.* **9**, 552–560.

Dollfus, R. Ph. (1936). Invertébrés marins et Thalassoides. *In Cestodes—Faune de France*, (Ch. Joyeux and J. G. Baer, eds), **30**, 509–539.

Dollfus, R. Ph. (1942). Etudes critiques sur les tétrarhynques du muséum de Paris. *Archs Mus. natn. Hist. nat., Paris, sér. 6*, **19**, 7–466.

Dollfus, R. Ph. (1946). Notes diverses sur des Tétrarhynques. *Mém. Mus. natn. Hist. nat., Paris, nouv. sér.* **22**, 179–220.

Dollfus, R. Ph. (1958). Copépodes, isopodes et helminthes parasites de céphalopodes de la Méditerranée et de l'Atlantique européen. *Faune mar. Pyrén.-orient.*, no. 1, 61–72.

Dollfus, R. Ph. (1964a). Sur le cycle évolutif d'un cestode diphyllide. Identification de la larve chez *Carcinus maenas* (L. 1758) hôte intermédiare. *Annls Parsit. hum. comp.* **39**, 235–241.

Dollfus, R. Ph. (1964b). Enumération des cestodes du plancton et des invertébrés marins. (6e contribution). *Annls Parasit. hum. comp.* **39**, 329–379.

Dollfus, R. Ph. (1967). Enumération des cestodes du plancton et des invertébrés marins. (7e contribution). *Annls Parasit. hum. comp.* **42**, 155–178.

Dollfus, R. Ph. (1969). Quelques espèces de cestodes tétrarhynques de la côte Atlantique des Etats-Unis, dont l'une n'était pas connu à l'état adulte. *J. Fish. Res. Bd Can.* **26**, 1037–1061.

Dollfus, R. Ph. (1974). Enumération des cestodes du plancton et des invertébrés marins. (8e contribution). *Annls Parasit. hum. comp.* **49**, 381–410.

Dollfus, R. Ph. (1976). Enumération des cestodes de plancton et des invertébrés marins. (9e contribution). *Annls Parasit. hum. comp.* **51**, 207–220.

Dubinina, M. N. (1965). [Cestoda: Ligulidae in the Fauna of the USSR]. Avtoreferat diss., Leningrad. (*from* Kuperman, B. I. 1973. *Vide infra*).

Essex, H. E. (1928). The structure and development of *Corallobothrium*. *Ill. biol. Monogr.* **11**, 1–74.

Euzet, L. (1955). Quelques cestodes de *Myliobatis aquila* L. *Recl Trav. Labs Bot. Géol. Zool. Univ. Montpellier, Sér. zool., fasc. I*, pp. 19–27.

Euzet, L. (1959). Recherches sur les cestodes tétraphyllides des sélaciens des côtes de France. Thèse. Université de Montpellier. Causse, Graille et Castelneau, Montpellier.

Euzet, L. and Combes, C. (1965). Contribution à l'étude de *Tetragonocephalum uarnak* (Shipley et Hornell, 1906). *Bull. Soc. neuchâtel. Sci. nat.* **88**, 101–122.

Feigenbaum, D. and Carnuccio, J. (1976). Comparison between the trypanorhynchid cestode infections of *Penaeus duorarum* and *Penaeus brasiliensis* in Biscayne Bay, Florida. *J. Invert. Path.* **28**, 127–130.

Fields, W. G. (1965). The structure, development, food relations, reproduction, and life history of the squid *Loligo opalescens* Berry. *Fish Bull. Calif.* **136**, 1–108.

Fischer, H. (1967). The life history of *Proteocephalus fluviatilis* Bangham (Cestoda) from smallmouth bass, *Micropterus dolomieui* Lacépède. M.Sc. Thesis, University of Toronto.

Fischer, H. (1968). The lifecycle of *Proteocephalus fluviatilis* Bangham (Cestoda) from smallmouth bass, *Micropterus dolomieui* Lacépède. *Can. J. Zool.* **46**, 569–579.

Fischer, H. (1972). Studies on the life cycles and ecology of *Proteocephalus ambloplitis*, *P. fluviatilis* and *P. pearsei* (Cestoda) from Lake Opeongo, Algonquin Park. Ph.D. Thesis, University of Toronto.

Fischer, H. and Freeman, R. S. (1973). The role of plerocercoids in the biology of *Proteocephalus ambloplitis* (Cestoda) maturing in the smallmouth bass. *Can. J. Zool.* **51**, 133–141.

Fisher, W. K. (1946). Echiuroid worms of the North Pacific Ocean. *Proc. U.S. natn. Mus.* No. 3198, **96**, 215–292.

Freeman, R. S. (1964). On the biology of *Proteocephalus parallacticus* MacLulich (Cestoda) from smallmouth bass, *Micropterus dolomieui* Lacépède. *Can. J. Zool.* **46**, 569–579.

Freeman, R. S. (1973). Ontogeny of cestodes and its bearing on their phylogeny and systematics. *In* "Advances in Parasitology", (Ben Dawes, ed.), Vol. 11, pp. 481–557. Academic Press, London and New York.

Freeman, R. S. (1981). Unified terminology and the logic of cestode ontogeny. "Review of Advances in Parasitology" (Proc. ICOPA IV) (in press).

Freeman, R. S. (1982a). How did tapeworms get that way? *Bull. Can. Soc. Zool.* **13**, 5–8.

Freeman, R. S. (1982b). Do any *Anonchotaenia*, *Cyathocephalus*, *Echeneibothrium* or *Tetragonocephalum* (= *Tylocephalum*) (Eucestoda) have hookless oncospheres or coracidia? *J. Parasit.* **68**, 737–743.

Freze, V. I. (1965). Proteocephalata in fish, amphibians and reptiles. *In* "Essentials of Cestodology", (K. I. Skrjabin, ed.), Vol. V. (Transl. R. Berick and A. Birron, 1970. Clearinghouse, Springfield, Va).

Fuhrmann, O. (1931). Dritte Klasse des Cladus Platyhelminthes. Cestoidea. *In* "Handb. Zool." (Kükenthal and Krumbach, eds), Vol. 2, pp. 141–416. Berlin and Leipzig.

Gruber, F. A. (1878). Ein neuer Cestoden-Wirt. *Zool. Anz.* **1**, 74–75.

Guttowa, A. (1961). Experimental investigations on the systems "procercoids of *Diphyllobothrium latum* (L.)—*Copepoda*". *Acta parasit. pol.* **9**, 371–408.

Guttowa, A. (1967). Experimental coinfection of *Copepoda* naturally infected with *Proteocephalus* sp. with the larvae of *Diphyllobothrium latum* (L.). *Acta parasit. pol.* **14**, 399–404.

Guttowa, A. (1971). [Host-parasite interrelationships in the system Pseudophyllidea procercoids—Copepoda in the light of physiological investigations]. *Trudy Vsesoyuznogo Instituta Gel'mintologii im K. I. Skrjabina*, **17**, 33–35. (See Abstract No. 3339, *Helminth. Abstr.* (Ser. A) (1973) **42**, 715).

Hall, M. C. (1929). Arthropods as intermediate hosts of helminths. *Smithson. misc. Collns*, **81**, 1–77.

Hamilton, K. A. and Byram, J. E. (1974). Tapeworm development: the effects of urea on a larval tetraphyllidean. *J. Parasit.* **60**, 20–28.

Herde, A. (1938). Early development of *Ophiotaenia perspicua* LaRue. *Trans. Am. miscrosc. Soc.* **57**, 282–291.

Herdman, W. A. and Hornell, J. (1906). Pearl production. *In* "Report to the government of Ceylon on the pearl oyster fisheries of the Gulf of Manaar", (W. A. Herdman, ed.), Part V, pp. 1–42. Royal Society, London.

Heyneman, D. and Voge, M. (1971). Host response of the flour beetle, *Tribolium confusum* to infections with *Hymenolepis diminuta, H. microstoma*, and *H. citelli* (Cestoda: Hymenolepididae). *J. Parasit.* **57**, 881–886.

Humes, A. G. (1950). Experimental copepod hosts of the broad tapeworm of man *Dibothriocephalus latus* (L.). *J. Parasit.* **36**, 541–547.

Hunter, G. W., III (1929). Life-history studies on *Proteocephalus pinguis* LaRue. *Parasitology*, **21**, 487–496.

Hutton, R. F., Sogandares-Bernal, F., Edred, B., Ingle, R. M. and Woodburn, K. D. (1959). Investigations on the parasites of and diseases of saltwater shrimps (Penaeidae) of sports and commercial importance to Florida. *Fla. St. Bd Conserv., Tech. Ser. No.* **26**, 1–38.

Jameson, H. L. (1912). Studies on pearl oysters and pearls. I. The structure of the shell and pearls of the Ceylon pearl-oyster (*Margaritifera vulgaris* Schumacher): with an examination of the cestode theory of pearl production. *Proc. zool. Soc. Lond.* **1912**, 266–358.

Janicki, C. von and Rosen, F. (1917). Le cycle évolutif du *Dibothriocephalus latus* L. *Bull. Soc. neuchâtel. Sci. nat.* **42**, 19–53.

Jarecka, L. (1959). On the life-cycle of *Bothriocephalus claviceps* (Goeze, 1782). *Acta parasit. pol.* **7**, 527–532.

Jarecka, L. (1964). Cycle évolutif à un seul hôte intermédiare chez *Bothriocephalus claviceps* (Goeze, 1782) cestode de *Anguilla anguilla* L. *Annls Parasit. hum. comp.* **39**, 149–156.

Joyeux, Ch. and Baer, J. G. (1961). Classe des Cestodes. *In* "Traité de Zoologie", (P. P. Grassé, ed.), Vol. IV, pp. 347–560. Masson, Paris.

Katkansky, S. C. and Warner, R. W. (1969). Infestation of the rough-sided little-neck clam, *Protothaca laciniata*, in Morro Bay, California, with larval cestodes (*Echeneibothrium* sp.). *J. Invert. Path.* **13**, 125–128.

Katkansky, S. C., Warner, R. W. and Poole, R. L. (1969). On the occurrence of larval cestodes in the Washington clam, *Saxidomus nuttalli*, and the gaper clam, *Tresus nuttalli*, from Drakes Estero, California. *Calif. Fish Game*, **55**, 317–322.

Kay, M. W. (1942). A new species of *Phyllobothrium* van Beneden from *Raja binoculata* (Girard). *Trans. Am. microsc. Soc.* **61**, 261–266.

Kennedy, C. R. (1965). The life-history of *Archigetes limnodrili* (Yamaguti) (Cestoda: Caryophyllaeidea) and its development in the invertebrate host. *Parasitology*, **55**, 427–437.

Kennedy, C. R. (1972). The effect of the cestode *Caryophyllaeus laticeps* upon the reproduction and respiration of its intermediate host. *Parasitology*, **64**, 485–499.

Keymer, A. E. (1980). The influence of *Hymenolepis diminuta* on the survival and fecundity of the intermediate host *Tribolium confusum*. *Parasitology*, **81**, 405–421.

Kisielewska, K. (1959). Types of Copepoda and *Drepanidotaenia lanceolata* (Bloch) host–parasite systems established experimentally. *Acta parasit. pol.* **7**, 371–392.

Klekowski, R. Z. and Guttowa, A. (1968). Respiration of *Eudiaptomus gracilis* infected with *Diphyllobothrium latum*. *Expl. Parasit.* **22**, 279–287.

Kotel'nikov, G. A. (1963). [Reservoir parasitism in hymenolepidids.] Problemy

Parazitologii (*Trudy IV Nauchnoĭ Konferentsii Parazitologov UKrSSR*), pp. 64–67. (See Abstract No. 4166, *Helminth. Abstr.* (Ser. A) (1972) **41**, 571).

Kruse, D. N. (1959). Parasites of the commercial shrimps, *Penaeus aztecus* Ives, *P. duorarum* Burkenroad and *P. setiferus* (Linnaeus). *Tulane Stud. Zool.* **7**, 123–144.

Kuperman, B. I. (1973). ["Tapeworms of the Genus *Triaenophorus*. Parasites of Fishes"]. *Izdatel'stvo "Nauka", Leningrad.* (Transl. B. R. Sharma, 1981, pp. 149–156; Amerind Publishing Co. Pvt. Ltd, New Delhi).

Lackie, A. M. (1976). Evasion of the haemocytic defense reaction of certain insects by larvae of *Hymenolepis diminuta* (Cestoda). *Parasitology*, **73**, 97–107.

Lackie, A. M. (1981). Humoral mechanisms in the immune response of insects to larvae of *Hymenolepis diminuta* (Cestoda). *Parasite Immun.* **3**, 201–208.

Léger, N. and Cavier, R. (1970). A propos de l'évolution d'*Hymenolepis nana* var. *fraterna*, chez des hôtes intermédiares inhabituels. *Annls Parasit. hum. comp.* **45**, 195–210.

Leid, R. W. Jr and Williams, J. F. (1979). Helminth parasites and the host inflammatory system. *In* "Chemical Zoology", (M. Florkin and B. T. Scheer, eds), Vol. 11, pp. 229–271. Academic Press, London and New York.

le Riche, W. H. (1967). World incidence and prevalence of the major communicable diseases. *In* "Ciba Foundation Symposium on Health of Mankind", (G. Wolstenholme and M. O'Connor, eds), pp. 1–46. J. and A. Churchill Ltd, London.

Lethbridge, R. C. (1971). The hatching of *Hymenolepis diminuta* eggs and penetration of the hexacanths in *Tenebrio molitor* beetles. *Parasitology*, **62**, 445–456.

Lethbridge, R. C. (1980). The biology of the oncosphere of cyclophyllidean cestodes. *Helminth. Abstr. (Ser. A)*, **49**, 59–72.

Li, H. C. (1929). The life histories of *Diphyllobothrium decipiens* and *D. erinacei. Am. J. Hyg.* **10**, 527–550.

MacKenzie, K. (1975). Some aspects of the biology of the plerocercoid of *Gilguinia squali* Fabricius 1794 (Cestoda: Trypanorhyncha). *J. Fish Biol.* **7**, 321–327.

Mackiewicz, J. S. (1972). Caryophyllidea (Cestoidea): A review. *Expl. Parasit.* **31**, 417–512.

Malek, E. A. and Cheng, T. C. (1974). "Medical and Economic Malacology", Academic Press, New York and London.

Marshall, A. G. (1967). The cat flea, *Ctenocephalides felis* (Bouché, 1835) as an intermediate host for cestodes. *Parasitology*, **57**, 419–430.

Marshall, S. M., Nicholls, A. G. and Orr, A. P. (1934). On the biology of *Calanus finnmarchicus.* V. Seasonal distribution, size, weight and chemical composition in Loch Striven in 1933, and their relation to phytoplancton. *J. mar. biol. Ass. U.K., N.S.* **19**, 793–827.

Meggitt, F. J. (1914). The structure and life history of a tapeworm (*Ichthyotaenia filicollis* Rud.) parasitic in the stickleback. *Proc. zool. Soc. Lond.* **8**, 113–138.

Mehlhorn, H., Becker, B., Andrews, P. and Thomas, H. (1981). On the nature of the proglottids of cestodes: a light and electron microscopic study on *Taenia, Hymenolepis* and *Echinococcus. Z. ParasitKde.* **65**, 243–259.

Michajlow, W. (1932). Les adaptations graduelles de Copépodes comme premiers hôtes intermédiares de *Triaenophorus nodulosus* Pall. *Annls Parasit. hum. comp.* **10**, 334–344.

Michajlow, W. (1953). O stosunkach weronatrzgatunkowych w populacjach procerkoidow *Triaenophorus lucii* (Müll.). *Acta parasit. pol.* **1**, 1–28.

Miller, R. B. (1943). Studies on cestodes of the genus *Triaenophorus* from fish of Lesser Slave Lake, Alberta. II. The eggs, coracidia and life in the first intermediate host of *Triaenophorus crassus* Forel and *T. nodulosus* (Pallas). *Can. J. Res. Series D*, **21**, 284–291.

Morozova, M. E. (1955). [Biology of the early phases of development of broad tapeworm in Karelian-Finnish SSR]. Thesis. [Karelian-Finnish Branch, Acad. Sci. USSR]. (*from* Kuperman, B.I. (1973) *Vide supra*).

Mudry, D. R. and Dailey, M. D. (1971). Postembryonic development of certain tetraphyllidean and trypanorhynchan cestodes with a possible alternative life cycle for the order Trypanorhyncha. *Can. J. Zool.* **49**, 1249–1253.

Mueller, J. F. (1959). The laboratory propagation of *Spirometra mansonoides* (Mueller, 1935) as an experimental tool. II. Culture and infection of the copepod hosts and harvesting the procercoid. *Trans. Am. microsc. Soc.* **78**, 245–255.

Mueller, J. F. (1966). The laboratory propagation of *Spirometra mansonoides* (Mueller, 1935) as an experimental tool. VII. Improved techniques and additional notes on the biology of the parasite (Cestoda). *J. Parasit.* **52**, 437–443.

Nakajima, K. and Egusa, S. (1969a). [Studies on a new trypanorhynchan larva, *Callotetrarhynchus* sp., parasitic on cultured yellowtail. II. On the source and route of infection]. *Bull. Jap. Soc. scient. Fish.* **35**, 351–357.

Nakajima, K. and Egusa, S. (1969b). [Studies on a new trypanorhynchan larva, *Callotetrarhynchus* sp., parasitic on cultured yellowtail. III. On the anchovy worm]. *Bull. Jap. Soc. scient. Fish.* **35**, 723–729.

Nakajima, K. and Egusa, S. (1969c). [Studies on a new trypanorhynchan larva, *Callotetrarhynchus* sp., parasitic on cultured yellowtail. IV. On the development of the scolex]. *Bull. Jap. Soc. scient. Fish.* **35**, 730–736.

Nakajima, K. and Egusa, S. (1972). [Studies on a new trypanorhynchan larva, *Callotetrarhynchus* sp., parasitic on cultured yellowtail. XII. Free proglottid]. *Bull. Jap. Soc. scient. Fish.* **38**, 1333–1340.

Nappi, A. J. (1975). Parasite encapsulation in insects. *In* "Invertebrate Immunity. Mechanisms of Invertebrate Vector–Parasite Reactions", (K. Maramarasch and R. E. Shope, eds), pp. 293–326. Academic Press, New York and London.

Overstreet, R. M. (1973). Parasites of some penaeid shrimps with emphasis on reared hosts. *Aquaculture*, **2**, 105–140.

Overstreet, R. M. (1978). Marine maladies? Worms, germs, and other symbionts from the Northern Gulf of Mexico. *Mississippi–Alabama Sea Grant Consortium*, 78–021, 140 p. Blossman Printing, Inc. Ocean Springs, Miss.

Overstreet, R. M. (1979). Crustacean health research at the Gulf Coast Research Laboratory. *Proc. Second bienn. Crustacean Health Workshop*, pp. 300–314. Texas A & M University, College Station, Tx.

Overstreet, R. M. (1982). Metazoan symbionts of Crustaceans. *In* "Biology of Crustacea", Academic Press. (by permission of the author) (in press).

Pesson, B. and Léger, N. (1975). *Hymenolepis nana* var. *fraterna* (Cestoda:

Hymenolepididae) chez *Leucophaea maderae* (Dictyoptera: Blattidae): la traversée de la paroi intestinale. *Annls Parasit. hum. comp.* **50**, 425–437.

Pesson, B. and Léger, N. (1977). La destinée d'*Hymenolepis nana* var. *fraterna* (Cestode) chez un hôte inhabituel: *Leucophaea maderae* (Dictyoptère). *Annls Parasit. hum. comp.* **52**, 78–80.

Pesson, B. and Léger, N. (1978). *Hymenolepis nana* var. *fraterna* (Cestoda: Hymenolepididae) chez *Leucophaea maderae* (Dictyoptera: Blattidae): l'expression du conflit hôte–parasite après inhibition expérimentale de la réaction hemocytaire. *Annls Parasit. hum. comp.* **53**, 147–154.

Pesson, B., Léger, N. and Bouchet, P. (1978). Le développement du cysticercoïde d'*Hymenolepis nana* var. *fraterna* (Cestoda: Hymenolepididae) dans la cavité général de *Tenebrio molitor* (Coleoptera: Tenebrionidae) et de *Leucophaea maderae* (Dictyoptera: Blattidae). *Annls Parasit. hum. comp.* **53**, 155–161.

Poinar, G. O. Jr (1969). Arthropod immunity to worms. *In* "Immunity to Parasitic Animals", (G. J. Jackson, R. Herman and I. Singer, eds), Vol. I, pp. 173–210. Appleton-Century-Crofts, New York.

Read, C. P. (1972). "Animal Parasitism", Prentice-Hall, Inc., Englewood Cliffs, NJ.

Reichenbach-Klinke, H. H. (1956). Die Entwicklung der Larven bei der Bandwürmerordnung Tetraphyllidea Braun, 1900. *Abh. braunschw. wiss. Ges.* **8**, 61–73.

Rifkin, E. and Cheng, T. C. (1968). The origin, structure, and histochemical characterization of encapsulating cysts in the oyster *Crassostrea virginica* parasitized by the cestode *Tylocephalum* sp. *J. Invert. Path.* **10**, 54–64.

Rifkin, E., Cheng, T. C. and Hohl, H. R. (1969). An electron microscope study of the constituents of encapsulating cysts in *Crassostrea virginica* formed in response to *Tylocephalum* metacestodes. *J. Invert. Path.* **14**, 211–226.

Rifkin, E., Cheng, T. C. and Hohl, H. R. (1970). The fine structure of the tegument of *Tylocephalum* metacestodes: with emphasis on a new type of microvilli. *J. Morph.* **130**, 11–24.

Riser, N. W. (1951). The procercoid larva of *Lacistorhynchus tenuis* (van Ben., 1858). *J. Parasit.* **37**, (Suppl.), p. 26 (Abstr.).

Riser, N. W. (1956a). Observations on the plerocercoid larva of *Pelichnibothrium speciosum* Monticelli 1889. *J. Parasit.* **42**, 31–33.

Riser, N. W. (1956b). Early larval stages of two cestodes from elasmobranch fishes. *Proc. helminth. Soc. Wash.* **23**, 120–124.

Roberts, R. J. (ed.) (1978). "Fish Pathology", Bailliere Tindall, London.

Ruszkowski, J. S. (1934). Etudes sur le cycle évolutif et sur la structure des cestodes de mer. IIIe partie. Le cycle évolutif du Tétrarhynche *Grillotia erinaceus* (van Beneden 1858). *Mém. Acad. pol. Sci., Classe Sci. Math. nat. Sér.* B, pp. 1–9.

Ryšavý, B. (1961). [The problem of reservoir parasitism in Hymenolepididae]. *Helminthologia*, **3**, 288–293.

Salt, G. (1963). The defence reactions of insects to metazoan parasites. *Parasitology*, **53**, 527–642.

Salt, G. (1970). "The Cellular Defence Reactions of Insects", Cambridge University Press, New York.

Sandeman, I. M. and Burt, M. D. B. (1972). Biology of *Bothrimonus* (=*Diplo-*

cotyle) (Pseudophyllidea: Cestoda): ecology, lifecycle and evolution; a review and synthesis. *J. Fish. Res. Bd Can.* **29**, 1381–1395.

Schiller, E. L. (1959). Experimental studies on morphological variation in the cestode genus *Hymenolepis.* I. Morphology and development of the cysticercoid of *H. nana* in *Tribolium confusum. Expl. Parasit.* **8**, 91–118.

Schmidt, G. D. (1970). "How to Know the Tapeworms", Wm. C. Brown Co. Publishers, Dubuque, Iowa.

Schom, C., Novak, M. and Evans, W. S. (1981). Evolutionary implications of *Tribolium confusum—Hymenolepis citelli* interactions. *Parasitology*, **83**, 77–90.

Scott, K. J. and Bullock, W. L. (1974). *Psammonyx nobilis* (Amphipoda: Lysianassidae), a new host for *Bothrimonus sturionis* (Cestoda: Pseudophyllidae [sic]). *Proc. helminth. Soc. Wash.* **41**, 256–257.

Shipley, A. E. and Hornell, J. (1904). The parasites of the pearl oyster. *In* "Report to the Government of Ceylon on the Pearl Oyster Fisheries of the Gulf of Manaar", (W. A. Herdman, ed.), Part II, pp. 77–106. Royal Society, London.

Shipley, A. E. and Hornell, J. (1905). Further report on parasites. *In* "Report to the Government of Ceylon on the Pearl Oyster Fisheries of the Gulf of Manaar", (W. A. Herdman, ed.), Part III, pp. 49–56. Royal Society, London.

Sparks, A. K. (1963). Infection of *Crassostrea virginica* (Gmelin) from Hawaii with a larval tapeworm, *Tylocephalum. J. Insect Path.* **5**, 284–288.

Sparks, A. K. and Chew, K. K. (1966). Gross infestation of the littleneck clam (*Venerupis staminea*) with a larval cestode (*Echeneibothrium* sp.). *J. Invert. Path.* **8**, 413–416.

Sparks, A. K. and Fontaine, C. T. (1973). Host response in the white shrimp, *Penaeus setiferus*, to infection by the larval trypanorhynchid cestode *Prochristianella penaei. J. Invert. Path.* **22**, 213–219.

Stark, G. T. C. (1965). *Diplocotyle* (Eucestoda), a parasite of *Gammarus zaddachi* in the estuary of the Yorkshire Esk, Britain. *Parasitology*, **55**, 415–420.

Stephen, D. (1978). First record of the larval cestode *Tylocephalum* from the Indian backwater oyster, *Crassostrea madrasensis. J. Invert. Path.* **32**, 110–111.

Thomas, L. J. (1937). Environmental relations and the life history of *Bothriocephalus rarus* Thomas, a tapeworm infecting the newt *Triturus viridescens. J. Parasit.* **23**, 133–152.

Thomas, L. J. (1941). The life cycle of *Ophiotaenia perspicua* LaRue, a cestode of snakes. *Revta Med. trop. Parasit. Habana*, **7**, 74–78.

Thurston, J. P. (1967). The morphology and life-cycle of *Cephalochlamys namaquensis* (Cohn, 1906) (Cestoda: Pseudophyllidea) from *Xenopus muelleri* and *X. laevis. Parasitology*, **57**, 187–200.

Tripp, M. R. (1969). General mechanisms and principles of invertebrate immunity. *In* "Immunity to Parasitic Animals", (G. J. Jackson, R. Herman and I. Singer, eds), Vol. I, pp. 111–128. Appleton-Century-Crofts, New York.

Ubelaker, J. E., Cooper, N. B. and Allison, V. F. (1970). Possible defensive mechanism of *Hymenolepis diminuta* cysticercoids to hemocytes of the beetle *Tribolium confusum. J. Invert. Path.* **16**, 310–312.

Valkounová, J. (1973). Reservoir parasitism in cestodes of the family Hymenolepididae (Ariola, 1899) parasitic in domestic and wild ducks. *Věst.čsl. Spol. zool.* **37**, 71–75 (read in abstract).

Villella, J. B., Iversen, E. S. and Sindermann, C. J. (1970). Comparison of the parasites of pond-reared and wild pink shrimp (*Penaeus duorarum* Burkenroad) in South Florida. *Trans. Am. Fish. Soc.* **99**, 789–794.

Voge, M. and Graiwer, M. (1964). Development of oncospheres of *Hymenolepis diminuta*, hatched *in vivo* and *in vitro*, in the larvae of *Tenebrio molitor*. *J. Parasit.* **50**, 267–270.

Voge, M. and Heyneman, D. (1957). Development of *Hymenolepis nana* and *Hymenolepis diminuta* (Cestoda: Hymenolepididae) in the intermediate host *Tribolium confusum*. *Univ. Calif. Publs Zool.* **59**, 549–579.

Vogel, H. (1929). Studien zur Entwicklung von *Diphyllobothrium*. II Teil. Die Entwicklung des Procercoids von *Diphyllobothrium latum*. *Z. ParasitKde.* **2**, 629–644.

Wagner, E. D. (1954). The life history of *Proteocephalus tumidocollus* Wagner, 1953 (Cestoda), in rainbow trout. *J. Parasit.* **40**, 489–498.

Wagner, O. (1917). Über den Entwicklungsgang und Bau einer Fischtaenie (*Ichthyotaenia torulosa* Batsch). *Jena Z. Naturw.* **55**, 1–66.

Wardle, R. A. (1935). Fish-tapeworm. *Bull. Biol. Bd Can.* **45**, 1–25.

Wardle, R. A. and McLeod, J. A. (1952). "The Zoology of Tapeworms", University of Minnesota Press, Minneapolis.

Wardle, R. A., McLeod, J. A. and Radinovsky, S. (1974). "Advances in the Zoology of Tapeworms, 1950–1970", University of Minnesota Press, Minneapolis.

Warner, R. W. and Katkansky, S. C. (1969a). Infestation of the clam *Protothaca staminea* by two species of tetraphyllidean cestodes (*Echeneibothrium* spp.). *J. Invert. Path.* **13**, 129–133.

Warner, R. W. and Katkansky, S. C. (1969b). A larval cestode from the Pismo clam, *Tivela stultorum*. *Calif. Fish Game*, **55**, 248–251.

Watson, N. H. F. and Price, J. L. (1960). Experimental infections of cyclopid copepods with *Triaenophorus crassus* Forel and *T. nodulosus* (Pallas). *Can. J. Zool.* **38**, 345–356.

Willey, A. (1907). Report on the window-pane oysters (*Placuna placenta*, "Muttuchchippi") in the back waters of the Eastern Province. (June, 1907). *Spolia zeylan.* **5**, 33–57.

Williams, H. H. (1966). The ecology, functional morphology and taxonomy of *Echeneibothrium* Beneden, 1849 (Cestoda: Tetraphyllidea), a revision of the genus and comments on *Discobothrium* Beneden, 1870, *Pseudoanthobothrium* Baer, 1956, and *Phormobothrium* Alexander, 1963. *Parasitology*, **56**, 227–285.

Williams, H. H. (1968). The taxonomy, ecology and host-specificity of some Phyllobothriidae (Cestoda: Tetraphyllidea), a critical revision of *Phyllobothrium* Beneden, 1849 and comments on some allied genera. *Phil. Trans. R. Soc., Ser. B, Biol. Sci.* No. 786; **253**, 231–307.

Williams, H. H. (1969). The genus *Acanthobothrium* Beneden, 1849. *Nytt Mag. Zool.* **17**, 1–56.

Wiśniewski, L. W. (1932). Zur postembryonalen Entwicklung von *Cyathocephalus truncatus* Pall. *Zool. Anz.* **98**, 213–218.

Wolf, P. H. (1976). Occurrence of larval stages of *Tylocephalum* (Cestoda: Lecanicephaloidea) in two oyster species from northern Australia. *J. Invert. Path.* **27**, 129–131.

Wundsch, H. H. (1912). Neue Plerocercoïde aus marinen Copepoden. *Arch. Naturgesch.* 78, *Abt.* A, *Heft*, **9**, 1–20.

Yamaguti, S. (1959). "Systema Helminthum. Vol. II. The Cestodes of Vertebrates", Interscience Publishers, New York.

Chapter 13

Pathology of Cestode Infections in the Vertebrate Host

C. Arme, J. F. Bridges and D. Hoole

I. INTRODUCTION

To attempt a comprehensive survey of the pathology of cestode infections in vertebrates is a daunting task; to achieve this objective within a limited space is impossible. We therefore seek the understanding of readers for the selectivity of our approach. Fortunately there exist several reviews of specialized and older literatures, consultation of which will go some way to remedying the deficiencies of this contribution (Rees, 1967; Smyth and Heath, 1970; Kennedy, 1976; Brown and Voge, 1982).

We have restricted our treatment of adult cestodes to three types, *Hymenolepis diminuta*, *H. microstoma* and *Diphyllobothrium latum*. These were selected because much is known concerning the interactions with their hosts. With the exception of *Bothriocephalus* spp. (see Loganov and Kolarova, 1979; Scott and Grizzle, 1979), information on other adult cestodes is generally sparse. Within the metacestodes, we have chosen to deal at length only with the plerocercoids of some Pseudophyllidea and a variety of cyclophyllidean types. The former, like the section on adult cestodes, have been dealt with from the point of view of pathophysiology;

BIOLOGY OF THE EUCESTODA Vol. 2
ISBN 0-12-062102-9

in the latter we have emphasized histopathology and clinical features. These differences of approach reflect both our own interests and the emphasis within the available literature.

II. ADULT CESTODES

A. *Hymenolepis microstoma*

Cysticercoids of *Hymenolepis microstoma* excyst in the small intestine of the definitive host, after which the worms migrate into the bile duct and attach to the biliary epithelium. As the parasites increase in size, their strobilae may protrude from the bile duct into the duodenum.

There have been many accounts of the pathology of *H. microstoma* infections in rodents. In the interests of economy of space, and clarity, comments will be restricted to only a few of these. The presence of the parasite in the bile duct of rodents is associated with a variety of host responses. The most obvious of these, and one that was first reported almost half a century ago, is an enlargement of the bile duct, its diameter increasing some twenty-five times during the first six weeks of infection. The histogenesis of the host response has been characterized at the ultrastructural level by Lumsden and Karin (1970). They described a typical inflammatory reaction in which an initial leucocytic invasion of the peribiliary connective tissues was followed by fibroplasia. In established infections, the characteristic cells found in connective tissues were eosinophils and plasma cells; these latter were found to incorporate tritiated leucine into a protein that might possibly be an antibody.

Although host bile duct hypertrophy is the most noticeable consequence of parasitization, other components of the host response have been described. The most comprehensive studies of changes that occur in a number of host organs are those of Pappas and co-workers. Pappas (1976) used 6-week-old CF-1 mice infected with 15 cysticercoids. A most interesting result from this work was the demonstration that leucocyte infiltration into the lamina propria of the common bile duct occurred *before* the worms had migrated from the small intestine. This is reminiscent of fascioliasis in which an increase in bile proline concentration and hyperplasia of the bile duct wall occur prior to the migration of the flukes (Campbell *et al.*, 1981). Four days following infection of mice with *H. microstoma* some erosion of the bile duct epithelium had occurred. On day 5, the bile duct wall had increased in thickness from the 30–50 μm of control animals to 100–150 μm. Seven days post-infection this value had increased to 0.5 mm and a maximum thickness of 1.0 mm was observed by day 12. Although the thickness of the bile duct wall decreased 15 days after infection this was not due to reversal of the histopathological changes described above. Rather, it was a consequence of an increase in the diameter of the bile duct with an associated stretching and thinning of its wall. In contrast to the bile duct, few changes in the histology of the duodenum were observed, even when the latter was distended by the presence of the posterior portions of the strobilae of the parasites. This finding was subsequently confirmed by Pappas and Schroeder (1977). Using scanning electron microscopy they observed that, whereas erosion of the biliary mucosa was evident eight days post-infection, no marked changes in the structure of intestinal villi occurred, even in infections of much longer standing.

Effects were also noted in other organs with which the tapeworm had no contact. Thus, one week after infection, cellular infiltration and fibrosis had occurred in intrahepatic bile ducts, and lesions later appeared on the surface of the liver (Pappas, 1976). No change in the histology of spleens was detected, but Lumsden and Karin (1970) observed pancreatic cellular infiltration after the same time interval.

All the host organs examined by Pappas (1976) increased in wet weight relative to host total body weight. It was demonstrated that this effect was not due solely to an increase in organ water content and Pappas (1978a) attempted to determine the chemical basis of increased organ weight. Liver, bile duct, small intestine and spleen of parasitized and non-parasitized CF-1 mice were analysed quantitatively for glucose, glycogen, RNA, DNA, total protein and hydroxyproline, the latter serving as a measure of collagen content. This study demonstrated that, although the weights of the organs studied increased, this increase was not, in most cases, due to an increase in the amount of any specific substance, and the relative biochemical composition of the organs varied little between parasitized mice and controls. This strongly suggested that organ growth (i.e. hyperplasia) rather than hypertrophy had occurred, and also that fibrosis, although prominent histologically, did not contribute disproportionately to the growth process.

The facts that (a) pathological changes occur in organs not in contact with the parasite, (b) cellular infiltration of the bile duct commences before the migration of the parasite and (c) host oxygen consumption is raised by 58% 2 days post-infection (Mayer and Pappas, 1976), i.e. before the most marked histopathological changes had developed, suggest that toxins produced by the parasite might be initiators of the observed pathological consequences of infection. Simpson and Gleason (1975) provided evidence that a toxic metabolite was present in Tyrode's saline in which parasites had been incubated for 24 h. Intraperitoneal injection or oral administration of these incubation media induced liver lesions similar to those occurring in parasitized mice. Preliminary attempts to characterize the toxin indicated that it was present in lipid extracts of the Tyrode's saline, and that its molecular weight was less than 10 000. Unfortunately, no further studies have been undertaken on this intriguing problem. In addition to determining further the biochemical properties of the toxic substance, it would be of interest to determine whether *in vitro* incubation of other parasites yielded media with similar toxic properties. Also, if a toxin is indeed produced, then the basis for its somewhat selective effects on host organs warrants further study.

B. *Diphyllobothrium latum*

The pathophysiology of vitamin B_{12} malabsorption has exercised physiologists over many years, and it is clear that there are a variety of etiologies of malabsorption syndromes. Here only one of these will be discussed: the vitamin B_{12} deficiency in man associated with the presence in the intestine of the tapeworm, *D. latum*.

Vitamin B_{12} has a complex structure comprising a central cobalt atom surrounded by four reduced pyrrole rings (the so-called corrin nucleus), and attached to the corrin ring is 5,6-dimethylbenzimidazole. A number of residues are attached to different parts of the molecule, including a cyanide group linked to the central cobalt atom—hence the synonym for vitamin B_{12}, cyanocobalamin.

The dietary requirement for vitamin B_{12} is estimated to be 2 μg per day. In food, the vitamin is bound to protein (Farquharson and Adams, 1976), and the first stage in its utilization by man is the severance of these bonds by the action of pepsin and/or low pH in the stomach. Free cyanocobalamin then binds to gastric intrinsic factor, a glycoprotein secreted by the parietal cells of the gastric mucosa. The principal site for vitamin B_{12} absorption is the lower half of the ileum, and the first stage of uptake is the binding of the vitamin-intrinsic factor complex to the ileal brush border. The apparent K_m (K_t) for this process is approximately 5×10^{-10} M (Cooper, 1964) indicating a high affinity between substrate and adsorption site. Because only intrinsic factor is able to stimulate the ileal uptake of vitamin B_{12}, it has been proposed that within the glycoprotein there exist separate binding sites for both cyanocobalamin and ileal brush border receptors (Herbert, 1959). Intrinsic factor is not absorbed by ileal cells (Allen, 1975) but vitamin B_{12} is, eventually appearing in the portal circulation bound to a polypeptide carrier molecule, transcobalamin II.

D. latum is a pseudophyllidean cestode for which man may serve as a final host. Once the plerocercoid larvae become established in the host intestine they grow and mature rapidly—as much as 5–20 cm per day (Petrushevsky and Tarasov, 1933), and the infection becomes patent within approximately 4 weeks. During observations on several *Diphyllobothrium* spp., Andersen (1975) concluded that the bothria of adult worms attached firmly to one or two intestinal villi. In *D. latum* and *D. ditremum*, a layer of secretory substance was present between host and parasite tissues, possibly serving an adhesive role.

A variety of symptoms may be shown by infected individuals. These have been comprehensively described by Saarni *et al.* (1963) and range from constipation to salt craving. However, undoubtedly the most interesting pathophysiological feature of the disease is the anaemia that develops in a proportion of infected individuals. Most publications dealing with the subject describe this condition as "pernicious anaemia", although von Bonsdorf (1977) has suggested that "tapeworm pernicious anaemia" is a more appropriate term. Tapeworm pernicious anaemia is a macrocytic, megaloblastic anaemia, with reduced platelet and white cell counts, increased haemolysis and reduced plasma vitamin B_{12} concentrations. In some infected individuals, serum vitamin B_{12} concentrations can be as much as 50% lower than in non-parasitized controls (Nyberg *et al.*, 1961).

It is well established that adult *D. latum* may contain up to 3 μg of vitamin B_{12}/g/dry wt (Nyberg, 1952; Rausch *et al.*, 1967). Using radiocobalt-labelled vitamin B_{12}, Nyberg (1958a, 1958b) demonstrated that *in vivo* intestinal cyanocobalamin absorption could be some 50% less in infected humans than in parasite-free individuals. It was also shown that radiolabelled vitamin B_{12} was readily absorbed by *D. latum* but not by a cyclophyllidean tapeworm parasite of the intestine, *Taenia saginata*. The biochemical basis of this difference between cestode species was not understood at the time, but more recent studies have provided a plausible explanation (see below). Scudamore *et al.* (1961) found that vitamin B_{12} concentrations were not uniform along the strobila of *D. latum*, but that greater amounts were present in the anterior, and presumably metabolically more active, proglottides.

In vitro, *D. latum* is apparently partially able to abolish the ability of intrinsic factor to bind with free vitamin B_{12} and also to dissociate the vitamin-intrinsic factor complex (Nyberg, 1960a, 1960b). However, the nature of the releasing factor that brings about this dissociation is not known. Mettrick and Podesta (1974) have suggested that proton secretion by the parasite may result in a lowering of intestinal pH and that this would enhance the dissociation of the protein-vitamin complex. In support of this view is the observation that vitamin B_{12} malabsorption is associated with pancreatic exocrine deficiency and the Zollinger-Ellison syndrome. In both of these conditions it has been suggested that a lowering of intestinal pH, below that optimal for cyanocobalamin absorption, was responsible for the vitamin deficiency syndromes. Certainly in pancreatic exocrine deficiency diseases administration of sodium bicarbonate corrected the malabsorption condition. However, Veeger *et al.* (1962) demonstrated that in patients with pancreatic insufficiency, and who were malabsorbing vitamin B_{12}, ileal pH did not differ from controls, and it has also been shown that malabsorption could be corrected by the administration of trypsin alone (Toskes *et al.*, 1973).

Whatever the mechanism involved, it is clear that *D. latum* is able to dissociate the intrinsic factor—vitamin B_{12} complex and then absorb the free vitamin. The preferred intestinal site for *D. latum* is the proximal small intestine. In this region there are few, if any, intestinal receptors for the intrinsic factor-vitamin complex. Thus the parasite can, by the action of its releasing factor, liberate vitamin B_{12} from the complex without experiencing any competition for the latter from the host. Cyanocobalamin is then absorbed by the anterior portion of the strobila as noted above. On the other hand, in the ileum, where host intrinsic factor: vitamin B_{12} receptors are abundant, the host is able to compete with the parasite for the complex.

A proportion of individuals with tapeworm pernicious anaemia also exhibit an impairment of intrinsic factor secretion (Salokannel, 1970) and this defect is not rectified following anthelmintic treatment. The relative roles of impaired intrinsic factor secretion and the effects of parasite releasing factor in the pathophysiology of *D. latum* infections remain to be determined.

Although vitamin B_{12} deficiency is the principal physiological consequence of *D. latum* infections, it has been postulated that interference with host folate metabolism also occurs (Reynolds, 1976). There is, however, no evidence that the cestode absorbs significant amounts of folate. Thus, the disturbances in folate metabolism observed in some *D. latum* carriers may be the result of alteration in absorption mechanisms that are of a non-specific nature and which may also occur in association with the presence of other intestinal parasites (Brasitus, 1979).

Von Bonsdorf (1977) concluded his review of diphyllobothriasis in man by asking:

> Why does *D. latum*, in contrast to other intestinal helminths, require such large amounts of vitamin B_{12}. The only suggestion that may be offered is that this parasite surpasses other comparable worm species in growth rate and production of eggs.

Recent studies however have suggested other explanations.

It is now well established that members of several cestode orders contain high concentrations of vitamin B_{12}; an exception is the order Cyclophyllidea (Barrett, 1981). Detailed studies on cyanocobalamin in *Spirometra mansonoides* by Tkachuck and co-workers (1976a,b, 1977a,b) have provided valuable insights into its metabolic role in this parasite and it is possible that some of their findings may be applicable to *D. latum*.

Although it has been clearly demonstrated that *D. latum* absorbs vitamin B_{12} *in vitro*, the characteristics of the uptake mechanisms involved have not been fully elucidated. In contrast, the initial rates of vitamin B_{12} uptake in *S. mansonoides* have been shown to be non-linear with respect to solute concentration and an apparent K_t of 250 nM was derived from absorption data. Uptake was not dependent on prior binding to intrinsic factor and cyanocobalamin transport could be inhibited by its structural analogues. The binding of the vitamin to presumed membrane carriers was dependent upon certain structural relationships associated with the benzimidazole moiety, and substitution of the cyanide group either by methyl or hydroxyl groups did not affect uptake.

Adult *S. mansonoides* can convert cyanocobalamin to adenosylcobalamin, and Nyberg *et al.* (1970) described a substance isolated from *D. latum* that migrated in the manner of adenosylcobalamin during electrophoresis. Tkachuck *et al.* (1977a) remarked that the only enzyme of Metazoa known to require adenosylcobalamin as a coenzyme is methylmalonyl-CoA mutase, which catalyses the reversible isomerization of methylmalonyl-CoA to succinyl-CoA. Adult *S. mansonoides* has been shown to excrete large amounts of propionate, and it is probable that this is derived via the decarboxylation of methylmalonyl-CoA by the enzyme propionyl carboxylase, a reaction that may be linked to a net production of ATP. The requirement in some parasites for a high internal concentration of vitamin B_{12} may reflect the fact that the methylmalonyl-CoA mutase of these organisms binds vitamin B_{12} only very weakly.

C. *Hymenolepis diminuta*

There is a wealth of information on the pathophysiology of *Hymenolepis diminuta* infections in the rat, largely resulting from the work of Mettrick and co-workers over the past 20 years. Space limitations do not permit justice to be done to the breadth and detail of their studies, and the reader is directed to recent reviews (Mettrick and Podesta, 1974; Mettrick, 1980) for comprehensive surveys of the work of this group.

In rats infected with 10 cysticercoids and fed *ad libitum*, Mettrick (1971) reported large reductions (up to 95%) in the total number of bacteria in the small intestine and colon, 16 days post-infection. The largest decrease occurred in the small intestine where the pH was lowest (see below), and the worm biomass greatest. Coliform bacteria were most affected, although in certain regions of the intestine significant reductions in the numbers of lactobacilli, yeasts and streptococci were also observed. Of the anaerobes, only *Clostridia* spp. were uniformly present in the intestines of infected and non-infected animals; anaerobic enterococci and yeasts were completely absent in parasitized rats and anaerobic streptococci and micrococci were present only in the ileum. Micro-organisms contribute significantly to the luminal pool of vitamins, and the reduced microbial fauna might be expected to result in host vitamin deficiencies under certain circumstances. However, Mettrick and Jackson (1979) demonstrated an enhanced mucosal uptake by parasitized rats of thiamine, riboflavin and folic acid at low concentrations (10^{-5} mM).

The pH of the proximal regions of the small intestine of non-parasitized rats is slightly acidic (about 6.8), becoming alkaline in the lower ileum (about 7.3) (Mettrick, 1971). In parasitized animals, jejunal and duodenal pH is lowered by varying degrees dependent upon worm biomass and the quality and quantity of ingested food. The principal mechanism whereby acidification is effected appears to be proton secretion by the parasite rather than production of organic acid excretory products (Podesta, 1978). The characteristics of the process, its relationship with luminal pCO_2 and bicarbonate ions, and certain physiological and biochemical consequences of low intestinal pH have been fully discussed by Podesta *et al.* (1976), Podesta (1978), Mettrick (1980) and Ovington and Bryant (1981). The lowering of luminal pH has different effects on host and parasite. For example, in the former the mucosal transport of glucose, salt and water is reduced, whereas the converse is true for *H. diminuta* (Podesta and Mettrick, 1974a,b).

In the intestine of non-infected rats fed *ad libitum*, oxidation-reduction potentials (Eh) were negative along the entire length (—28 mV to —195 mV). In contrast, in parasitized animals Eh values were more positive (+75 mV to —76 mV) and reflected the distribution of worm biomass (Mettrick, 1975). The factors that affect the equilibrium between oxidized and reduced luminal metabolites (i.e. Eh) are complex, and may act directly or indirectly. They include pH, intestinal micro-organisms and luminal carbon dioxide and oxygen (see Mettrick, 1980).

Oxygen tensions in the gut of parasitized rats were found to be up to 20% higher than in non-parasitized controls (Podesta and Mettrick, 1974c). In perfusion experiments the rate of change of the partial pressure of oxygen was greater in parasitized animals. Podesta and Mettrick (1974c) suggested several possible explanations for this. For example, the observed reduction in the weight of the mucosa as a proportion of total intestinal weight in infected rats might reduce the effectiveness of the epithelium as a diffusion barrier. Also considered were the possible effects of lowered intestinal pH on the number of intestinal micro-organisms and the enhanced bacteriacidal effect of bile in acid conditions. Both of these phenomena would tend to reduce the intestinal microbial flora with a consequent reduction in the amount of reducing agents in the intestine.

There exist in the rat intestine, luminal gradients in the concentration of protein, amino acids and carbohydrates (Mettrick, 1970). Animals were starved for 18 h, allowed to feed on a known quantity of an experimental diet and then killed at various times post-feeding. Protein nitrogen gradients in different regions of the intestine varied with the time elapsed after feeding and also whether the dietary protein was casein or egg albumen; these differences were related to the slower rates of casein digestion. The molar ratios of amino acids in the lumen also changed following feeding and both the above suggested that endogenous recruitment of protein nitrogen was insufficient to maintain homeostasis in the intestine (see Arme and Read, 1969; Nasset and Ju, 1961; Gitler, 1964). The nature of the protein component of the diet also affected luminal carbohydrate gradients. In further studies Mettrick (1971) determined nutritional gradients in non-parasitized and parasitized rats fed *ad libitum*. Both groups of animals consumed similar amounts of food so that the marked differences observed between infected and non-infected rats were considered to be related directly to the presence of the parasite. Thus, the amount of TCA-soluble and TCA-insoluble nitrogen, and of lipid, was significantly higher in non-parasitized animals, perhaps indicating their utilization by *H. diminuta*. In non-infected rats the greatest amounts of carbohydrates were found in the duodenum and the lowest in the ileum, but this gradient was reversed in parasitized hosts. The absolute and relative concentrations of amino acids were also altered in the presence of the tapeworm. Mettrick (1980) has commented on possible relationships between nutritional gradients and worm migration (see also Chapter 2; Mettrick and Cho, 1981).

Although much is known concerning the interactions between cestodes and host digestive enzymes (see Chapter 7) there have been few estimates of the activity of intestinal enzymes in normal and parasitized rats. Pappas (1978b) found that tryptic and total protease activity did not vary with infection when the intestine was treated as a whole or analysed as three separate regions. These data suggest that trypsin inactivation does not occur *in vivo*. However, it is possible that a small, but for the parasite a highly significant, amount of trypsin is inactivated at the worm surface. This would escape detection in the assay system because of the masking effect of the large amounts of trypsin present. Mead (1976), using histochemical techniques, found no variation in the distribution of amylase in parasitized and non-parasitized rats and no differences in the rates of carbohydrate digestion following solid test meals were detected by Mead and Roberts (1972).

The presence of *H. diminuta* is also associated with pathophysiological changes in sites other than the small intestine. In addition to the features directly related to host immune responses, Turton *et al.* (1975) recorded an increase in plasma viscosity in human infections. Kartasheva *et al.* (1980)

noted that in "hymenolepid" infections of man, changes occurred in the concentration of plasma amino acids, in various parameters of gastric function and in "the intensity of peroxidation of erythrocyte membranes". Dunkley and Mettrick (1977) studied changes in blood glucose levels in rats following carbohydrate meals. With a glucose meal, parasitized rats remained hypoglycaemic for 24 h post-feeding, whereas after a corn-starch meal, the period of hypoglycaemia extended to 5 h. A possible complex relationship between the presence of the parasite and host–pancreatic endocrine function was suggested.

III. METACESTODES

A. Plerocercoids

(1) *Ligula intestinalis*

Ligula intestinalis completes its life-cycle in three hosts: the definitive host is a fish-eating bird; the first intermediate host is a copepod in which the procercoid develops, and the second intermediate host is a fish in which the plerocercoid occurs. The plerocercoid is the dominant phase of the life-cycle in terms of both longevity and host involvement. A variety of fish are able to act as hosts (Orr, 1967), but in Britain the plerocercoid is restricted to the body cavity of cyprinids. Multiple infections are common, with the weight of parasites occasionally exceeding that of host-tissue. Under these conditions it is hardly surprising that parasitization is associated with a number of pathological changes in infected fish.

Ligulosis results in a retardation of growth in *Alburnus alburnus* (Harris and Wheeler, 1974), *Perca flavescens* (Pitt and Grundmann, 1957), *Acanthobrama marmid*, *Chalcalburnus mossulensis*, *Leuciscus cephalus orientalis* (Basaran and Kele, 1976) and *Abramis brama* (Brylinski, 1969; Jarzynowa, 1971a; Garadi and Biro, 1975). No decrease in growth rate was observed by Sweeting (1976) in parasitized *Rutilus rutilus*; he suggested that this might have been due to the ready availability of food in the water studied.

With the exception of *Gobio gobio*, infected fish often exhibit a pronounced distension of the ventral body wall. In *R. rutilus*, Arme and Owen (1968) and Sweeting (1977) noted that this ventral pouching tended to reduce the degree of overlap of adjacent scales. Sweeting (1977) also observed that the "body wall index" (the weight of a defined area of body wall expressed as a percentage of the body weight) decreased as the parasitization index (weight of parasite/weight of intact fish × 100) increased. With an index of more than 30% the musculature of the body wall was "completely absent".

There are conflicting reports on the effects of *Ligula*-plerocercoids on the shape of *A. brama*. Arme and Owen (1968) and Richards and Arme (1981) noted that there was no obvious distension of the host ventral body wall in fish samples from a variety of British waters. In contrast Brylinski (1970) and Jarzynowa (1971a) observed this phenomenon. Jarzynowa (1974) found that fillet weight decreased as the weight and number of plerocercoids increased. Richards and Arme (1981), investigating the musculature of the body wall of infected *A. brama*, found no evidence of stretching. Although there was a reduction in the number of muscle fibres, atrophy was never detected (cf. Sweeting, 1977), and they suggested that parasitization was associated with an inhibition of muscle fibre development during fish growth.

Compression, displacement and changes in the weight of the organs of the perivisceral coelom of the host have frequently been described. For example, liver and gonad weights are reduced in parasitized fish (Arme and Owen, 1968; Brylinski, 1972; Mahon, 1976; Sweeting, 1977). Histological examination of the gonads of infected *R. rutilus*, and other host species, has revealed that reduced gonad weights are due to an inhibition of gametogenesis, an inhibition that occurs at stages of gamete development known to be dependent upon pituitary gonadotrophins (Arme, 1968, 1975). Thus, the effects of *L. intestinalis* on host gonads is not one of parasitic castration, but represents an interaction between the parasite and the pituitary-gonadal axis of the host. This view is supported by the observation that the presence of the plerocercoid is associated with an apparent suppressive effect on fish pituitary gonadotrophins. Similar effects were also demonstrated in the gonadotrophs of *Xenopus laevis*, an abnormal host infected by surgical implantation of parasites into the dorsal lymph sac. Here the observed cytological changes resembled those that followed testosterone implantation into the toad. This led Arme (1968) to suggest that the mechanism whereby *Ligula* exerted its effect on the host pituitary-gonadal axis might be related to the production of an anti-gonadotrophin by the parasite, possibly a steroid. However, recent studies, using a variety of investigative techniques, have failed to demonstrate steroid production by *L. intestinalis* (Arme *et al.*, 1982).

Disturbances in host metabolism are a frequently recorded consequence of infection. Kosareva (1961) found that liver glycogen was reduced in parasitized *R. rutilus*, *A. brama* and *Blicca bjoernka*, and this was confirmed for *R. rutilus* by Strazhik and Davydov, 1975 (quoted by Dabrowski, 1980). A decrease in "fat reserves" associated with parasitization has been observed in *A. alburnus* (Harris and Wheeler, 1974) *Leucaspius delineatus* (Shpolyanskaya, 1953), *A. brama* (Kosheva, 1956) and *R. rutilus* (Shpolyanskaya, 1953; Dabrowski, 1980). In contrast, Jarzynowa (1971b) found no significant differences between the fat and protein content of the muscle and viscera in parasitized and non-parasitized *A. brama*.

The concentrations of amino acids and other ninhydrin-positive compounds in normal and parasitized *R. rutilus* have been determined by Dabrowski (1980) and Soutter *et al.* (1980). The latter observed that the concentrations of a number of blood amino acids were significantly lowered as a consequence of parasitism. Whereas in non-infected fish, amino acid concentrations in blood and perivisceral fluid were generally similar, in parasitized fish a number of differences were found. Significantly lower concentrations of threonine, glutamic acid, proline, glycine, alanine, valine, isoleucine, leucine, lysine and histidine were present in blood when compared with coelomic fluid. It was suggested that the effects of *Ligula* on host amino acids were comparable to those accompanying starvation.

Changes in the cellular composition of the blood have also been associated with *Ligula* infections. Sadkovskaya (1953) and Shpolyanskaya (1953) have observed an increase in the monocyte and polymorphonuclear leucocyte counts in *Gobio gobio* and *Carassius carassius* respectively. The latter author also noted a decrease in haemoglobin; a similar fall has been observed in infected *A. brama* by Kosheva (1956) and infected *R. rutilus* by Arme and Owen (1968).

In *Ligula*-infected *A. brama*, the concentrations of serum albumin and α-globulin were decreased and that of β- and γ-globulin, increased (Guttowa and Honowska, 1973). Comparable data were obtained from sera of *C. carassius* by Ljubina, 1970 (quoted by Guttowa and Honowska, 1973). Strazhnik and Davydov (1971) demonstrated that ligulosis can be associated with changes in host vitamin metabolism. In particular, the thiamine content of the liver of infected *R. rutilus* was significantly lower than that of non-infected fish. Oxygen consumption in *Ligula*-infected fish in autumn decreases with increasing intensity of infection (Dabrowski and Szpilewski (1980). In summer, however, the oxygen consumption of parasitized fish was 12% higher than in non-infected individuals. Jara *et al.* (1977) found no differences in the localization of alkaline and acid phosphatases in the intestinal wall of parasitized and non-parasitized *A. brama*. However, using biochemical and histochemical techniques, Witala (1975) showed that in infected fish alkaline phosphatase activity decreased in the liver, mesonephros and ovaries, while acid phosphatase activity increased in the spleen and decreased in the ovaries.

With the exception of *G. gobio*, *Ligula*-infections in British cyprinids are associated with a host tissue response. Light microscopical observations by Arme and Owen (1968, 1970) have shown that the host response is evident in fish aged 0+ soon after infection. In the early stages of the response in fry of *R. rutilus*, the parasite is surrounded by a cellular exudate comprising macrophages, fibroblasts and polymorphonuclear leucocytes. In infections of longer standing, plerocercoids become enmeshed in sheets of connective tissue, and rarely calcification may occur.

The ultrastructural characteristics of the host tissue response in *R. rutilus* have been described by Hoole and Arme (1982, 1983). In established infections a layer of leucocytes is present, adjacent to the microtriches of the parasite, external to which is a connective tissue layer containing fibroblasts, collagen fibres and cell remnants. Three cell types are present within the leucocyte layer, believed to be macrophages, neutrophils and monocytes. The former phagocytose portions of the microtriches. In fry, the connective tissue layer is less prominent than in established infections, and lymphocytes, together with an unusual cell type characterized by granules which contain an asymmetrical electron dense component, are present. No ultrastructural evidence of dead or dying plerocercoids has been noted in natural infections of British cyprinids. Thus, despite the massive host cellular response with phagocytosis of microtriches, the parasite is able to survive. Why a host-response does not occur in *G. gobio* is not known. Hoole and Arme (unpublished observations) have demonstrated that *Ligula* from *G. gobio* is capable of stimulating a host response when surgically implanted into the body cavity of *R. rutilus*, and that *G. gobio* is able to mount a cellular response to the presence of inert materials and implanted *L. intestinalis* derived from *R. rutilus* but not to plerocercoids from *G. gobio*. It has not been established whether a humoral component is involved in the host response to *L. intestinalis*. Molnar and Berczi (1965), using Ouchterlony plates, detected specific antibodies in serum from parasitized *A. brama*. Sweeting (1977), using the same technique, failed to detect precipitating antibodies in the serum of infected *R. rutilus*, although an increase in γ-globulins was demonstrated electrophoretically.

(2) *Schistocephalus solidus*

The general life-cycle is similar to that described for *L. intestinalis*. The plerocercoid is commonly found in the body cavity *Gasterosteus aculeatus* and most studies on the pathology of infections have been made on this host–parasite system.

Parasitized fish frequently exhibit a characteristic distension of the body, both anterior and posterior to the ventral bony plate. This pouching affects flexibility so that propulsive movements are restricted to the pectoral fins and post-anal region. Pennycuick (1971) observed that the Le Cren's condition-factor was lower in infected *G. aculeatus* and that these hosts had a reduced growth rate and weight when compared with non-infected fish. In contrast, *Schistocephalus* sp. infections of *Pungitius pungitius* did not result in significant changes in host condition-factors (Curtis, 1981).

The effect of *S. solidus* on the growth and condition of *G. aculeatus* has been investigated using a bio-energetics approach. Comparison of the gross efficiencies (growth/food intake × 100) of infected and non-infected fish have been made by Walkey and Meakins (1970) and Meakins (1974a). The mean gross efficiency of the combined host–parasite unit was higher (8.21) than the mean gross efficiency of unparasitized fish (6.08). When the data were corrected for that proportion of calorific expenditure used for worm growth, the mean gross efficiency of parasitized fish was considerably reduced (3.68). Although in infected fish there was a negative imbalance in the energy budget, in non-infected animals there was a positive imbalance. It was concluded that a heavy parasite burden, particularly when this consisted of small rapidly growing plerocercoids, seriously depleted endogenous food reserves, resulting in loss of host weight. These views have received support from the dietary-stress study of Pascoe and Mattey (1977). Parasitized fish needed less food (per unit total weight) to maintain their total body weight than do non-parasitized individuals. However, it was suggested that, within the symbiosis, the parasite was the more efficient partner. Thus, although the dietary level required to maintain body weight in non-parasitized fish was 17% higher than that for parasitized fish, the latter may actually be losing weight, even though the host–parasite unit taken as a whole was gaining weight. Certainly, when parasitized and non-parasitized fish are starved, or kept on a restricted diet, survival time is longer in the latter (Walkey and Meakins, 1970; Meakins, 1974a; Pascoe and Mattey, 1977).

Walkey and Meakins (1970) and Lester (1971) demonstrated that both the routine or standard respiratory rate (oxygen consumption of fish swimming in still water) and the active respiratory rate (oxygen consumption of fish swimming against a current) are higher in infected fish. Lester (1971) suggested that these data might well result from the greater water resistance offered by the swollen body of parasitized animals, and that the different oxygen demands of infected and non-infected fish may affect the spatial distribution of fish in natural waters. Meakins (1974a) showed that at 15°C, parasitized fish consume at least 2.5 μl O_2/mg dry wt/h more than uninfected individuals. Meakins and Walkey (1975) confirmed these results and showed that, whereas standard and active respiratory rates were affected by parasitism, the minimum respiratory rate (the lowest oxygen consumption occurring after a period of active respiration) apparently was not. A seasonal variation in respiratory rates was also described. In February, the rate in parasitized fish was higher than that in non-parasitized animals, while in August this pattern was reversed. It was suggested that increased respiratory rates associated with parasitization may result from carbon dioxide and fermentation organic acids, secreted by the plerocercoid, reducing blood pH. This would, in turn, reduce the oxygen-carrying capacity of the blood, and hence lead to an increase in oxygen consumption. Arme and Owen (1967) also noted that the packed-cell volume of erythrocytes decreased in parasitized animals.

Pathological effects on the liver and gonads of *G. aculeatus* have been noted by many workers (e.g. Arme and Owen, 1967; Pennycuick, 1971; Meakins, 1974a,b). Arme and Owen (1967) noted that in some parasitized fish, at the beginning of the breeding season, fully developed yolky oocytes were not present, and often those oocytes that developed later in the year were not shed. The latter may result in the ovary-weight of parasitized fish exceeding that of non-parasitized individuals in the post-spawning period. No histological changes in the pituitary glands of infected fish were noted (Kerr, 1948; Arme and Owen, 1967), so that it would appear that interference with host reproduction in *S. solidus* infections has a different physiological basis from that described above for *Ligula*-infected cyprinids. Calorific investigations by Meakins (1974a,b) have shown that plerocercoids of *S. solidus* produce a substantial metabolic drain on the host, especially in winter, when energy that would normally be used to support oogenesis is diverted from the host. Spermatogenesis is also delayed in parasitized fish but to a lesser extent than oogenesis (Kerr, 1948), and Pennycuick (1971) noted an absence of secondary sexual characteristics in parasitized male fish. This latter phenomenon was not observed by Arme and Owen (1967), even in the most heavily parasitized male *G. aculeatus* examined. In none of the above studies was any effort made to determine water quality, particularly with respect to heavy metal pollutants. References to fish survival may now have to be re-evaluated since Pascoe and co-workers (Pascoe and Cram, 1977; Pascoe and Woodworth, 1980) have shown that cadmium can reduce substantially the survival time of infected fish.

(3) *Spirometra mansonoides*

Plerocercoids (spargana) of the genus *Spirometra* have been described from man and numerous other animal species in many parts of the world. A characteristic symptom of human infection is localized pain and a swelling at the site of the parasite. Encapsulation may occur with a pronounced host–cellular invasion. Although the medical importance of sparganosis in man should not be underestimated, it is from a laboratory model, in which plerocercoids of *S. mansonoides* are grown sub-cutaneously in rodents, that much has been learned concerning host–parasite interactions in this remarkable symbiosis.

Mueller (1963) first reported that the growth rate of mice harbouring *S. mansonoides* was greater than that of controls. Subsequently, some 50 papers have been written investigating the phenomenon of parasite-induced weight-gain (see Arme, 1975; Mueller, 1974, 1980; Odening, 1979). It has been suggested that the parasite produces a growth-promoting substance, PGF (=plerocercoid growth factor). Growth enhancement occurs in a variety of animals including parasitized lizards (*Anolis carolinenesis*), deer mice, mice, hamsters and intact and endocrine deficient (e.g. thyroidectomized, hypophysectomized, alloxan-diabetic) rats. PGF is found in media in which plerocercoids have been incubated, and it is thought to be a protein of approximately 70 000 molecular weight. Comparison of the properties of PGF with those of various mammalian hormones has been the subject of extensive research. Many of the properties of PGF resemble mammalian growth hormone, although there is no cross-reaction between these two substances using immunological techniques. Insulin-like and prolactin-like properties of PGF have also been described.

Since the review of Arme (1975), particular attention has been paid to the effects of PGF on host lipid metabolism. Ruegamer and Phares (1974) noted that, in parasitized female and castrated male rats, there was a large increase in serum triglyceride and total lipid concentration, rather than a reduction as is obtained following administration of growth hormone. These workers suggested that, since the fatty acid composition of the host and parasite was similar, the plerocercoid might alter host lipid biochemical pathways to provide appropriate substrates for its own metabolism. In hamsters, Phares and Carroll (1977) noted similar increases in host serum lipids following infection. The livers of the infected animals contained 17% more cholesterol, 12% more lipid phosphorus and 24% less triglyceride than controls. A significant increase in the incorporation of [2-^{14}C] acetate into the lipid fraction of the liver (47%) and serum (77%) was also observed. In alloxan diabetic rats, Phares and Carroll (1978) showed that those animals infected with *S. mansonoides* had heavier epididymal fat pads, and higher concentrations of liver and serum triglycerides than alloxan non-infected controls. Infected rats also exhibited a slightly increased concentration of serum cholesterol and a significantly decreased amount of liver cholesterol. By comparison, bovine growth hormone treatment led to a smaller gain in weight of epididymal fat pads, and a lowering of both serum and liver cholesterol concentrations; no effect was observed on serum and liver triglycerides.

In contrast to the differences demonstrated between PGF and growth hormone, particularly with respect to lipid metabolism, a number of recent studies have emphasized similarities between the two substances. In a radioreceptor assay for growth hormone Tsushima *et al.* (1974) showed that PGF and growth hormone competed for the same binding sites on rabbit liver membranes. Phares and co-workers have demonstrated that both PGF and growth hormone have similar effects on lymphoid tissue. Phares *et al.* (1976) noted that incorporation of deoxycytidine, thymidine and uridine into DNA by isolated thymocytes from infected, immunologically depressed rats (hypophysectomized), was higher than in controls. Additionally, thymidine kinase, an enzyme the concentration of which is usually considered to be elevated in rapidly growing tissues, had a far greater activity in the spleens of infected treated-animals. Increased metabolic activity in lymphoid tissue in diabetic, hypophysectomized rats infected with *S. mansonoides*, has also been described (Phares and Cook, 1978).

Cook and Phares (1975) observed that in both growth-hormone treated and infected, hypophysectomized rats there was a decreased rate of hepatic drug metabolism. Veech *et al.* (1976) investigated the effects of growth hormone on the glycolytic pathway and TCA cycle of hypophysectomized rats. Few differences between test and control animals were noted although, in the former, some inhibitory effects on pyruvate utilization and stimulation of NADPH involvement in reductive biosynthesis were observed. The effects of PGF on intermediary carbohydrate metabolism were similar to those described above for growth hormone. In infected animals, however, a significant decrease in the concentration of glucose-6-phosphate was detected. The livers of hypophysectomized growth hormone- and PGF-treated animals differed with respect to their [NAD+]: [NADH] ratios and glutamate concentrations.

The insulin-like activity of PGF has been less well studied than its growth hormone-like properties. Phares and Carroll (1977, 1978) showed that epididymal fat pads increased in weight and the concentration of serum and hepatic triglycerides were raised in infected hamsters and diabetic, hypophysectomized rats. Both of these features were claimed to be indicative of insulin-like activity. However, since PGF failed to stimulate acetate incorporation into the fat pads, Phares and Carroll (1977) concluded that there was no direct insulin-like effect on adipose tissue.

Whether the diverse effects produced by the plerocercoids of *S. mansonoides* are the result of one or several compounds is not known. It is difficult to envisage that *S. mansonoides* could evolve several substances whose actions mimic thyroxine, pituitary growth hormone and, to some extent, insulin and prolactin. However, in a closely related species, *Spirometra erinacei*, Hirai *et al.* (1978) have suggested that two factors occur, a growth hormone-like substance which was derived from the medium in which plerocercoids were incubated and an insulin-like substance obtained from worm homogenates.

Whether PGF is simply a by-product of parasite metabolism whose effects on the host are accidental, or whether there is a more intricate relationship between the two partners of the symbiosis is not known. However, some African and Far Eastern strains of *S. mansonoides* are able to develop without producing any marked growth promoting effects on hypophysectomized rats (Mueller, 1972; Opuni *et al.*, 1974).

B. Cysticerci

The cysticercus is the metacestode or bladderworm stage of some taeniid species. In the intermediate host they are characteristically spherical or oval milky-white cysts containing fluid and a single invaginated scolex. Infection of intermediate hosts occurs by direct ingestion of embryonated eggs and, where one species can serve as both definitive and intermediate host, autoinfection is possible (e.g. *Taenia solium*). Oncospheres are liberated in the gut of the host and, following penetration of the intestinal wall, are distributed throughout the body via the blood and lymphatic systems.

According to Smyth and Heath (1970), three stages in the pathology associated with the developing cysticercus can be recognized: the pathway of migration of the oncosphere before reaching the predilection site; the pathology at the predilection site; and pathology due to toxins released and infections caused by degenerating parasites.

(1) Pathology due to migration to predilection site

The migration of the oncosphere may cause some generalized malaise and pyrexia but often passes unnoticed (Hughes, 1974). Oncospheres of most cestode species enter a villus and penetrate a venule of sufficient diameter to allow their passive transport to the liver via the hepatic portal vein; large oncospheres may only be able to penetrate the lacteal. Heath (1971) showed that penetration of the host gut was achieved by the combined action of hooks and enzymes produced by the penetration gland of the larva. Entry was effected near the apex of a villus and resulted in destruction of the columnar epithelium. Cells of the lamina propria were also lysed before the oncosphere entered the blood or lymphatic system. Primary infections of oncospheres resulted in little or no lymphocytic reaction in the villus, but the rate of migration of oncospheres in challenge infections was reduced (Banerjee, 1972).

(a) *Cysticercus tenuicollis*

C. tenuicollis (metacestode of *T. hydatigena*) is of great economic importance in lambs under one year of age, a resistance to infection developing after the first few months of life (Edwards and Herbert, 1980). When parasite burdens are low, infection with *C. tenuicollis* is typically asymptomatic, although acute disease and death have been recorded in a number of species of intermediate host (Edwards and Herbert, 1980). In heavy infections, clinical symptoms (including pyrexia, diarrhoea, anaemia, jaundice and depression of growth rate) usually appear at 3–5 days and intensify up to 7–10 days post-infection. Where death occurs, it is usually within 9–13 days following infection (Shepelev, 1959). According to Jensen and Piersen (1975), cysticercosis is not generally the primary cause of death, but infection favours the development of other potentially lethal conditions such as acidosis, salmonellosis, coccidiosis and polioencepholomalacia.

Pullin (1955) and Sweatman and Plummer (1957) found that four stages could be recognized in the development of cysticerci. Following ingestion of eggs, the oncospheres remain in the gut/portal blood for up to one week; a period of one to two weeks is then spent in the liver followed by migration to the peritoneal cavity (days 18–21 following infection), and finally to the development sites on the omentum. Jensen and Pierson (1975) noted that migrating parasites created tortuous channels in the liver parenchyma and perforations in the capsule. Histopathological changes varied with the age of the lesion. Immediately behind the migrating cyst the channel consisted of a blood-filled cavity containing necrotic hepatic cells and a wall of fibrocytes and reticulo-endothelial cells. In older channels (yellow in colour), the cavity contained haemolysed blood and neutrophils while the wall contained macrophages, giant cells, lymphocytes, eosinophils and connective tissue. Completely healed channels were grey, and were occluded by connective tissue and macrophages.

Biochemical studies by Edwards and Herbert (1980) showed that the concentrations of serum glutamate-oxaloacetate transaminase (SGOT) and serum glutamate-pyruvate transaminase (SGPT) were elevated in infected lambs. Enzyme concentrations in lambs given "trickle"-infections were lower than those administered in a single acute dose. Lambs just prior to death had vastly elevated enzyme levels. In the same study it was shown that clinical signs developed in only 6% of infected pigs, and no changes in the concentrations of the above serum enzymes occurred. Feder *et al.* (1974) measured the mineral content of plasma and liver in infected pigs and sheep. Heavy infections ($>100\,000$ embryospheres) resulted in host death. In lighter infections the concentration of blood Fe^{++}, Cu^{++} and Zn^{++} fell slowly, reaching their lowest level 9–15 days post-infection. The degree and timing of the fall in ion concentrations was largely independent of the number of parasites administered. Electrolyte deposition (Na^{+}, K^{+}, Ca^{++}, Mg^{++}, Fe^{++}, Cu^{++}, Zn^{++} and Cl^{-}) in the liver reflected the severity of the infection but did not parallel the fall in their plasma concentrations.

From a diagnostic point of view, changes in serum enzyme levels correlate well with the degree of liver damage in heavily infected lambs but are not sufficiently specific to detect low levels of infection. Serological tests have also proved to be of little value in diagnosing the disease (Edwards and Herbert, 1980).

(b) *Cysticercus pisiformis*

C. pisiformis (metacestode of *T. pisiformis*) utilizes rats, rabbits, squirrels etc. as intermediate hosts, the life-cycle being completed when infected prey are eaten by a suitable carnivorous definitive host. Following migration through the liver, parasites usually encyst in the peritoneal cavity although limited development can occur in the lungs, mesentery and mesenteric lymph nodes (Flatt and Moses, 1975; Soulsby, 1968; Worley, 1974). Haemorrhagic tunnels are produced in liver tissue 24 h post-infection (Flatt and Campbell, 1974); migration in the liver continues for 15–30 days after which the parasite leaves the liver and enters the peritoneum (Soulsby, 1968; Flynn, 1973; Worley, 1974). Occasionally parasites become entrapped in the liver by the host inflammatory response and can be demonstrated at the centre of granulomas up to 70 days following infection. Nemeth (1970) suggested that the host immune mechanism operated prior to the post-hepatic migration of oncospheres, and that this contributed to their entrappment in the liver. The formation and healing of liver lesions produced by migrating *C. pisiformis* have been described by Flatt and Moses (1975) and broadly resemble those noted above for *C. tenuicollis.*

Clinically, the symptoms associated with *C. tenuicollis* and *C. pisiformis* infections are similar. In infections of the latter species in rabbits, Liebermann and Boch (1960) and Chevrier *et al.* (1971) detected a mild eosinophilia, two weeks post-infection, accompanied by a decrease in serum albumin and an increase in γ-globulin concentrations. Such changes are often associated with migration of many species of larval parasites and are therefore valueless in differential diagnosis.

(2) Pathology at the predilection site

(a) *C. tenuicollis* and *C. pisiformis*

Although Flatt and Moses (1975) have recorded migration of *C. pisiformis* to the lung and lymph nodes, most cysts develop in the peritoneal cavity; the same site is preferred by *C. tenuicollis.* The cysts have space in which to grow so that pathological effects due to pressure on host organs are few. Where encystment in organs occurs, vascular damage, thrombosis and infarction can result. There have been no detailed studies of the possible effects of degenerating cysticerci of the above species on the intermediate host.

(b) *Cysticercus cellulosae*

C. cellulosae (metacestode of *T. solium*) infections of pig result in measly pork, and the parasite causes a serious financial and public health problem in many parts of the world. The pig is the usual intermediate host with cysticerci being found commonly, but not exclusively, in voluntary muscle and the brain.

Man can also act as intermediate host and the clinical symptoms of the disease have been most studied in humans. During the development of the parasite there may be mild and transient manifestations of the disease, e.g. headache, fever, malagic pain and eosinophilia (Reeder and Palmer, 1981a). In skin and subcutaneous tissue, cysticerci may cause single or multiple painless swellings, up to 2 cm diameter. In the eye, infection is often associated with periorbital pain, light-flashes, grotesque shapes in the visual field and blurring or loss of vision. Brain cysticerci may result in meningitis, hydrocephalus, Jacksonian epilepsy and increased intra-cranial pressure. Involvement of the heart, skeletal muscles, lungs, liver and kidneys is usually asymptomatic, even in heavy infections (Sparks *et al.*, 1976).

Except in the eye and brain, cysticercosis is associated with a mild inflammatory response and the cyst eventually becomes encapsulated (Adams, 1975). When cysticerci degenerate there is an initial infiltration of neutrophils, histiocytes and eosinophils followed by granuloma formation and calcification. Cysticerci in the brain become surrounded by neuroglia (Reeder and Palmer, 1981a). Both blood and cerebrospinal fluid may exhibit a pronounced eosinophilia and elevated albumin concentrations, however, such changes are not specific to *C. cellulosae*. Positive pathological diagnosis can only be made by removal and subsequent identification of the parasite, and/or by radiology. The serological method of choice is the indirect haemagglutination test (Rydzewski *et al.*, 1975).

(c) *Cysticercus bovis*

C. bovis (metacestode of *T. saginata*) does not apparently infect man. Domestic cattle are the usual intermediate host, but many other species may be infected (Nelson *et al.*, 1965; Abuladze, 1964). A detailed review of the life-cycle patterns of this parasite may be found in Hird and Pullen (1979).

The preferred sites for *C. bovis* are the heart and masticatory muscles, but it may also occur at other sites throughout the body (Viljoen, 1937). In massive experimental cysticercosis, parasites were found in the lungs in 78.5% of animals, liver 42.8%, parotid gland 28.5%, kidney 14.2%, and wall of the large intestine 7%, (Blazek and Schramlova, 1981).

As with *C. cellulosae*, light infections of *C. bovis* are largely asymptomatic, but heavy infections are often associated with pyrexia, anorexia, muscular weakness and emaciation and destruction of heart and skeletal muscle (Viljoen, 1937). Under these circumstances fatalities occur. Blazek *et al.* (1981) described the histopathological changes associated with the developing cyst. In the early stages of invasion (14–21 days post-infection), extensive tissue necrosis and vigorous activation of eosinophils, macrophages, fibroblasts and lymphocytes occurred. Twenty-one to twenty-three days post-infection the initial host cellular response diminished and fibrosis and granulization commenced. At approximately 40–50 days following infection the cysticerci became located in an unusual cyst formed from a dilated lymphatic duct that had either grown to reach it or had regenerated from one destroyed at the beginning of the invasion. Granulation of the lymphatic vessel wall only occurred in patches. Macrophages and eosinophils were again activated 50–60 days post-infection. This vigorous activation of the lymphohistiocystic system in both the early and late stages of the infection were considered by Blazek *et al.* (1981) to be due to the metabolic and antigenic activity of the worm.

If these reactions occurred in the liver, eosinophilous hepatitis resulted; in the lungs, fibrinous alveolitis, granulomatous lymphangitis and phlebitis occurred; in the parotid gland, hyperplasia of the duct epithelium was seen. The process of cyst degeneration is similar to that of *C. cellulosae* and does not appear to result in severe pathology (Viljoen, 1937). Biochemically, Cena (1976) has shown that in infected calves there is a maximum decrease in the concentration of serum proteins 2 weeks after infection. This change was associated with reversible liver damage and was non-specific. Thus, no specific biochemical diagnostic procedures are available for the detection of *C. bovis*. Serodiagnostic techniques are also unsatisfactory (Gemmell and Johnstone, 1977), although an ELISA method may be of value in the future (Walls *et al.*, 1977).

(d) *Cysticercus ovis*

C. ovis (metacestode of *T. ovis*) occurs in sheep of all ages (McCleery and Wiggins, 1960) and occupies a variety of sites including the heart (in 100% of animals) diaphragm (76–7%), skeletal muscle (26–6%), brain and eye (10%) (Macarie *et al.*, 1979). The histopathology associated with infection resembles that of *C. bovis*.

(e) *Cysticercus fasciolaris*

C. fasciolaris is the metacestode of *T. taeniaeformis* and occurs in the livers of rats and mice. It has also been reported in pheasant (Rysavy, 1973) and man (Sterba *et al.*, 1977).

Following infection and encapsulation in the liver a neoplastic-like growth may develop (Schwabe, 1955; Marcial-Rojas, 1971). Macroscopic changes caused by the cysticercus are similar irrespective of the species of intermediate host (Sterba *et al.*, 1977). According to Banerjee (1972), 24 h after infection parasites were located in the lumina of liver sinusoids, and caused congestion. No further pathological changes occurred for 9 days until the cysts were visible on the surface of the liver. Each cyst was surrounded by host lymphocytes, macrophages and eosinophils, and changes in liver parenchyma extended 60–140 μm beyond the cyst wall. Focal areas of necrosis were observed and hepatic cells surrounding the cysts were depleted of glycogen. Similar alterations in glycogen and glycoprotein levels, and differences in phosphohydrolase distribution, have been observed by Lewert and Lee (1955). Thirty days post-infection, parasites became encapsulated; calcification was common (Sterba *et al.*, 1977) and resulted in death of the cysticercus. Light infections caused no pathological changes; heavy infections may cause extensive destruction of the liver and liver sarcoma (Schwabe, 1955).

(f) *Mesocestoides corti*

Although details of the life-cycle of this genus are sparse, the cysticercus-type metacestode, known as a tetrathyridium, can be readily passaged in mice and it is this host–parasite system from which details of the pathology of infections have been derived.

In mice, *M. corti* proliferates in the peritoneal cavity, initiating a severe inflammatory response (Mitchell and Handman, 1977) before invading and migrating through the liver and lung (Specht and Voge, 1965; Specht and Widmer, 1972; Pollaco *et al.*, 1978; Todd *et al.*, 1978; White *et al.*, 1982). In heavy infections invasion of other organs also occurs (Todd *et al.*, 1978). Extensive liver damage occurs prior to encapsulation of the parasite (Specht and Widmer, 1972). Todd *et al.* (1978) showed that five days post-infection, infiltration of polymorphonuclear leucocytes and portal collagenization had occurred; after 13 days foci of necrosis were observed and these increased with time. In some livers, amyloidosis and granuloma formation occurred one year after infection. In the lung, there was a time-related increase in peribronchiolar lymphocytic infiltration, inflammation, foci of bronchopneumonia and granuloma formation.

Destruction of liver parenchyma was accompanied by biochemical changes in serum (White *et al.*, 1982). Concentrations of alanine-aminotransferase (ALT) and aspartate-aminotransferase (AST) gradually increased and reached a maximum 35 days post-infection, when encapsulation of tetrathyridia appeared complete. Subsequent liver regeneration was accompanied by a restoration of ALT and AST concentrations to normal. Serum alkaline phosphatase and albumin concentrations also fell during the course of the infection, a phenomenon noted in other parasitic diseases (Sadun *et al.*, 1965). Slight decreases in the amount of α-1-globulins and slight increases in α-2-globulins were observed, although progressive increases in the concentration of β-1- and β-2-globulins and γ-globulins occurred. These results may indicate antibody synthesis as a result of infection (Mitchell and Handman, 1977).

C. Hydatid cyst

Hydatidosis is caused by metacestodes of the genus *Echinococcus*. There are many species within the genus (see Rausch *et al.*, 1981; Smyth and Heath, 1970), but only two principal types of disease exist in intermediate hosts. There is a difference in the clinical and pathological features of these two types and they will be discussed separately below. The more common and better understood form of the disease is caused by unilocular cysts of *E. granulosus*; the metacestode of *E. multilocularis* is a multilocular or alveolar cyst, made up of a series of proliferating vesicles embedded in a dense fibrous stroma.

Following ingestion of eggs, liberated oncospheres penetrate villi in a manner similar to that described for other taeniids. Predilection for the lungs and liver is thought to be related to the size of oncospheres relative to that of the venules and lymphatic lacteals; the filtration role played by the lung and liver is reflected in the records of cyst distribution in man and other intermediate hosts (Thompson, 1977). An average of 60% of cysts of *E. granulosus* develop in the liver, with fewer being found in other sites (lung, 20%; kidneys, 4%; muscles, 4%; spleen, 3%; soft tissue, 3%; brain, 3%; bones, 2%; other sites, 1%). At least 90% of alveolar cysts of *E. multilocularis* occur in liver.

(1) Unilocular hydatidosis

When the oncosphere reaches the predilection site it becomes surrounded by host inflammatory cells within a few hours (Webster and Cameron, 1961). These authors suggest that the intensity of this initial host response influences the development of the parasite. An intense response results in granuloma formation; a less intense reaction allows the oncosphere to form a hydatid cyst, but it may deprive the parasite of nutrients, causing death or arrested growth. Such cysts are usually sterile and are found characteristically in unsuitable hosts. A weak host-response permits the development of a fertile cyst. The nature of these host reactions varies with time, strain of parasite, species of host and type of tissue invaded.

There is an extensive literature concerning the pathology of unilocular hydatidosis in man (e.g. Deve, 1949; Lupascu and Panaitesco, 1968; Poole and Marcial-Rojas, 1971; Carcassonne *et al.*, 1973; Delahaye and Laaban, 1973; Sparks *et al.*, 1976; Grove *et al.*, 1976; Thompson, 1977; Reeder and Palmer, 1981b). Pathology may be largely due to the mechanical problems caused by an enlarging cyst or to the production by the parasite of toxins or allergens. Of the cysts that develop in the liver, most are asymptomatic and clinical features do not present until they reach about 10 cm in diameter. Developing cysts may cause jaundice or portal hypertension and complications include cholangitis and rupture. Cysts in lung may provoke cough, haemoptysis, dyspnoea and chest pain. In the brain, the earliest symptom is an increase in intra-cranial pressure. Cysts in the kidneys compress parenchyma and cause haematuria and albuminaemia. Involvement of bone is relatively rare, but when occurring the common site is the lower vertebrae often resulting in "hydatid-Pott's disease" with compression of the spinal cord and paraplegia. Cyst rupture may occur in any site, resulting in a variety of conditions, including death (Schantz, 1977).

Biochemically, some disturbances in bile, cholesterol and glycogen functions of the liver have been recorded. Evranova and Mosina (1966) observed a depression in rates of glycogen synthesis in liver and skeletal muscle of infected sheep and Moustafa *et al.* (1965) also observed a decrease in host liver glycogen in infected camels. Vessal *et al.* (1972) found that reserves of liver fatty acids were depleted and Vosokoboinik and Vasil'ev (1972) noted a rise in the concentrations of serum cholesterol and phospholipid in infected sheep. Ranucci and Grol-Ranucci (1978) recorded changes in SGOT, glutamate dehydrogenase, leucine aminopeptidase, γ-glutamyl-transferase, total serum protein and electrolytically fractionated serum proteins in sheep; however no changes occurred in the amounts of serum alkaline phosphatase, SGPT, sorbitol dehydrogenase, bilirubin or cholesterol. Although some of these biochemical data are conflicting, the general indication is that of a disruption of normal liver function.

Vitamin levels in blood, liver and muscle have been analysed in infected animals. Izmagilova (1968) found a lowering of the concentrations of vitamins A and C in infected sheep; Podgornova and Donskova (1972) noted a reduction in the content of vitamin C in blood of infected cattle.

Matossian *et al.* (1976) found persistant hypergammaglobulinaemia in patients with hydatidosis. This was due to increased IgG (present in all cases) and increased IgA and IgM (present only in pulmonary cases). None of the above changes are sufficiently specific to form the basis of a diagnostic test for hydatidosis, and the diagnostic methods of choice remain those with an immunological basis (Matossian *et al.*, 1972; Huldt *et al.*, 1973; Grove *et al.*, 1976; Araj *et al.*, 1977; Tassi *et al.*, 1981; Dada *et al.*, 1981).

(2) Alveolar (multilocular) hydatidosis

Metacestodes of *E. multilocularis* develop in the liver in 90% of all cases. Growth is invasive and destructive; the metacestode resembles a malignant neoplasm and is difficult to eradicate surgically. In the liver exogenous budding occurs, but when the cyst reaches a certain size it begins to undergo a central necrosis, forming an abscess, while more cysts continue to develop peripherally. Metastatic foci may be produced in other organs.

The pathogenesis of *E. multilocularis* has been widely studied in both natural and experimental hosts. Ali-Khan and Siboo (1980) consider that two types of development can occur in *E. multilocularis*, characterized by either progressive or restrictive growth. The type of growth is dependent upon the species or strains of host used (Lubinsky, 1964; Rausch and Schiller, 1956; Yamashita *et al.*, 1958; Ohbayashi *et al.*, 1971; Ali-Khan, 1974). The restrictive pattern of growth is found in partially susceptible or refractory hosts that restrict or abort the larval cyst mass by producing a massive inflammatory response. The progressive pattern is found in hypersusceptible hosts in which there is an insignificant tissue reaction. Ali-Khan (1978a,b,c) suggested that the factor which governs the pattern of cyst mass growth is whether or not immunodepression of the specific cellular response occurs. In hypersusceptible hosts, growth of the cyst mass is associated with a depressed cell-mediated immune response and T-cell depletion which favour unrestricted parasite growth.

The histopathology of infection in hypersusceptible hosts has recently been reviewed (Ali-Khan and Siboo, 1980). During the first three days of the infection cellular infiltration occurred in the liver which became hyperaemic with minute white foci observable macroscopically (Rausch, 1954).

The walls of the interlobular veins became congested and contained neutrophil perivascular infiltrates. Each parasite vesicle was surrounded by leucocytes beyond which lay degenerating liver tissue and fibroblasts. Two to three weeks post-infection, secondary vesicles were formed by endogenous budding of the germinal membrane. These young cysts lacked laminated layers and were encircled by neutrophils. By 34 days scoleces could be discerned in the cysts; liver tissue continued to be destroyed and hepatomegaly ensued.

In man, infection with alveolar hydatid cysts runs a chronic course often with a fatal outcome. Eosinophilia of over 5% is seen in some 18% of patients, although the majority of liver function tests yield normal values except in cases of jaundice (Delahaye and Laaban, 1973; Sparks *et al.*, 1976; Wilson and Rausch, 1980; Reeder and Palmer, 1981b).

Increased amounts of total serum protein are frequently recorded, with an elevation in the number of electrophoretically detectable fractions and additional peaks in the transferrin zone (Euchuk *et al.*, 1978). Ustinov (1978) found that alveolar hydatidosis was accompanied by disaminoacidaemia, the degree of disturbance depending on the stage of the disease and degree of liver involvement. Increases in the concentration of serum aromatic amino acids during treatment indicated an unfavourable prognosis. Serological methods remain the diagnostic techniques of choice, especially the indirect haemagglutination technique and immunoelectrophoresis (Reeder and Palmer, 1981b).

D. Coenuri

Coenuri are metacestodes of the genus *Multiceps*. Many authors have accorded the *Multiceps* group generic status, but Esch and Self (1965) have proposed that these parasites should be regarded as a single species of the genus *Taenia*, (*T. multiceps*). The coenurus is distinguished from a cysticercus by virtue of containing numerous scoleces not included in brood capsules. Coenuri develop in a variety of sites, and cysts are often recovered from the central nervous system. As with cysticercosis described above, the first expression of pathology at the predilection site is an inflammatory response, followed by changes due to pressure as the cyst grows. The degenerating coenurus also provokes an inflammatory reaction that culminates in fibrosis and calcification. Clinical features of brain coenurosis in domesticated animals are staggering, ataxia, grinding of teeth, head tremors and abnormal ear movements (Fankhauser *et al.*, 1959). In man, subcutaneous coenuri form painless nodules; in the brain, parasites tend to develop in the sub-arachnoid space and cause basal arachnoiditis. Common symptoms include headache and vomiting induced by increased intracranial pressure. Palsies may occur, especially in the sixth cranial nerve, and Jacksonian epilepsy may develop. In intra-ocular cases, localization is most often intravitreal in the anterior chamber; in subretinal areas the coenurus may resemble a neoplasm or granuloma (Manschot, 1976).

IV. CONCLUSION

It is clear from this brief survey that in all the cestode species discussed infection is associated with some degree of host pathology. There has, however, been some controversy in the past concerning the pathogenicity of tapeworms. Whereas there seems to be little disagreement concerning those cyclophyllidean metacestodes that often cause overt clinical disease or death of the host, a number of authors have subscribed to the view that infections with adult worms result in little or no host pathology. Thus, Rees (1967) states, "adult cestodes have little effect on the host and are very rarely pathogenic"; Insler and Roberts (1976) concluded that infection of rats with *H. diminuta* "appeared not to affect nutrient utilization or consumption in a healthy unstressed host, at least on a gross level . . .", and Ingham and Arme (1973) found that the presence of *Eubothrium crassum* and *Proteocephalus* sp. had no effect on the growth of farm-reared *Salmo gairdneri* or the absorption by the host of L-leucine and D-glucose.

It is perhaps premature to attempt to arrive at any general conclusion on the pathogenicity of adult cestode infections, since the majority of species have not been subjected to sufficient study. However, for one parasite–host system, *H. diminuta* in the laboratory rat, there is a wealth of relevant information. Nevertheless, the available data present a paradox. Whole organism studies reveal that parasitized animals grow normally and show few if any signs of clinical disease and yet, largely through the studies of Mettrick and co-workers, we know that many physiological changes occur in the rat host. The fact that those changes result in few gross effects in the rat may be related to a long evolutionary relationship between host and parasite and/or to a compensatory effect of alternative physiological mechanisms brought into play in parasitized animals (Mettrick, 1980). This latter is truly a rewarding area for future study.

V. REFERENCES

Abuldaze, K. (1964). Taeniata of animals and man and diseases caused by them. *In* "Essentials of Cestodology", (K. Skrjabin, ed.), pp. 154. Academy of Sciences of the USSR. Helminthological Laboratory, Israel. Program for Scientific Translations, 1970.

Adams, R. (1975). Inflammatory diseases—cysticercosis. *In* "Pathology of Muscle Diseases", (R. Adams, ed.), 3rd edition, pp. 329–331. Harper and Row, Maryland.

Ali-Khan, Z. (1974). Host–parasite relationships in Echinococcosis I. Parasite biomass and antibody response in three strains of inbred mice against graded doses of *Echinococcus multilocularis* cysts. *J. Parasit.* **60**, 231–235.

Ali-Khan, Z. (1978a). Cellular changes in the lymphoreticular tissues of C57L/J mice infected with *Echinococcus multilocularis* cysts. *Immunology*, **34**, 831–839.

Ali-Khan, Z. (1978b). Pathological changes in the lymphoreticular tissue of swiss mice infected with *Echinococcus granulosus*. *Z. ParasitKde.* **58**, 47–54.

Ali-Khan, Z. (1978c). Cell mediated immune response in early and chronic alveolar murine hydatidosis. *Expl. Parasit.* **46**, 157–165.

Ali-Khan, Z. and Siboo, R. (1980). Pathogenesis and host response in subcutaneous alveolar hydatidosis 1. Histogenesis of alveolar cysts and a qualitative analysis of the inflammatory infiltrates. *Z. ParasitKde.* **62**, 241–254.

Allen, R. H. (1975). Human vitamin B_{12} transport proteins. *Prog. Hemat.* **9**, 57–84.

Andersen, K. (1975). The functional morphology of the scolex of *Diphyllobothrium* Cobbold (Cestoda, Pseudophyllidea). A scanning electron and light microscopical study on scoleces of adult *D. dendriticum* (Nitzch), *D. latum* (L) and *D. ditremum* (Creplin). *Int. J. Parasit.* **5**, 487–494.

Araj, G. F., Matossian, R. M. and Frayha, G. J. (1977). The host response in secondary hydatidosis of mice. 1. Circulating antibodies. *Z. ParasitKde*, **52**, 23–30.

Arme, C. (1968). Effects of the plerocercoid larva of a pseudophyllidean cestode, *Ligula intestinalis*, on the pituitary gland and gonads of its host. *Biol. Bull. mar. biol. Lab. Woods Hole*, **134**, 15–25.

Arme, C. (1975). Tapeworm–host interactions. *In* "Symbiosis", (D. H. Jennings and D. L. Lee, eds), pp. 505–532. Cambridge University Press, Cambridge.

Arme, C. and Owen, R. W. (1967). Infections of the three-spined stickleback, *Gasterosteus aculeatus* L., with the plerocercoid larvae of *Schistocephalus solidus* (Muller, 1776), with special reference to pathological effects. *Parasitology*, **57**, 301–314.

Arme, C. and Owen, R. W. (1968). Occurrence and pathology of *Ligula intestinalis* infections in British fishes. *J. Parasit.* **54**, 272–280.

Arme, C. and Owen, R. W. (1970). Observations on a tissue response within the body cavity of fish infected with the plerocercoid larvae of *Ligula intestinalis* (L.) (Cestoda: Pseudophyllidea). *J. Fish Biol.* **2**, 35–37.

Arme, C. and Read, C. P. (1969). Fluxes of amino acids between the rat and a cestode symbiote. *Comp. Biochem. Physiol.* **24**, 1135–1147.

Arme, C., Griffiths, D. V. and Sumpter, J. P. (1982). Evidence against the hypothesis that the plerocercoid larva of *Ligula intestinalis* (Cestoda: Pseudophyllidea) produces a sex steroid that interferes with host reproduction. *J. Parasit.* **68**, 169–171.

Banerjee, D. (1972). Histopathological changes in the liver of albino rats with cysticercus fasciolaris infection. *J. Commun. Dis.* **4**, 156–161.

Barrett, J. (1981). "Biochemistry of Parasitic Helminths", Macmillan, London.

Basaran, A. and Kele, A. (1976). [The effects and infection percentage of plerocercoid of *Ligula intestinalis* (L.) on the fish species living in the lake of Devegecidi Down.] *Biyolegii Dergisi Ist.* **26**, 45–56.

Blazek, K. and Schramlova, J. (1981). [The organ reaction during the localization of *C. bovis* in the internal organs of cattle.] *Vet. Med. (Praha)*, **26**, 37–47.

Blazek, K., Schramlova, J. and Kursa, J. (1981). [Pathological changes in the skeletal muscles and heart of cattle during the development of *C. bovis* larva.] *Vet. Med. (Praha)*, **26**, 23–35.

Brasitus, T. A. (1979). Parasites and malabsorption. *Am. J. Med.* **67**, 1058–1065.

Brown, W. J. and Voge, M. (1982). "Neuropathology of Parasitic Infections", Oxford Medical Publications, Oxford.

Brylinski, E. (1969). [Effect of *Ligula intestinalis* (L) plerocercoids on growth of bream (*Abramis brama* L.)] *Roczn, Nauk roln. (Rocz)*, **91**, 345–360.

Brylinski, E. (1970). [Ligulosis and mortality in invaded bream population.] *Roczn. Nauk roln. (Rocz)*, **92**, 35–39.

Brylinski, E. (1972). [The influence of tapeworm *Ligula intestinalis* (L.) on the fertility and procreation of bream—*Abramis brama* (Linnaeus 1758).] *Roczn. Nauk roln. (Rocz)*, **94**, 7–16.

Campbell, A. J., Sheers, M., Moore, R. J., Edwards, S. R. and Montague, P. E. (1981). Proline biosynthesis by *Fasciola hepatica* at different developmental stages *in vivo* and *in vitro*. *Molec. Biochem. Parasit.* **3**, 91–101.

Carcassonne, M., Aubrespy, P., Dor, V. and Choux, M. (1973). Hydatid cysts in childhood. *In* "Progress in Paediatric Surgery", (P. P. Rickman, W. A. Heeker and J. Prevot, eds), Vol. 5, pp. 1–35. Urban and Schwarzenberg, Munich, Berlin and Vienna.

Cena, H. (1976). On the influence of the infection with *Taenia saginata* oncospheres upon the plasma proteins in calves. *Vet. Arh.* **46**, 207–214.

Chevrier, L., Calamel, M. and Soule, C. (1971). [Experimental infection of domestic rabbits with *Taenia pisiformis*.] *Revue Méd. vet.* **122**, 521–528.

Cook, D. E. and Phares, C. K. (1975). Initial studies on the effects of the growth factor produced by plerocercoids of the tapeworm, *Spirometra mansonoides*, on liver drug metabolism. *Biochem. Pharmac.* **24**, 1919–1922.

Cooper, B. A. (1964). The uptake of Co^{57}—labelled vitamine B_{12} by everted sacs of intestine *in vitro*. *Medicine*, **43**, 689–696.

Curtis, M. A. (1981). Observations on the occurrence of *Diplostomum spatheceum* and *Schistocephalus* sp. in ninespine sticklebacks (*Pungitius pungitius*) from the Belcher Islands, Northwest Territories, Canada. *J. Wildl. Dis.* **17**, 241–246.

Dabrowski, K. R. (1980). Amino-acid composition of *Ligula intestinalis* (L.) (Cestoda) plerocercoids and of the host parasitized by these cestodes. *Acta parasit. pol.* **27**, 45–48.

Dabrowski, K. and Szpilewski, K. (1980). Studies on the roach *Rutilus rutilus* (L.), infected with *Ligula intestinalis* (L.) plercocercoids (Cestoda, Pseudophyllidea). *Acta parasit. pol.* **27**, 37–44.

Dada, B. J. O., Adegboye, D. S. and Mohammed, A. N. (1981). Experience in Northern Nigeria with countercurrent immunoelectrophoresis, double diffusion and indirect haemagglutination tests for diagnosis of hydatid cysts in camels. *J. Helminth.* **55**, 197–203.

Delahaye, R. P. and Laaban, J. (1973). Hydatidosis. *In* "Radiologie Tropicale et Parasitaire", (R. P. Delahaye, ed.), pp. 141–221. Delachaux and Nieste S.A., Neuchâtel, Switzerland.

Deve, F. (1949). "L'echinococcose primitive", pp. 362. Masson and Cie, Paris.

Dunkley, L. C. and Mettrick, D. F. (1977). *Hymenolepis diminuta*: migration and the rat hosts' intestinal and blood plasma glucose levels following dietary carbohydrate intake. *Expl. Parasit.* **41**, 213–228.

Edwards, G. T. and Herbert, I. V. (1980). The course of *Taenia hydatigena* infections in growing pigs and lambs: Clinical signs and post-mortem examination. *Br. vet. J.* **136**, 256–264.

Esch, G. W. and Self, J. T. (1965). A critical study of the taxonomy of *Taenia pisiformis* (Bloch, 1780); *Multiceps multiceps.* (Leske, 1780); and *Hydatigena taeniaeformis* (Batsch, 1786). *J. Parasit.* **51**, 932–937.

Euchuk, G. N., Makarevich, N. I. and Mitrafonov, L. N. (1978). [The study of the serum protein spectrum by disc-electrophoresis of patients infected with *Echinococcus multilocularis*.] *Medskaya. Parazit.* **47**, 33–36.

Evranova, V. G. and Mosina, S. K. (1966). [Histochemical changes in liver and lungs in experimental hydatidosis in sheep and cysticercosis in cattle.] *Uchen. Zap. kazan. vet. Inst.* **96**, 218–223.

Fankhauser, R., Hintermann, J. and Valette, H. (1959). [Coenurosis.] *Schweizer Arch. Tierheilk*, **101**, 15–32.

Farquharson, J. and Adams, J. F. (1976). The forms of vitamin B_{12} in foods. *Br. J. Nutr.* **36**, 127–136.

Feder, H., Enigk, K. and Dey-Hazra, A. (1974). [Mineral content of blood plasma and liver in the pig and sheep during infection with the larva of *Taenia hydatigena*.] *Zentbl. Vet. Med.* B, **21**, 259–270.

Flatt, R. E. and Campbell, W. W. (1974). Cysticercosis in rabbits: Incidence and lesions of naturally occurring disease in young domestic rabbits. *Lab. Anim. Sci.* **24**, 914–918.

Flatt, R. E. and Moses, R. W. (1975). Lesions of experimental cysticercosis in domestic rabbits. *Lab. Anim. Sci.* **25**, 162–167.

Flynn, R. J. (1973). "Parasites of Laboratory Animals", The Iowa State University Press.

Garadi, P. and Biro, P. (1975). The effect of ligulosis on the growth of bream (*Abramis brama* L.) in Lake Balaton. *Annal. Biol. Tihany*, **42**, 165–173.

Gemmell, M. and Johnstone, P. (1977). Experimental immunization of six-month-old calves against infection with the cysticercus stage of *Taenia saginata. Trop. Anim. Hlth Prod.* **8**, 233–242.

Gitler, C. (1964). Protein digestion and absorption in non-ruminents. *In* "Mammalian Protein Metabolism", (H. H. Munro and J. B. Allison, eds), Vol. 1, pp. 35–69. Academic Press, London and New York.

Grove, D. I., Warren, K. S. and Mahmoud, A. A. F. (1976). Algorithms in the diagnosis of exotic diseases X. Echinococcosis. *J. Infect. Dis.* **133**, 354–358.

Guttowa, A. and Honowska, M. (1973). Changes in the serum protein fractions in the course of *Ligula intestinalis* L. plerocercoids infestation in the bream *Abramis brama* (L.). *Acta parasit. pol.* **21**, 107–114.

Harris, T. and Wheeler, A. (1974). *Ligula* infestation of bleak *Alburnus alburnus* (L.) in the tidal Thames. *J. Fish Biol.* **6**, 181–188.

Heath, D. D. (1971). The migration of oncospheres of *Taenia pisiformis*, *T. serialis* and *Echinococcus granulosus* within the intermediate host. *Int. J. Parasit.* **1**, 145–152.

Herbert, V. (1959). Mechanism of intrinsic factor action in everted sacs of rat small intestine. *J. Clin. Invest.* **38**, 102–109.

Hirai, K., Nishida, H., Shiwaku, K. and Okuda, H. (1978). Studies on the plerocercoid growth factor of *Spirometra erinacei* (Rudolphi, 1819), with special reference to the effect on lipid mobilization *in vitro*. *Jap. J. Parasit.* **27**, 527–533.

Hird, D. W. and Pullen, M. M. (1979). Tapeworms, meat and man: a brief review and update of cysticercosis caused by *Taenia saginata* and *Taenia solium*. *J. Food Protection* **42**, 58–64.

Hoole, D. and Arme, C. (1982). Ultrastructural studies on the cellular response of roach, *Rutilus rutilus* L., to the plerocercoid larva of the pseudophyllidean cestode, *Ligula intestinalis*. *J. Fish Dis.* **5**, 131–144.

Hoole, D. and Arme, C. (1983). *Ligula intestinalis* (Cestoda: Pseudophyllidea): an ultrastructural study on the cellular response of roach fry, *Rutilus rutilus*. *Int. J. Parasit.* **13**, 359–364.

Hughes, J. J. (1974). Microbial diseases—Cysticercosis. *In* "The Pathology of Muscle", (J. J. Hughes, ed.), Vol. 4 in the series *Major Problems in Pathology*, Saunders and Co., Philadelphia and London.

Huldt, G., Johansson, S. G. O. and Lantto, S. (1973). Echinococcosis in Northern Scandinavia: immune reactions to *Echinococcus granulosus* in Kautokeino Lapps. *Archs environ. Hlth*, **26**, 36–40.

Ingham, L. and Arme, C. (1973). Intestinal helminths in rainbow trout, *Salmo gairdneri* (Richardson): absence of effect on nutrient absorption and fish growth. *J. Fish Biol.* **5**, 309–313.

Insler, G. D. and Roberts, L. S. (1976). *Hymenolepis diminuta*: lack of pathogenicity in the healthy rat host. *Expl. Parasit.* **39**, 351–357.

Izmagilova, R. G. (1968). [Changes in the vitamin A and C balance in helminthiasis.] *Materiali Seminara—Soveshchaniya po Bothe S Gel 'mintzamisel'-Khoz. Zhivotnith v Chimkente, Alma-Ata.* 73–74.

Jara, Z., Olech, W. and Witala, B. (1977). Localization of alkaline and acid phosphatase activity in the intestine of healthy breams (*Abramis brama* L.) and those infected with plerocercoid of tapeworm *Ligula intestinalis* (Linne 1758). *Acta histochem.* **58**, 232–241.

Jarzynowa, B. (1971a). [The effect of tapeworm *Ligula intestinalis* (L.) on the growth rate and biometric features of bream (*Abramis brama* L.) from the lakes of the Lubelskie Voivodship.] *Roczn. Nauk roln. (Rocz)*, **93**, 31–50.

Jarzynowa, B. (1971b). [The influence of tapeworm *Ligula intestinalis* (L.) on the chemical composition of the body of bream—*Abramis brama* (Linnaeus).] *Roczn. Nauk roln. (Rocz)*, **93**, 35–46.

Jarzynowa, B. (1974). [The consumption output of bream—*Abramis brama* (Linnaeus, 1758) infected with ligulosis.] *Roczn. Nauk roln. (Rocz)*, **96**, 47–58.

Jensen, R. and Piersen, R. E. (1975). Cysticercosis from *Taenia hydatigena* infection in feedlot lambs. *J. Am. Vet. med. Assoc.* **166**, 1183–1186.

Kartasheva, L. D., Prokofieva, M. S., Lysakova, L. A. and Petrova, T. A. (1980). [The functional activity of the gastrointestinal tract and the intensity of peroxidation of erythrocyte lipids in hymenolepidosis.] *Med. Parasitol. Parasit. Dis.* **3**, 32–37.

Kennedy, C. R. (1976). "Ecological Aspects of Parasitology", North Holland, Amsterdam.

Kerr, T. (1948). The pituitary in normal and parasitised roach *Leuciscus rutilus* (Flem). *Q. J. microsc. Sci.* **89**, 129–137.

Kosareva, N. A. (1961). [Metabolic disturbances in cyprinids as a result of their invasion by *Ligula* and *Digramma.*] *Dokl. Akad. Nauk. SSSR*, **139**, 510–512.

Kosheva, A. F. (1956). [Influence of the parasites *Ligula intestinalis* and *Digramma interruptu* on their fish hosts.] *Zool. Zh.* **35**, 1629–1632.

Lester, R. J. G. (1971). The influence of *Schistocephalus* plerocercoids on the respiration of *Gasterosteus* and a possible resulting effect on the behaviour of the fish. *Can. J. Zool.* **49**, 361–366.

Lewert, R. M. and Lee, C. L. (1955). Studies on the passage of helminth larvae through host tissues III. The effects of *Taenia taeniaformis* on the rat liver as shown by histochemical techniques. *J. inf. Dis.* **97**, 177–186.

Liebermann, H. and Boch, J. (1960). Untersuchungen an *Cysticercus pisiformis* befallen Kaninchen. *Berl. Munch. tierarztl. Wschr.* **73**, 123–125.

Loganov, L. and Kolarova, V. (1979). [The pathology and pathogenesis of *Bothriocephalus gowkongensis* in carp.] *Obs. Srav. Patol.* **7**, 127–134.

Lubinsky, G. (1964). Growth of the vegetatively propogated strain of larval *Echinococcus multilocularis* in some strains of Jackson mice and their hybrids. *Can. J. Zool.* **42**, 1099–1103.

Lumsden, R. D. and Karin, D. S. (1970). Electron microscopy of the peribiliary connective tissues in mice infected with the tapeworm *Hymenolepis microstoma. J. Parasit.* **56**, 1171–1183.

Lupascu, G. and Panaitesco, D. (1968). "Hydatidoza Bucaresti", Editura Academiei Republicii Socaliste Romania, Bucharest.

Macarie, I., Lungu, T. R., Fromuda, V., Oproiu, V. and Olaru, H. (1979).

Morphological contributions to sheep cysticercosis. *Lucrari stintifice, I.A.N.B.*, ser. C, 20/21, 49–54.

Mahon, R. (1976). Effect of the cestode *Ligula intestinalis* on spottail shiners, *Notropis hudionius. Can. J. Zool.* **54**, 2227–2229.

Manschot, W. A. (1976). *Coenurus* infestation of eye and orbit. *Arch. Ophthalmol., Chicago*, **94**, 961–964.

Marcial-Rojas, R. A. (ed.) (1971). "Pathology of Protozoal and Helminthic Disease with Clinical Correlation", Williams and Wilkins, Baltimore.

Matossian, R. M., Alami, S. Y., Salti, I. and Araj, G. F. (1976). Serum immunoglobulin levels in human hydatidosis. *Int. J. Parasit.* **6**, 367–371.

Matossian, R. M., Kane, G. J., Chantler, S. M., Batty, I. and Sarhadian, H. (1972). The specific immunoglubulin in hydatid disease. *Immunology*, **22**, 423–430.

Mayer, L. P. and Pappas, P. W. (1976). *Hymenolepis microstoma*: Effect of the mouse bile duct tapeworm on the metabolic rate of CF-1 mice. *Expl. Parasit.* **40**, 48–51.

McCleery, E. F. and Wiggins, G. S. (1960). A note on the occurrence of *Cysticercus ovis* in sheep derived from sources within the U.K. *Vet. Rec.* **72**, 847–848.

Mead, R. W. (1976). Histochemical study on the distribution of amylase activity within the intestine of the rat and the effect of cestode (*Hymenolepis diminuta*) infection. *Trans. Am. microsc. Soc.* **95**, 183–188.

Mead, R. W. and Roberts, L. S. (1972). Intestinal digestion and absorption of starch in the intact rat: effects of cestode (*Hymenolepis diminuta*) infection. *Comp. Biochem. Physiol.* **41**, 749–760.

Meakins, R. H. (1974a). The bioenergetics of the *Gasterosteus Schistocephalus* host-parasite system. *Polskie Archwm Hydrobiol.* **21**, 455–466.

Meakins, R. H. (1974b). A quantitative approach to the effects of the plerocercoid of *Schistocephalus solidus* Muller 1776 on the ovarian maturation of the three-spined stickleback *Gasterosteus aculeatus* L. *Z. ParasitKde*, **44**, 73–79.

Meakins, R. H. and Walkey, M. (1975). The effects of parasitism by the plerocercoid of *Schistocephalus solidus* Muller 1776 (Pseudophyllidea) on the respiration of the three-spined stickleback *Gasterosteus aculeatus* L. *J. Fish Biol.* **7**, 817–824.

Mettrick, D. F. (1970). Protein, amino acid and carbohydrate gradients in the rat intestine. *Comp. Biochem. Physiol.* **37**, 517–541.

Mettrick, D. F. (1971). *Hymenolepis diminuta*: the microbiota, nutritional and physico-chemical gradients in the small intestine of uninfected and parasitized rats. *Can. J. Physiol. Pharmacol.* **49**, 972–984.

Mettrick, D. F. (1975). *Hymenolepis diminuta*: effect of oxidation—reduction potential in the mammalian gastrointestinal canal. *Expl. Parasit.* **37**, 223–232.

Mettrick, D. F. (1980). The intestine as an environment for *Hymenolepis diminuta. In* "Biology of the Tapeworm *Hymenolepis diminuta*", (H. P. Arai, ed.), pp. 281–356. Academic Press, London and New York.

Mettrick, D. F. and Cho, C. H. (1981). Migration of *Hymenolepis diminuta* (Cestoda) and changes in 5HT (serotonin) levels in the rat host following parenteral and oral 5HT administration. *Can. J. Physiol. Pharmacol.* **59**, 281–286.

Mettrick, D. F. and Jackson, D. J. (1979). Vitamin absorption in the *in vivo* intestine of normal and infected (*Hymenolepis diminuta*: Cestoda) rats. *J. Helminth.* **53**, 213–222.

Mettrick, D. F. and Podesta, R. B. (1974). Ecological and physiological aspects of helminth–host interactions in the mammalian gastrointestinal canal. *In* "Advances in Parasitology", (B. Dawes, ed.), Vol. 12, pp. 183–279. Academic Press, London and New York.

Mitchell, G. F. and Handman, E. (1977). Studies on the immune response to larval cestodes in mice. A simple mechanism of non-specific immunosuppression in *Mesocestoides corti* infected mice. *Aust. J. exp. Biol. med. Sci.* **55**, 615–622.

Molnar, K. and Berczi, I. (1965). [Demonstration of parasite-specific antibodies in fish-blood by agar—gel-diffusion precipitation test.] *Ztschr. Immuno. Allergie—Forsch.* **129**, 263–267.

Moustafa, A. M. B., Moustafa, I. H. and Soliman, M. K. (1965). Histochemical studies on glycogen deposition in normal camel livers and those infected with *Echinococcus granulosus* cysts. *J. Vet. Sci. United Arab Rep.* **2**, 83–91.

Mueller, J. F. (1963). Parasite-induced weight gain in mice. *Ann. N.Y. Acad. Sci.* **113**, 217–233.

Mueller, J. F. (1972). Failure of oriental spargana to immunize the hypophysectomized rat against the spargunum growth factor of *Spirometra mansonoides. J. Parasit.* **58**, 872–875.

Mueller, J. F. (1974). The biology of *Spirometra. J. Parasit.* **60**, 3–14.

Mueller, J. F. (1980). A growth factor produced by a larval tapeworm and its biological activity. In "Growth and Growth Factors", (K. Shizume and K. Takano, eds), pp. 193–201. University of Tokyo Press, Tokyo.

Nasset, E. S. and Ju, J. S. (1961). Mixture of endogenous and exogenous protein in the alimentary tract. *J. Nutr.* **74**, 461–465.

Nelson, G., Pester, F. and Rickman, R. (1965). The significance of wild animals in the transmission of cestodes of medical importance in Kenya. *Trans. R. Soc. trop. med. Hyg.* **59**, 507–524.

Nemeth, I. (1970). Immunological study of rabbit cysticercosis II. Transfer of immunity to *Cysticercus pisiformis* (Bloch, 1780) with parenterally administered immune serum of lymphoid cells. *Acta vet. hung.* **20**, 69–79.

Nyberg, W. (1952). Microbiological investigations on anti-pernicious anaemia factors in the fish tapeworm. *Acta med. scand. Suppl.* **271**, 1–68.

Nyberg, W. (1958a). The uptake and distribution of CO^{60}-labelled vitamin B_{12} by the fish tapeworm *Diphyllobothrium latum. Expl. Parasit.* **7**, 178–190.

Nyberg, W. (1958b). Absorption and excretion of vitamin B_{12} in subjects infected with *Diphyllobothrium latum* and in non-infected subjects following oral administration of radioactive B_{12}. *Acta haemat.* **19**, 90–98.

Nyberg, W. (1960a). The influence of *Diphyllobothrium latum* on the vitamin B_{12}-intrinsic factor complex, II *In vitro* studies. *Acta med. scand.* **167**, 189–192.

Nyberg, W. (1960b). The influence of *Diphyllobothrium latum* on the vitamin B_{12}-intrinsic factor complex. I. *In vivo* studies with Schilling test techniques. *Acta med. Scand.* **167**, 185–187.

Nyberg, S., Wuobela, X. and Grasbeck, R. (1970). Corinoids in *Diphyllobothrium latum*. *In* "H. D. Srivastava Commeration Volume", (K. S. Singh and B. K. Tandam, eds), Indian Veterinary Research Institute, Izatnagar, India.

Nyberg, W., Grasbeck, R., Saarni, M. and von Bonsdorff, B. (1961). Serum vitamin B_{12} levels and incidence of tapeworm anemia in a population heavily infected with *Diphyllobothrium latum. Am. J. clin. Nutr.* **9**, 606–612.

Odening, von K. (1979). [Investigation on the "sparganum growth factor" of *Spirometra.*] *Angew. Parasitol.* **20**, 185–192.

Ohbayashi, M., Rausch, R. L. and Fay, F. H. (1971). On the ecology and distribution of *Echinococcus spp.* (Cestoda, Taeniidae), and the characteristics of their development within the intermediate host. II. Comparative studies of the development of larval *E. multilocularis* (Leukart 1863) in the intermediate host. *Jap. J. vet. Res.* **19**, Suppl. 1–53.

Opuni, E. K., Muller, R. and Mueller, J. F. (1974). Absence of sparganum growth factor in African *Spirometra* spp. *J. Parasit.* **60**, 375–376.

Orr, T. S. C. (1967). Distribution of the plerocercoid of *Ligula intestinalis. J. Zool., Lond.* **153**, 91–97.

Ovington, K. S. and Bryant, C. (1981). The role of carbon dioxide in the formation of end-products by *Hymenolepis diminuta. Int. J. Parasit.* **11**, 221–228.

Pappas, P. W. (1976). *Hymenolepis microstoma*: correlation of histolpathological host response and organ hypertrophy. *Expl. Parasit.* **40**, 320–329.

Pappas, P. W. (1978a). Biochemical alterations in organs of mice infected with *Hymenolepis microstoma*, the mouse bile duct tapeworm. *J. Parasit.* **64**, 265–272.

Pappas, P. W. (1978b). Tryptic and protease activities in the normal and *Hymenolepis diminuta*-infected rat small intestine. *J. Parasit.* **64**, 562–564.

Pappas, P. W. and Schroeder, L. L. (1977). Biliary and intestinal pathology in mice infected with *Hymenolepis microstoma*, as determined by scanning electron microscopy. *J. Parasit.* **63**, 762–764.

Pascoe, D. and Cram, P. (1977). The effect of parasitism on the toxicity of cadmium to three-spined stickleback, *Gasterosteus aculeatus* L. *J. Fish Biol.* **10**, 467–472.

Pascoe, D. and Mattey, D. (1977). Dietary stress in parasitized and non-parasitized *Gasterosteus aculeatus* L. *Z. ParasitKde*, **51**, 179–186.

Pascoe, D. and Woodworth, J. (1980). The effects of joint stress on sticklebacks. *Z. ParasitKde*, **62**, 159–163.

Pennycuick, L. (1971). Quantitative effects of three species of parasites on a population of three-spined sticklebacks, *Gasterosteus aculeatus. J. Zool., Lond.* **165**, 143–162.

Petrushevsky, G. K. and Tarasov, V. A. (1933). Versuche über die Ansteckung des Menschen mit verschiedenen Fischplerozerkoiden. *Arch. Schiffs—u. Tropenhyg.* **37**, 370–372.

Phares, C. K. and Carroll, R. M. (1977). A lipogenic effect in intact male hamsters infected with plerocercoids of the tapeworm, *Spirometra mansonoides. J. Parasit.* **63**, 690–693.

Phares, C. K. and Carroll, R. M. (1978). Comparison of the effects of the growth factor produced by *Spirometra mansonoides* and growth hormone in diabetic-hypophysectominzed rats: lipid composition. *J. Parasit.* **64**, 401–405.

Phares, C. K. and Cook, D. E. (1978). Comparison of the effects of the growth factor produced by *Spirometra mansonoides* and growth hormone in diabetic-hypophysectomized rats: lymphoid tissue. *J. Parasit.* **64**, 406–410.

Phares, C. K., Hofert, J. F. and Pettinger, C. L. (1976). Growth stimulation of lymphatic tissue by plerocercoid larvae of the tapeworm, *Spirometra mansonoides. Gen. comp. Endocr.* **28**, 103–106.

Pitt, C. E. and Grundmann, A. W. (1957). A study into the effects of parasitism on the growth of the yellow perch produced by the larvae of *Ligula intestinalis* (Linnaeus, 1758) Gmelin 1790. *Proc. helminth. Soc. Wash.* **24**, 73–80.

Podesta, R. B. (1978). Characterization *in vitro* of H^+ secretion and $H^+:Na^+$ exchange by an organism normally inhabiting a CO_2-rich environment: *Hymenolepis diminuta* (Cestoda) in the rat intestine. *Can. J. Zool.* **56**, 2344–2354.

Podesta, R. B. and Mettrick, D. F. (1974a). Components of glucose transport in the host–parasite system, *Hymenolepis diminuta* (Cestoda) and the rat intestine. *Can. J. Physiol. Pharmacol.* **52**, 183–197.

Podesta, R. B. and Mettrick, D. F. (1974b). The effect of bicarbonate and acidification on water and electrolyte absorption by the intestine of normal and infected (*Hymenolepis diminuta*: Cestoda) rats. *Am. J. dig. Dis.* **19**, 725–735.

Podesta, R. B. and Mettrick, D. F. (1974c). Pathophysiology of cestode infections: effect of *Hymenolepis diminuta* on oxygen tensions, pH and gastrointestinal function. *Int. J. Parasit.* **4**, 277–292.

Podesta, R. B., Mustafa, T., Moon, T. W., Hulbert, W. C. and Mettrick, D. F. (1976). Anaerobes in an aerobic environment: role of CO_2 in energy metabolism of *Hymenolepis diminuta. In* "Biochemistry of Parasites and Host–Parasite Relationships", (H. van den Bossche, ed.), pp. 81–88. Elsevier/North Holland, Amsterdam.

Podgornova, G. P. and Donskava, T. I. (1972). [Vitamin C content in the organs of hosts infected with hydatid.] *J. Voprosy Morfologii, Ekologii i Parazitologii Zhivotnykh, volograd*, USSR: Pedagog Instituta. 137–140.

Pollacco, C., Nicholas, W. L., Mitchell, G. F. and Stewart, A. (1978). T-cell dependent collagenous encapsulating response in the mouse liver to *Mesocestoides corti* (Cestoda). *Int. J. Parasit.* **8**, 457–462.

Poole, J. B. and Marcial-Rojas, R. A. (1971). Echinococcosis. *In* "Pathology of Protozoal and Helminthic Diseases", (R. A. Marcial-Rojas, ed.), pp. 635–657. Williams and Wilkins, Baltimore.

Pullin, J. W. (1955). Observations on liver lesions in lambs experimentally infected with *Cysticercus tenuicollis. Can. J. comp. Med.* **19**, 17–25.

Ranucci, S. and Grol-Ranucci, H. (1978). Blood chemistry of sheep with parasitic lesions in the liver. *Clinica vet.* **10**, 324–333.

Rausch, R. L. (1954). Studies on the helminth fauna of Alaska XX. The histiogenesis of the alveolar larva of *Echinococcus* species. *J. Infect. Dis.* **94**, 178–186.

Rausch, R. L. and Schiller, E. L. (1956). Studies on the helminth fauna of Alaska XXV. The ecology and public health significance of *Echinococcus sibiricensis.* (Rausch and Schiller, 1954) on St Lawrence Island. *Parasitology*, **46**, 395–419.

Rausch, R. L., D'Alessandro, A. and Rausch, V. R. (1981). Characteristics of the larval *Echinococcus vogeli* (Rausch and Bernstein, 1972) in the natural and intermediate host, the paca, *Cuniculus paca L.* (Rodentia, Dasyproctidae). *Am. J. trop. med.* **30**, 1043–1052.

Rausch, R. L., Scott, E. M. and Rausch, V. R. (1967). Helminths in Eskimos in western Alaska, with particular reference to *Diphyllobothrium* infection and anaemia. *Trans. roy. Soc. trop. Med. Hyg.* **61**, 351–357.

Reeder, M. M. and Palmer, P. S. (1981a). Taeniasis (Tapeworm) Cysticercocis,

Sariocystis. *In* "The radiology of Tropical Diseases with Epidemiological, Pathobiological and Clinical Correlation", (M. M. Reeder and P. S. Palmer, eds), pp. 911–933. Williams and Wilkins, Baltimore.

Reeder, M. M. and Palmer, P. S. (1981b). Hydatid disease (Echinococcosis). *In* "The Radiology of Tropical Disease with Epidemiological, Pathobiological and Clinical Correlation", (M. M. Reeder and P. S. Palmer, eds), pp. 157–221. Williams and Wilkins, Baltimore.

Rees, G. (1967). Pathogenesis of adult cestodes. *Helminth. Abstr.* **36**, 1–23.

Reynolds, E. H. (1976). Neurological aspects of folate and vitamin B_{12} metabolism. *Clinics in Haemat.* **5**, 661–696.

Richards, K. S. and Arme, C. (1981). The effects of the plerocercoid larva of the pseudophyllidean cestode *Ligula intestinalis* on the musculature of bream (*Abramis brama*). *Z. ParasitKde*, **65**, 207–215.

Ruegamer, W. R. and Phares, C. K. (1974). Effects of age on growth and food efficiency response in rats infected with tapeworm larvae (38175). *Proc. Soc. exp. Biol. Med.* **146**, 698–702.

Rydzewski, A., Chisholm, E. and Kagan, I. (1975). Comparison of serologic tests for human cysticercosis by indirect haemagglutination, indirect immunofluorescent antibody and agar gel precipitin tests. *J. Parasit.* **61**, 154–155.

Rysavy, B. (1973). Unusual finding of larval stages of the cestode *Hydatigera taeniaformis* (Batsch, 1786) in the pheasant. *Folia parasit. (Praha)*, **20**, 15.

Saarni, M., Nyberg, W., Grasbeck, R. and von Bonsdorff, B. (1963). Symptoms in carriers of *Diphyllobothrium latum* and in non-infected controls. *Acta med. scand.* **173**, 147–154.

Sadkovskaya, O. D. (1953). [Changes in the leucocyte blood formula of the common gudgeon during *Ligula* infection.] *In* "K. I. Skryabin's 75th Birthday Presentation Papers", Moscow, Igdatelvo Akad. Nauk. SSSR, pp. 617–619.

Sadun, E. H., Williams, J. S., Meroney, F. C. and Hutt, G. (1965). Pathophysiology of *Plasmodium berghei* infection in mice. *Expl. Parasit.* **17**, 277–286.

Salokannel, J. (1970). Intrinsic factor in tapeworm anaemia. *Acta med. scand.* Suppl. **517**, 1–51.

Schantz, P. M. (1977). *Echinococcus granulosus*: Acute systematic allergic reactions to hydatid cyst fluid in infected sheep. *Expl. Parasit.* **43**, 268–285.

Schwabe, C. W. (1955). Helminth parasites and neoplasia. *Am. J. vet. Res.* **16**, 485–491.

Scott, A. L. and Grizzle, J. M. (1979). Pathology of cyprinid fishes caused by *Bothriocephalus gowkongensis* Yeh, 1955 (Cestoda: Pseudophyllidea). *J. Fish Dis.* **2**, 69–73.

Scudamore, H. H., Thompson, J. H. Jr and Owen, C. A. (1961). Absorption of Co^{60} labelled vitamin B_{12} in man and uptake by parasites, including *Diphyllobothrium latum*. *J. Lab. clin. Med.* **57**, 240–246.

Shepelev, D. (1959). [Experimental *T. hydatigena* infection in lambs and kids] *Sb. Rab. vologod. nauchno—issled. vet. opyt. Sta.* **4**, 181–186.

Shpolyanskaya, A. Y. (1953). [Changes in the leucocyte formula of fish blood under the influence of *Ligula*.] *Dokl. Akad. Nauk. SSSR*, **90**, 319–320.

Simpson, G. F. and Gleason, L. N. (1975). Lesion formation in the livers of mice caused by metabolic products of *Hymenolepis microstoma*. *J. Parasit.* **61**, 152–154.

Smyth, J. D. and Heath, D. D. (1970). Pathogenesis of larval cestodes in mammals. *Helminth. Abstr.* **39**, 1–23.

Soulsby, E. J. L. (1968). "Helminths, Arthropods and Protozoa of Domestic Animals", 6th edition, pp. 120–122. Williams and Wilkins, Baltimore.

Soutter, A. M., Walkey, M. and Arme, C. (1980). Amino acids in the plerocercoid of *Ligula intestinalis* (Cestoda: Pseudophyllidea) and its fish host, *Rutilus rutilus. Z. ParasitKde*, **63**, 151–158.

Sparks, A. K., Connor, D. H. and Neafie, R. C. (1976). Echinococcosis. *In* "Pathology of Tropical and Extraordinary Diseases—An Atlas", (C. H. Binford and D. H. Connor, eds), Vol. 2, pp. 530–533. Armed Forces Pathology, Washington D.C., USA. Castle House Publications Ltd, UK.

Sparks, A. K., Neafie, R. C. and Connor, D. H. (1976). Diseases caused by Cestodes 3. Cysticercosis. *In* "Pathology of Tropical and Extraordinary Diseases—An Atlas", (C. H. Binford and D. H. Connor, eds), Vol. 2, pp. 539–543. Armed Forces Pathology, Washington D.C., USA. Castle House Publications Ltd, UK.

Specht, D. and Voge, M. (1965). Asexual multiplication of *Mesocestoides* tetrathyridia in laboratory animals. *J. Parasit.* **52**, 268–272.

Specht, D. and Widmer, E. A. (1972). Response of mouse liver to infection with tetrathyridia of *Mesocestoides* (Cestoda). *J. Parasit.* **58**, 431–437.

Sterba, J., Blazek, K. and Barus, V. (1977). Contribution to the pathology of strobilocercus (*Strobilocercus fasciolaris*) in the liver of man and some animals. *Folia parasit. (Praha)*, **24**, 41–46.

Strazhnik, L. V. and Davydov, O. N. (1971). Relative thiamine content of the tissues of certain tapeworms of fishes. *Gidrobiol. Zh.* **7**, 81–83.

Sweatman, G. K. and Plummer, P. J. G. (1957). Biology and pathology of *Taenia hydatigena* in domestic hosts. *Can. J. Zool.* **35**, 93–109.

Sweeting, R. A. (1976). Studies on *Ligula intestinalis* (L.): effects on a roach population in a gravel pit. *J. Fish Biol.* **9**, 515–522.

Sweeting, R. A. (1977). Studies on *Ligula intestinalis*: some aspects of the pathology in the second intermediate host. *J. Fish Biol.* **10**, 43–50.

Tassi, C., Dottorini, S., Scalise, G. and Geranio, N. (1981). *Echinococcus granulosus*. Diagnosis of human hydatid disease by the indirect haemagglutination reaction with antigens from hydatid fluid and scoleces. *Int. J. Parasit.* **11**, 85–88.

Thompson, R. C. A. (1977). Hydatidosis in Great Britain. *Helminth. Abstr.* **46**, 837–845.

Tkachuk, R. D., Weinstein, P. P. and Mueller, J. F. (1976a). Comparison of the uptake of vitamin B_{12} by *Spirometra mansonoides* and analogs affecting uptake. *J. Parasit.* **62**, 94–101.

Tkachuk, R. D., Weinstein, P. P. and Mueller, J. F. (1976b). Isolation and identification of a carbamide coenzyme from the tapeworm *Spirometra mansonoides. J. Parasit.* **62**, 948–950.

Tkachuk, R. D., Weinstein, P. P. and Mueller, J. F. (1977a). Metabolic fate of cyanocobalamin taken up by *Spirometra mansonoides* spargana. *J. Parasit.* **63**, 694–700.

Tkachuk, R. D., Saz, H. J., Weinstein, P. P., Finnegan, K. and Mueller, J. F. (1977b). The presence and possible function of methylmalonyl—CoA mutase and propionyl—CoA carboxylase in *Spirometra mansonoides*. *J. Parasit.* **63**, 769–774.

Todd, K. S., Simon, J. and Dipietro, J. A. (1978). Pathological changes in mice infected with tetrathyridia of *Mesocestoides corti*. *Lab. Anim.* **12**, 51–53.

Toskes, P. P., Deren, J. J. and Fruiterman, J. (1973). Specificity of the correction of vitamin B_{12} malabsorption of pancreatic extract and its clinical significance. *Gastroenterology*, **65**, 199–204.

Tsushima, T., Friesen, H. G., Chang, T. W. and Ruben, M. S. (1974). Identification of sparganum growth factor by a radioreceptor assay for growth hormone. *Biochem. biophys. Res. Commun.* **59**, 1062–1068.

Turton, J. A., Williamson, J. R. and Harris, W. G. (1975). Haematological and immunological responses to the tapeworm *Hymenolepis diminuta* in man. *Tropenmed. Parasit.* **26**, 196–200.

Ustinov, G. G. (1978). [Changes in the free amino acid spectrum in the blood plasma of patients with alveolar echinococcosis.] *Medskaya Parazit.* **47**, 36–40.

Veech, R. L., Hawkins, R. A., Nielson, R. C., Phares, C. K., Ruegamar, W. R. and Mehlman, M. A. (1976). A comparison of the metabolic effects of bovine growth hormone and growth factor from *Spirometra mansonoides* on rat liver *in vivo*. *J. Toxicol. and Envir. Health*, **1**, 793–806.

Veeger, W., Abels, J. and Hellemans, N. (1962). Effect of sodium bicarbonate and pancreatin on the absorption of vitamin B_{12} and fat in pancreatic insufficiency. *N. Eng. J. Med.* **267**, 1341–1344.

Vessal, M., Zekavat, S. Y. and Mohammadzadeh-K, A. A. (1972). Lipids of *Echinococcus granulosus* protoscolices. *Lipids*, **7**, 287–296.

Viljoen, N. F. (1937). Cysticercosis in swine and bovines with special reference to South African conditions. *Onderstepoort J. vet. Res.* **9**, 337–570.

Von Bonsdorf, B. (1977). "Diphyllobothriasis in man", Academic Press, London and New York.

Vosokoboinik, L. V. and Vasilev, A. A. (1972). [Clinical course of experimental hydatidosis in sheep and pigs and some questions of lipid metabolism.] *Byulletin Vsesoyuznogo Instituta Gelzmintologiya im. K. I. Skryabina*, **7**, 13–18.

Walkey, M. and Meakins, R. H. (1970). An attempt to balance the energy budget of a host–parasite system. *J. Fish Biol.* **2**, 361–372.

Walls, K., Allain, D., Arambulo, P., Bullock, S. and Dykes, A. (1977). The use of enzyme-linked immunospecific assay for the serodiagnosis of parasitic infections. *Abstr. Annual Meet. Am. Soc. Microbiol.* **77**, 37.

Webster, G. A. and Cameron, T. W. M. (1961). Observations on experimental infections with *Echinococcus* in rodents. *Can. J. Zool.* **39**, 877–891.

White, T. R., Thompson, R. C. A., Penhale, W. J., Pass, D. A. and Mills, J. N. (1982). Pathophysiology of *Mesocestoides corti* infection in the mouse. *J. Helminth.* **56**, 145–153.

Wilson, J. F. and Rausch, R. L. (1980). Alveolar hydatid disease in Eskimos. A review of clinical features of 33 indigenous cases of *Echinococcus multilocularis* infection in Alaskan Eskimos. *Am. J. trop. med. Hyg.* **29**, 1340–1355.

Witala, B. (1975). [Alkaline and acid phosphatases and hemosiderin in the internal organs of bream (*Abramis brama* L.) infected with plerocercoid *Ligula intestinalis* (Linne, 1758).] *Wiad. Parazyt.* **21**, 399–418.

Worley, D. E. (1974). Quantitative studies on the migration and development of *Taenia pisiformis* larvae in laboratory animals. *Lab. Anim. Sci.* **24**, 517–522.

Yamashita, J., Ohbayaski, M., Kitamura, Y., Suzuki, K. and Okugi, M. (1958). Studies on Echinococcosis VIII. Experimental *Echinococcus multicularis* in various rodents; especially on the difference of susceptibility among uniform strains of the mouse. *Jap. J. vet. Res.* **6**, 135–155.

Chapter 14

Immunity

M. D. Rickard

I. INTRODUCTION

Cestode infections are of considerable importance not only because of economic losses due to the unsightly lesions they produce in carcasses of meat animals, but also because of their severe effect on the health of humans infected with the larval stages of some species. Although substantial advances have been made in our understanding of the epidemiology of infection with these parasites and in methods for their control and treatment (Gemmell, 1978), there has been little evidence of an overall reduction in infection on a global basis, and in some cases infection may be becoming more widespread (Matossian *et al.*, 1977).

Immunity plays a significant role in the epidemiology of infection with cestodes utilizing mammalian intermediate hosts (Gemmell and Johnstone, 1977; Gemmell, 1978), but the importance of acquired resistance in the transmission of tapeworms using invertebrate or non-mammalian vertebrate hosts is not known. There are two major phases of the life-cycle in the tissues of the intermediate host: invasion by the oncosphere and development of the metacestode stage (i.e. cysticercus, cysticercoid or plerocercoid). The adult worms are found in the gut lumen or associated ducts of

BIOLOGY OF THE EUCESTODA Vol. 2
ISBN 0-12-062102-9

the definitive host, and the immune response has been shown to have some effect on this stage also.

This chapter will consider the manifestations and mechanisms of immunity to larval and adult cestodes, as well as prospects for the development of vaccines against them. Most research on immunity has been on members of the family *Taeniidae* in man and in domesticated and laboratory animals (*Taenia* spp. and *Echinococcus* spp.), and the family *Hymenolepididae* in laboratory animals (*Hymenolepis* spp.). Much less information is available concerning pseudophyllidean cestodes. Space does not permit discussion of important by-products of the immune response such as immunopathology and immunodiagnosis. Discussions of pathology and immunopathology can be found elsewhere in this volume as well as in Šlais (1970), Smyth and Heath (1970) and Rickard and Williams (1982), and immunodiagnosis is reviewed elsewhere (Kagan, 1976; Geerts *et al.*, 1977; Flisser *et al.*, 1979; Rickard, 1979; Williams, 1979, 1983).

II. IMMUNITY TO ADULT CESTODES

Whereas early experiments showed clearly that larval cestodes stimulate strong immunity in the intermediate host (see Section III), the question of immunogenicity of adult, lumen-dwelling stages has been much more controversial. Resistance to superinfection could not be demonstrated with *T. taeniaeformis* in cats (Miller, 1932) or with *Multiceps glomeratus* (Clapham, 1940) and *T. hydatigena* (Vukovic, 1949) in dogs. Chandler (1939, 1940) concluded that rats did not develop acquired immunity to infection with *H. diminuta*, and that the poor growth rate of worms in a second infection could be attributed to a "crowding effect". These early experiments were given added weight by the subsequent report by Heyneman (1962) that only the tissue-dwelling stages and not the lumen phase of *H. nana* were immunogenic in mice. Early workers ascribed this lack of immunity to the fact that lumen-dwelling adult cestodes were "outside" the body, and therefore isolated from the defence mechanisms of the host. Some early experiments on immunization against infection with *E. granulosus* in dogs (Turner *et al.*, 1933, 1936; Matov and Vasilev, 1955; Forsek and Rukavina, 1959; Gemmell, 1962) suggested that, in this case at least, partial protection could be stimulated in the host by injecting various crude preparations of larval or adult worm antigens. This apparent anomaly was generally explained by the fact that the scolex of *E. granulosus* had very intimate contact with the host's intestinal epithelium, to the extent of invading the lamina propria (Smyth *et al.*, 1970).

Recent investigations have now clearly established that true acquired immunity is a feature of infection with adult cestodes in many host–parasite systems, e.g. *H. diminuta* in mice (Weinmann, 1966; Hopkins *et al.*, 1972a,b) and rats (Harris and Turton, 1973; Hinsbo *et al.*, 1975; Chappell and Pike, 1976), *H. microstoma* in mice (Tan and Jones, 1968; Moss, 1971; Howard, 1976), *H. citelli* in mice (Hopkins and Stallard, 1974) and the white-footed deer mouse (Wassom *et al.*, 1973) and *H. nana* in mice (Ito, 1978, 1982).

There are several reviews of the early literature (Weinmann, 1966, 1970; Gemmell and Soulsby, 1968; Williams, 1971) and of more recent findings (Gemmell, 1976; Williams, 1979, 1983; Hopkins, 1980) concerning immunity to adult cestodes. The vast majority of information has been derived from studies of hymenolepidid cestodes in laboratory animals, and this will be presented separately from the little information that is available concerning immunity to adult cestodes in domesticated animals and man.

A. Laboratory animals

A number of cestodes belonging to the genus *Hymenolepis* occur in rats and mice and these have provided convenient models for experimental studies on immunity. Not only is a great deal of information available on the immune system of the various inbred strains of these hosts, especially mice, but the parasites are also readily maintained in the laboratory. All except *H. nana* have indirect life-cycles through a variety of arthropod intermediate hosts, and flour-beetles are frequently used for their maintenance in the laboratory. *H. nana* is exceptional in that the mammalian definitive host can also act as intermediate host, with cysticercoids maturing in the wall of the intestine; thus, self-augmentation of infection can occur with this parasite.

Mice reject a primary infection with *H. diminuta* some 10–14 days after infection (Hopkins *et al.*, 1972a), and the speed of rejection is, to some extent, dependent upon the number of cysticercoids administered (Befus, 1975). That this rejection is immunologically mediated is supported by a number of experimental observations on both this parasite and other hymenolepidids. Natural biological factors known to cause deficiencies in the immune response, such as pregnancy and lactation (Hopkins, 1980), immaturity of the immune system in young mice (Befus and Featherston, 1974), stress induced by fighting or predation (Weinmann and Rothman, 1967; Hamilton, 1974), and simultaneous infection with *Trypanosoma cruzi* (Machnicka and Choromanski, 1978) enhance survival of these cestodes in mice. Immunodeficient athymic nude mice do not reject infection with one to three cysticercoids of *H. diminuta* (Isaak *et al.*, 1975; Bland, 1976; Andreassen *et al.*, 1978, 1982), and similar results have been obtained with *H. nana* in mice (Isaak *et al.*, 1977). When athymic mice are reconstituted using thymus grafts, they reject infection in the same way as normal animals (Isaak *et al.*, 1975). Treatment with anti-lymphocyte serum (ALS) delays rejection of *H. diminuta* in mice (Hopkins *et al.*, 1972b), and immunosuppressive drugs such as cortisone (Hopkins *et al.*, 1972b; Hopkins and Stallard, 1976), methotrexate (Hopkins *et al.*, 1972b) and cyclophosphamide (Choromanski, 1978; Hopkins, 1980) enhance *H. diminuta* infections. Cortisone also enhances *H. microstoma* infections in mice (Moss, 1972). Sub-lethal X-irradiation suppresses rejection of *H. diminuta* in non-immune mice (Hopkins and Zajac, 1976).

None of these factors in itself constitutes conclusive proof that immunity operates in the rejection of hymenolepidids in mice, because in all cases other mechanisms which influence host resistance could also be deficient. However, taken together, with the demonstration that immunological "memory" persists in the absence of infection, they constitute a considerable body of evidence in favour of immune mechanisms. Enhanced rejection of a challenge infection has been demonstrated with *H. diminuta* (Weinmann, 1966; Hopkins *et al.*, 1972a) and *H. microstoma* (Tan and Jones, 1968; Howard, 1976) in mice, and in *H. diminuta* "memory" persists for at least six months (Hopkins, 1982).

Rats are thought to be the "normal" host of *H. diminuta*, and differences in the immune response of rats and mice make an interesting comparison. Worms which develop from infection with only a few cysticercoids persist for the life of the host and, because of this, it was assumed that the rat does not mount an effective immune response against this parasite. However, recent experiments have shown that as a certain threshold of infection is exceeded (approximately 10–20 worms), an increasing proportion of worms is rejected from the host (Harris and Turton, 1973; Hesselberg and Andreassen, 1975; Chappell and Pike, 1976). This could, of course, be interpreted in terms of a "crowding" effect, and Hopkins (1980) discusses this question at length. However, it has been shown recently that secondary infections are rejected more quickly and effectively than primary infections (Andreassen and Hopkins, 1980) and that cortisone suppresses the rejection of a 100-worm primary infection (Hinsbo *et al.*, 1975). Furthermore, immunological "memory" persists for a short time at least (Hopkins *et al.*, 1980). Rats produce antibodies to *H. diminuta* (Coleman *et al.*, 1968; Harris and Turton, 1973), and IgA, IgM and C3 are found on the surface of worms (Befus, 1977), although their role in protective immunity is not clear. Antithymocyte serum has been shown to cause a small delay in the rejection of *H. diminuta* from rats (Hinsbo *et al.*, 1976).

The immunogenicity of adult *H. nana* in mice has been a much more vexed question because of the complexity of analysing immunity stimulated by three different developmental phases, i.e. oncospheres, cysticercoids and adults. Rapid development of resistance following egg inoculation was described by early workers (Hunninen, 1935; Hearin, 1941), but Heyneman (1962) concluded that only the tissue phases were immunogenic and that the lumen phase was not. However, recent work has strongly suggested that the lumen phase is certainly immunogenic, and the immunogenicity of the various stages may be qualitatively different (Ito, 1978, 1980, 1982). Ito (1982) examined the kinetics of infection with *H. nana* in BALB/c and dd strains of mice; the fecundity and longevity of *H. nana* is much less in the former than in the latter. The immune response in BALB/c mice given mouse-derived cysticercoids was rapid enough to almost totally prevent auto-infection, whereas almost all dd mice given identical infections developed large burdens of worms. Eggs from adults of the initial infection of BALB/c mice developed into cysticercoids in the intestinal epithelium, but did not develop further into adults. The lumen phase of *H. nana* was clearly able to stimulate immunity against adult infection but did not prevent the development of eggs into cysticercoids.

B. Immunity in domesticated animals and birds

Literature concerning immunity in dogs to infection with taeniid cestodes is confusing to say the least. Several workers were unable to show immunity to superinfection with adult taeniids in dogs and cats (Miller, 1932; Clapham, 1940; Vukovic, 1949). Rickard *et al.* (1977d) gave five successive infections with six cysticerci each to beagle puppies and could detect no effect of the previous infections that was not attributable to increasing age of the animals. On the other hand, early experiments on immunization against *E. granulosus* in dogs showed evidence of an effect of immunity on the development of challenge infections (Turner *et al.*, 1933, 1936; Forsek and Rukavina, 1959; Gemmell, 1962; Movsesijan and Mladenović, 1970). Dogs injected with antigens collected from adult *E. granulosus* developing *in vitro* were able to reduce the numbers, growth and sexual development of worms from a challenge infection with *E. granulosus* (Herd *et al.*, 1975). However, in a later experiment (Herd, 1977), specific immunity could not be demonstrated and control dogs inoculated only with adjuvant were equally well protected. Ramazanov (1971) found that prior infection of dogs with *T. multiceps* rendered them more resistant to the establishment of *E. granulosus*, but Smyth *et al.* (1970) were unable to stimulate immunity to *E. granulosus* by parenteral injection with living oncospheres of *T. hydatigena*. Similarly, Rickard *et al.* (1975) found that *T. hydatigena* larvae developing in diffusion chambers implanted into the peritoneal cavity of dogs had no effect on subsequent challenge infection with *E. granulosus*.

Following the feeding of aliquot portions of protoscoleces of *E. granulosus*, it has been observed (Gemmell, personal communication) that the susceptibility of dogs to an initial infection is highly variable. This variability is not age-dependent. Some dogs acquire resistance to reinfection, and this is manifested by the complete or almost complete rejection of the second, third and subsequent challenge infections. Other dogs, however, remain susceptible to reinfection even after eight challenge infections over a 2-year period. Herd (1977) also came to the conclusion that natural resistance factors may operate in some dogs. These observations are strikingly similar to those made by Wassom *et al.* (1973) on *H. citelli* infection in white-footed deer mice. These authors found that only a few individuals in the population allowed the parasite to develop to sexual maturity, and this resulted in natural control of the infection rate in the population at large. Dogs develop reaginic antibodies during the course of infection with *E. granulosus* (Williams and Pérez-Esandi, 1971), but no protective role has been attributed to these.

Gray (1972, 1973) showed clearly that chickens develop immunity to *Raillietina cesticillus* which is manifested by gradual rejection of a primary infection and striking resistance to a challenge infection. This resistance was inhibited by dexamethasone. The reaction around attached scoleces was characterized by infiltration with mast cells and eosinophils, and antibody to the parasite could be detected in the serum of infected chickens. The early loss of tapeworms from lambs has also been attributed to immunological mechanisms (Seddon, 1931; Stoll, 1937).

There is no direct evidence of acquired immunity in man to adult cestodes, although epidemiological evidence concerning the frequency of single worm infections despite frequent opportunities for reinfection could suggest acquired resistance. Antibodies are found in infected persons (Machnicka-Roguska and Zwierz, 1970; Machnicka and Zwierz, 1974).

C. Mechanisms of immunity to adult cestodes

(1) Antigen

Suggestions by early workers that adult cestodes did not stimulate an immune response fitted the trend of thought at that time that antigens released into the lumen of the intestine were unlikely to be available for interaction with the immune system of the host. The success of early immunization experiments with *E. granulosus* in dogs was attributed to the very intimate contact between this parasite and the gut mucosa. However, it is now well established that many antigens are readily absorbed through the intestinal wall. For example, immunization against infection with larval taeniids has been achieved by oral dosing with non-living antigens (Ayuya and Williams, 1979; Lloyd, 1979; Rickard *et al.*, 1981b). Thus, antigen released from any part of a tapeworm, not just a scolex embedded in the lamina propria, could be absorbed and possibly stimulate the immune system of the host.

There is rapid turnover of the tegumental surface (Oaks and Lumsden, 1971) and antigen release could occur over the whole surface of the parasite. In addition, digestive and protective secretions released from the surface of the worm (Lumsden, 1975; Pappas and Read, 1975) could also be antigenic. Herd *et al.* (1975) found that antigens collected during *in vitro* cultivation of *E. granulosus* stimulated a degree of protective immunity in dogs; these antigens could have been either secretory products or breakdown products of tegument. Host antibodies have been demonstrated on the tegument of cestodes in mice and rats, but it is not known with certainty that these are specific antibodies directed against tegumental antigens (Befus, 1977).

The close association between the scolex and the intestinal epithelium has led to speculation that the scolex may be an important source of "functional" antigen. Rostellar glands occur in many cestodes (Farooqi, 1958; Smyth, 1964; Shield, 1969; Öhman-James, 1973); those of *E. granulosus* actively secrete material *in vivo* and *in vitro* (Smyth, 1969; Thompson *et al.*, 1979), and when the worms are incubated in serum from infected dogs a visible precipitate is formed with the secretions (Smyth, 1969). Andreassen *et al.* (1978) suggested that the functional antigens of *H. diminuta* may be related to the scolex. In support of this, Hopkins and Barr (1982) showed that when primary infections of *H. diminuta* were terminated by chemicals within three days, an effective immune response in mice was stimulated, and at this stage the worms consisted only of a scolex and small neck region. Further experiments, using heavily irradiated cysticercoids which survive but do not grow in the intestine (Christie and Moqbel, 1980), also strongly suggested that the scolex may be a prime candidate for a source of functional antigen. However, as Hopkins and Barr (1982) point out, this neither implicates the rostellar glands nor rules out the remainder of the strobila as an additional antigenic source. These workers showed that irradiated cysticercoids failed to stimulate any form of protection in rats, which suggested that the development of the strobila was necessary to reach the "threshold" level of antigenic stimulation required in this host. It is quite likely that scolex and strobilar antigens are qualitatively different.

(2) Immune damage to adult cestodes

When adult tapeworms are affected by the immune response they may either die and be expelled or undergo destrobilation (Hopkins *et al.*, 1972a; Gray, 1973). Initially, destrobilated worms are not permanently damaged, and will recover if transplanted into non-immune hosts (Hopkins *et al.*,

1972a; Gray, 1973), or if immune hosts are treated with cortisone (Hopkins and Stallard, 1976). Serum from artificially immunized animals (Heyneman and Welsh, 1959) or infected animals (Smyth, 1969; Herd, 1976) has been shown to have an adverse effect on adult cestodes *in vitro*. Also, Weinmann (1966) showed that mucosal extracts from the intestine of immune mice had a lethal effect on adult *H. nana*. Abnormalities described in these *in vitro* experiments were confined to macroscopic observations of darkening, "blebbing" of the tegument, rupture of the tegument with extrusion of the contents and the formation of precipitates around the worms.

Befus and Threadgold (1975) described the occurrence of opaque, darkened areas in the neck region of *H. diminuta* in mice which increased in number and appeared in other parts of strobila prior to destrobilation or worm expulsion. More immunogenic infections with increased numbers of cysticercoids resulted in more dark areas per worm. Similar dark areas could be induced by mechanical injury of the worm, and dark areas disappeared when worms were maintained for 4 h in Hank's balanced salt solution (Befus and Threadgold, 1975). When the dark areas were examined by electron microscopy, changes such as increase in electron density, abnormal mitochondria, reduced granular endoplasmic reticulum, Golgi complexes and discoidal secretory bodies, and accumulation of lipid droplets were apparent (Befus and Threadgold, 1975). These authors suggested that tegumental changes produced a net efflux of fluid from the cytoplasm, and that the dark areas were a result of this dehydration. They concluded that the initial tegumental damage was immunologically mediated. Dark areas also occurred in the tegument of *H. diminuta* in rats which do not reject the worms (Befus and Threadgold, 1975), but they were less in number than in mice. It was suggested that an equilibrium exists between damage by the immune system and repair by the worms, and that in heavy infections in rats this equilibrium is upset, resulting in rejection of the worms.

(3) Antibody

There have been many demonstrations of antibody in the sera of animals and humans infected with adult cestodes (Coleman *et al.*, 1968; Smyth, 1969; Moss, 1971; Slusarski and Zapart, 1971; Williams and Pérez-Esandi, 1971; Harris and Turton, 1973; Machnicka and Zwierz, 1974; Kondo *et al.*, 1977). Befus (1977) used immunofluorescent techniques to visualize antibody on the tegument of *H. diminuta* developing in the intestine of mice. He was able to demonstrate IgA, IgM, IgG_1 and IgG_2 on the parasite, but did not show whether this represented specific or non-specific absorption of antibody onto the surface. Although a role for antibodies in host protection, through mechanisms such as complement-mediated lysis and opsonization is well known, no such information is available for adult cestodes. In fact, immunoglobulins occur on the surface of *H. microstoma* in mice and *H. diminuta* in rats which are not rejected (Befus, 1977). Experiments attempting to demonstrate passive protection by transfer of immune serum have been uniformly unsuccessful, and no evidence is available from ultrastructural studies to show that cells which may be involved in antibody-mediated cellular effects are present on the surface of these parasites in the lumen of the gut. Hopkins (personal communication) has shown that bursectomized and irradiated chickens reject *R. cesticillus* as readily as do sham-operated chickens, which provides further evidence that antibody is not critical for rejection to take place.

Hopkins (1980) postulated that IgA may actually exert a host-protective role by combining with antigens released by tapeworms in the gut, thus preventing their absorption into the lamina propria where they could activate lytic systems and damage surrounding tissues. He further suggested that antibodies may enhance survival of the tapeworm by masking foreign antigens.

IgE and mast cells have been implicated in the rejection of several intestinal parasites. Andreassen *et al.* (1978) described a marked increase in the numbers of intestinal mast cells and globular leucocytes at the time of rejection of *H. diminuta* in mice, and Gray (1976) found that mast cells accumulated in the intestinal mucosa of chickens during primary and secondary infections with *R. cesticillus*. However, Hopkins (1980) was unable to show a demonstrable effect of promethazine, methysergide or cyproheptidine hydrochloride on the course of *H. diminuta* infections in mice. Furthermore, athymic mice, which did not show accumulation of mast cells, were able to reject infections with five worms (Andreassen *et al.*, 1978). It was suggested (Andreassen *et al.*, 1982) that T-cell dependent cells, such as mast cells, are not essential for worm damage but may be responsible for the accelerated responses seen in normal mice.

(4) Complement

Herd (1976) found that "normal" dog serum had a greater lethal effect on adult *E. granulosus* incubated *in vitro* than did immune serum. The lethal effect of normal serum was abolished by heat-inactivation to destroy complement activity and complement was consumed during the process of lysis. Lysis of *E. granulosus* protoscoleces in normal human serum has also been described by Kassis and Tanner (1976) and Rickard *et al.* (1977c), and shown to occur via the alternate pathway of complement activation (Herd, 1976; Rickard *et al.*, 1977c). Similar lysis of schistosome cercariae has also been demonstrated (Machado *et al.*, 1975). Leid and Williams (1979) have reviewed the interactions between parasites and the host inflammatory system, and the role played by complement activation *in vivo* is certainly not clear. C3 has been demonstrated on the surface of *H. diminuta* in mice, but it also occurs on *H. microstoma* in mice and *H. diminuta* in rats which are not rejected by the host. The presence of anticomplementary substances in a variety of cestodes will be discussed later in this chapter, as will be their potential role as a mechanism for evading the immune response.

D. Protective mechanisms of the parasite

Hopkins (1980) posed the question, "does *H. diminuta* survive in the rat because it evokes only a non-protective response or does it survive because it evades the potentially protective response it evokes?" He suggested that survival of a cestode in the intestinal lumen of the host may reflect the outcome of a balance between the resistance mechanisms of the host and the self-protection mechanisms of the parasite, and that this same balance may play a role in host-specificity. Apart from the *H. diminuta*/mouse/rat system, there are other reports of cestodes reaching various degrees of development before rejection in abnormal hosts (Beveridge and Rickard, 1975). It may be significant that immunosuppressive treatments have been successful in allowing establishment and development of parasites in unusual hosts, e.g. *T. solium* in golden hamsters (Verster, 1974).

The adult cestode has surface characteristics in common with the mature metacestode stage in the intermediate host. Speculation on the mechanisms that allow survival of the larval stage in an immunologically hostile environment will be presented later in this review, and it is highly likely that many of the same mechanisms could assist the survival of the adult stages, e.g. membrane turnover, blocking antibodies, molecular mimicry, anti-complementary substances. Apart from the known rapid turnover at the adult worm surface (Oaks and Lumsden, 1971), there is no specific evidence concerning any of these points. However, the proposals by Hopkins (1980) are attractive and warrant further investigations.

E. Immunization against infection

Section III of this chapter will emphasize the efficacy of the immune response against infection with larval cestodes and the success of vaccination studies. The results attained with adult cestodes are less promising. Early studies on immunization against *E. granulosus* in dogs by injecting crude antigens or dosing with irradiated protoscoleces showed promise (Turner *et al.*, 1933, 1936; Forsek and Rukavina, 1959; Gemmell, 1962; Movsesijan and Mladenović, 1970). Immunization of dogs with antigens collected during *in vitro* incubation of adult *E. granulosus* also resulted in a significant level of immunity manifested by decreased numbers and size of worms and reduced fecundity (Herd *et al.*, 1975). However, later experiments (Herd, 1977) produced disappointing results, and it was suggested that there may be an interaction between efficiency of vaccination and the innate resistance of the host. Such an interaction between vaccination, strain and sex of host has been described in experiments on immunization against infection with larval cestodes (Rickard *et al.*, 1981b, 1982). Smyth *et al.* (1970) and Rickard *et al.* (1975) were unable to detect any effect of immunizing dogs against *E. granulosus* using antigens from oncospheres of the related parasite, *T. hydatigena.*

Experiments on immunization of dogs against infection with *Taenia* spp. by feeding eggs (Rickard *et al.*, 1977d), or by injecting antigens collected during *in vitro* cultivation of adult worms (Heath *et al.*, 1980) have been unsuccessful. In our laboratory, we injected dogs twice with antigens collected during 24 h or 2 weeks of *in vitro* cultivation of *T. pisiformis* oncospheres, and challenged each dog with six cysticerci 4 weeks later. No effect of the immunization on establishment or development of worms recovered 6 weeks later was detected (unpublished observations).

Seddon (1931) protected lambs against infection with *Moniezia expansa* by feeding ground-up proglottides, whereas injection of the same material failed to stimulate resistance. The establishment and growth of *H. diminuta* in rats was not affected by a series of injections of powdered worm antigen (Chandler, 1940). Elowni (cited by Hopkins, 1980) showed that the addition of Freund's incomplete adjuvant to homogenates of *H. diminuta* did not enhance their immunogenicity. Furthermore, intraperitoneal implantation of live immature and mature worms was also without effect, but three intraperitoneal inoculations of 80 excysted cysticercoids produced a marginally statistically significant result. Hopkins (1980) points out that parenteral immunization has been notoriously poor at stimulating immunity against enteric infections, and Elowni (cited by Hopkins, 1980) achieved a 35% reduction in the size of a challenge infection with *H. diminuta* in mice by giving worm homogenates orally.

Although these results do not hold out great promise for the development of practical vaccines against infection with adult cestodes, as Hopkins (1980) states, "The field is certainly worthy of exploration, after all it is easier to vaccinate a pastoralist than his herd or a sheep dog than its flock".

III. IMMUNITY TO LARVAL CESTODES

Early work on immunity to the larval stages of *T. taeniaeformis* and *T. pisiformis* during the 1920s to 1940s showed clearly that these cestodes establish high levels of concomitant immunity in their intermediate hosts, i.e. almost absolute immunity to a challenge infection with eggs despite prolonged survival of mature metacestodes which developed from the initial infection. At that time it was also shown that mice rapidly became resistant after infection with *H. nana* eggs (Hunninen, 1935; Hearin, 1941). These early experiments paved the way for the tremendous advances that have been made during the last two decades. There are several excellent works describing the findings of the early workers (Larsh, 1951; Heyneman, 1963; Weinmann, 1966, 1970; Gemmell and Soulsby, 1968; Smyth, 1969; Gemmell and Macnamara, 1972; Gemmell, 1976; Gemmell and Johnstone, 1977), and the many recent advances have also been reviewed (Flisser *et al.*, 1979; Williams, 1979, 1983; Rickard and Howell, 1982; Rickard and Williams, 1982).

The efficacy of the immune response in preventing reinfection with larval taeniid cestodes has encouraged research workers to carry out a number of experiments on immunization against infection with some of the more important species. However, perhaps because of the logistics of working with larger economic animals or man, little is known concerning the mechanisms of immunity in these hosts, and we have to rely largely on extrapolating from experiments carried out with *T. taeniaeformis*, *T. pisiformis* and *H. nana* in laboratory rodents. There are some cestodes, e.g. *T. crassiceps*, *Mesocestoides corti*, *E. granulosus* and *E. multilocularis*, which can be maintained in the laboratory by serial passage of fully developed metacestode stages. These parasites have been widely used in experiments on immunity, and the results may yield important clues concerning mechanisms of survival of larvae and methods for immunotherapy of established infection.

A. Innate resistance

Innate resistance encompasses a vast field ranging from those factors which determine the host range or host specificity of a parasite, to the subtle influences which determine intra-specific variations in resistance/susceptibility such as strain (breed), sex, age, health etc. of the host. Aspects of innate resistance have been discussed in some detail by previous authors (Smyth, 1969; Weinmann, 1970; Wakelin, 1976; Rickard and Williams, 1982).

There is abundant evidence that inter- and intra-specific variations of susceptibility/resistance to a parasite may be governed in part by genetically determined anatomical, physical and physiological characteristics of both the host and parasite. Examples are the role of bile in determining the host range of *E. granulosus* (Smyth and Haselwood, 1963) and the lack of proteolytic enzyme in the gut of young rats which prevents hatching of *T. taeniaeformis* eggs (Musoke *et al.*, 1975). Weinmann (1970) has discussed much of the literature pertaining to this aspect of cestode infections. Despite the undoubted importance of these factors, it is also attractive to ascribe a role for the immune defence mechanisms of the host in determining host specificity. Sprent (1962) proposed that animals may react to the maximum of their immunological potential when confronted with a "new" parasite, but as a result of long-standing association between host and parasite they may achieve an immunological equilibrium, i.e. "adaptation tolerance". The frequent success of immunosuppressive drugs such as cortisone in allowing the establishment of parasites in foreign hosts gives some support to this notion, although such drugs may have wider ranging effects than simple immunosuppression of the host. Much of the experimental data concerning the importance of factors affecting intra-specific variation in innate resistance is conflicting, but this conflict only serves to highlight how poorly we understand such phenomena. Interactions between these factors also occur, such as between age and strain susceptibility to *T. taeniaeformis* infection in mice (Dow and Jarrett, 1960) and between sex and strain resistance of mice to *T. taeniaeformis* (Dow and Jarrett, 1960; Mitchell *et al.*, 1977a) and *E. multilocularis* (Yamashita *et al.*, 1963) infections.

Innate resistance influences not only the establishment but also the long-term survival of cestode larvae. The mechanisms responsible are as yet poorly understood, but some observations have been made:

decreased penetration of *T. taeniaeformis* oncospheres into the intestine of older mice (Turner and McKeever, 1976) and of *T. pisiformis* oncospheres in older rabbits (Heath, 1971);

the more rapid cellular response in the intestine of older mice (Turner and McKeever, 1976);

the rapid antibody response to *T. taeniaeformis* infection in innately resistant strains of mice (Mitchell *et al.*, 1980);

the fact that resistant strains of mice become susceptible if their antibody response is delayed by administration of cyclophosphamide (Mitchell *et al.*, 1977a);

enhanced resistance to *T. taeniaeformis* in mice following the injection of BCG (Thompson *et al.*, 1982) due, perhaps, to increased rates of cellular response and antibody production;

variations in complement function in different strains of mice (Mitchell *et al.*, 1977a);

reduction of innate resistance by injecting cortisone (Olivier, 1962).

Hormone status and age are known to be important factors in determining the readiness of the body's defence against the entry of many infectious agents (Ingram and Smith, 1965; Solomon, 1969).

Whatever the mechanisms of innate resistance, it is of undeniable practical importance. Williams *et al.* (1981) pointed out the extreme care that must be taken to define and standardize animals used in experiments on cestode immunity, and the difficulties that arise out of comparing work from different laboratories using different strains of hosts and/or parasites. While the objective of standardization is highly desirable, it is impossible to achieve when working with the outbred, larger domesticated animals. Innate resistance will certainly be important in determining the success of vaccination methods for preventing infection with larval cestodes under natural conditions. For example, interactions between immunization, and sex and strain of host (Rickard *et al.*, 1981b, 1982) may create difficulties in achieving a practical reduction in infection rates by vaccination.

B. Acquired immunity

Following ingestion of a cestode egg by its mammalian intermediate host, the oncosphere is released and becomes active, frees itself from the oncospheral membrane and penetrates into the intestinal wall. Oncospheres travel via veins or lymphatics to their site of election where they develop into the mature metacestodes. During this sequence of development there are marked transformations in the surface characteristics of the larva. The oncosphere is initially surrounded by delicate cytoplasmic folds, but these are rapidly replaced by a dense microvillar coat, and by one week after infection the larva has a microthrix layer comparable with that of the mature metacestode (Nieland, 1968; Engelkirk and Williams, 1982; Furukawa *et al.*, 1981; Rickard and Williams, 1982). Coincident with this morphological transition, the larva passes from a stage of being highly susceptible to attack by the host defences to a stage of almost complete insusceptibility (Heath, 1973a; Rickard, 1974; Musoke and Williams, 1975a; Ito, 1977; Mitchell *et al.*, 1977a). The metacestode stage usually survives in the tissues for a long period of time, apparently invulnerable to the host defences, but the onset of intense inflammatory reaction around the larva heralds damage to, or death of, the parasite (Showramma and Reddy, 1963; Verheyen *et al.*, 1978).

Thus, in the early stages at least, successful establishment of the parasite depends upon the outcome of a race between the rate of development of the larva and the rapidity of the host response to it. In a primary infection the parasite usually reaches the "resistant" stage before the host can kill it. In strains of animals with a high level of innate resistance the rapidity of the immune response may kill some, or all larvae before they reach this stage (Mitchell *et al.*, 1980). During a second infection, the rapid anamnestic response of the host kills most, if not all, invading oncospheres before they develop resistance to the host defences (Rickard, 1974). Detailed analysis of the structural and physiological changes taking place in the parasite membrane during these early stages could provide rewarding information concerning the mechanism of host attack upon the parasite, and the factors which allow larvae to evade the host defences.

(1) Antibody and complement

Antibody clearly plays a major role in the first line of defence against invading oncospheres. IgA in colostrum and gut secretions has been shown to passively protect mice (Lloyd and Soulsby, 1978) and rats (Musoke *et al.*, 1975) against infection with *T. taeniaeformis.* Because this antibody is not absorbed from the intestine, its action is probably directed against oncospheres in the lumen of the gut (Hammerberg *et al.*, 1977), and IgA in colostrum and gut secretions can be removed by absorption with oncospheres (Lloyd and Soulsby, 1978). Colostral immunity has also been demonstrated in sheep (Rickard and Arundel, 1974) and cattle (Lloyd and Soulsby, 1976), although the antibody class responsible in these species is not known. It is highly likely that systemically absorbed complement-fixing IgG antibodies are also involved in colostral immunity because Lloyd and Soulsby (1976) were able to passively protect neonatal calves against *T. saginata* infection by feeding immune serum.

Several studies have shown that reduced numbers of oncospheres penetrate the intestine of an immune host (Bailey, 1951; Bannerjee and Singh, 1969; Miyazato *et al.*, 1979), but some oncospheres still escape this initial barrier and enter the intestinal epithelium (Friedberg *et al.*, 1979; Furukawa *et al.*, 1981). *H. nana* oncospheres that penetrate the intestinal epithelium of immune mice are rapidly destroyed (Friedberg *et al.*, 1979; Furukawa *et al.*, 1981), and it is likely that this also occurs with taeniid oncospheres. A few oncospheres may survive the "gut barrier" and reach their site of election, but these parasites often succumb to the so-called "post-encystment" immunity, and the inflammatory reaction around these larvae may give rise to macroscopically visible lesions (Campbell, 1938a). It is possible that different immune mechanisms operate at this phase because Campbell (1938b) observed qualitative differences in the ability of worm antigen to absorb out the protective effect of serum collected at different times after infection. However, it is also likely that many parasites dying at this later stage in their development were sub-lethally damaged when they first entered the host, and that other host defence mechanisms, e.g. inflammatory cells, administered the *coup de grâce.*

When antibody is administered passively to rats, more larvae from an oral challenge infection are able to circumvent the "gut-barrier" and reach the liver than in those animals immunized by egg infection (Heath and

Pavloff, 1975; Musoke and Williams, 1975b). These larvae are eventually destroyed, but this observation suggests that natural infection via the gut stimulates additional mechanisms which enhance the action of serum antibody, e.g. local antibody and cellular defences.

Serum antibody is lethal to cestode oncospheres both *in vitro* (Silverman, 1955; Heath, 1970; Rickard and Outteridge, 1974; Heath and Lawrence, 1981) and *in vivo* (see reviews by Gemmell and Soulsby, 1968; Flisser *et al.*, 1979; Williams, 1979). Killing of oncospheres by incubation in immune serum *in vitro* is complement dependent, and studies using Cobra Venom Factor to deplete complement in the host (Musoke and Williams, 1975a; Mitchell *et al.*, 1977a) suggest that complement-fixing antibodies play a major role *in vivo* as well. Extensive analyses of sera collected during *T. taeniaeformis* infection in rats and mice have shown that IgG_{2a} plays a major role in rats, and IgG_{2a} and IgG_1 in mice (Leid and Williams, 1974a; Musoke and Williams, 1975a,b; Mitchell *et al.*, 1977a). The antibody classes responsible for passive transfer of immunity in ruminants have not been examined in detail, but Craig and Rickard (1982) have shown that sheep produce mainly IgG_1 and IgG_2 antibodies against *T. ovis* and *T. hydatigena* oncospheres.

Experiments with nude mice have suggested that the antibody response to *T. taeniaeformis* infection is T-cell dependent (Mitchell *et al.*, 1977a), as is probably also the case with *H. nana* infection in mice (Okamoto, 1968; Okamoto and Koizumi, 1972; Isaak *et al.*, 1977). The mode of action of antibody on oncospheres is yet to be determined. Furukawa *et al.* (1981) used electron microscopy to examine *H. nana* oncospheres in the intestinal wall of immune mice, and found that initial attack occurred as early as 8 h after infection and was characterized by changes in the epithelium of the oncosphere such as increased electron density, shrinking of the cytoplasm, and the formation of large empty vacuoles. These authors concluded that the changes were consistent with antibody-dependent complement-mediated lysis.

(2) Antibody-dependent cellular effects

IgE antibodies were demonstrated during infections of *T. taeniaeformis* in rats and *T. pisiformis* in rabbits (Leid and Williams, 1974b, 1975; Musoke *et al.*, 1978), and although antibodies were not essential for passive transfer of protection with serum from infected rats, they enhanced the degree of protection achieved and led to more rapid death of oncospheres in the recipients. Mast cells often accumulate at the site of infection with larval cestodes (Coleman and De Salva, 1963; Singh and Rao, 1967; Varute, 1971; Siebert *et al.*, 1979; Lindsay, 1981), as well as in the intestinal mucosa of rats with *T. taeniaeformis* larvae in the liver (Cook and Williams, 1981). It has been suggested (Leid and Williams, 1974b) that increased vascular permeability caused by IgE-mediated degranulation of mast cells in the vicinity of larvae may enhance the availability of complement-fixing antibody at the site. On the other hand, histamine or peritoneal anaphylactic diffusate injected directly into the intestine of rats had an inhibitory effect on the penetration of *T. taeniaeformis* oncospheres (Musoke *et al.*, 1978), which suggests that IgE and mast cells may also have a more direct role in preventing the establishment of cestode infections. Some authors have not mentioned an increase in the number of mast cells during challenge infections with *T. taeniaeformis* (Bannerjee and Singh, 1969) and *H. nana* (Friedberg *et al.*, 1979; Miyazato *et al.*, 1979; Furukawa *et al.*, 1981).

Antibody-dependent eosinophil-mediated killing of parasites has been prominent in recent literature (reviewed by Butterworth *et al.*, 1980), but a direct role for these cells in killing cestode larvae has not been substantiated. There have been many reports of increased numbers of eosinophils during cestode infections both at the site of infection and peripherally (see Ansari and Williams, 1976; Ansari *et al.*, 1976), and it has been suggested that antibody plays a role in stimulating the accumulation of these cells (Heath and Pavloff, 1975; Ansari *et al.*, 1976). Furukawa *et al.* (1981) showed by electron microscopy that eosinophils are closely applied to dying oncospheres of *H. nana* in the intestinal wall of mice, but there was no evidence of degranulation and further studies are needed to clarify the role of these cells.

The function of other cells, such as macrophages and neutrophils, is yet to be determined. These cells are known to accumulate around larval cestodes very early in infection (Engelkirk and Williams, 1982; Furukawa *et al.*, 1981), and they have been shown to cause *in vitro* killing of schistosomula (Bout *et al.*, 1981; Incani and McLaren, 1981).

(3) Antibody-independent cellular effects

Although lymphocytes are present near invading oncospheres of *H. nana* in mice (Friedberg *et al.*, 1979; Furukawa *et al.*, 1981), and splenic cells from normal and immune mice adhere to *H. nana* oncospheres *in vitro* (Furukawa, 1974), there is no evidence that these cells have any deleterious effect on the parasite. The many experiments using neonatal thymectomy, anti-thymocyte serum and cell transfer have been unhelpful in distinguishing between true cell-mediated immune effects and T-cell-dependent antibody effects (Okamoto, 1968; Nemeth, 1970; Okamoto and Koizumi, 1972). T-cell-deficient nude mice could provide a powerful tool for such analyses when the technology for preparing "pure" cell populations is sufficiently refined. Moreover, advances in methods for *in vitro* cultivation of cestode larvae (Heath, 1973b; Heath and Lawrence, 1976; Lawrence *et al.*, 1980) may ultimately make *in vitro* analyses of cell-mediated immune events possible.

(4) Mature larval stages

(a) Cyclophyllidea

It was stated earlier that larval cestodes rapidly become resistant to the defence mechanisms of the host, and then usually survive for long periods of time. If the larvae are damaged, or die, a vigorous inflammatory response ensues, producing either clinical signs in the host (*T. solium*) or unsightly lesions in the carcass (*T. ovis*, *T. saginata*). Mechanisms of immunity to mature metacestodes are difficult to study using egg-induced infections because the larvae are usually firmly encapsulated in host tissue, and any physical interference can damage them and result in their death. For this reason, most experimental observations have been made using polyembryonic larval forms that can be serially passaged without recourse to egg infection, e.g. *T. crassiceps*, *E. granulosus*, *E. multilocularis*, *M. corti*. However, there may be some danger in extrapolating from all results obtained in such studies; for example Musoke and Williams (1976) showed that intraperitoneal implantation of *T. taeniaeformis* strobilocerci into rats gave rise to an antibody response qualitatively different to that from an egg-induced infection.

Passive immunization with serum had no adverse effect on *T. taeniaeformis* strobilocerci transplanted into the peritoneal cavities of rats (Musoke and Williams, 1976), and antibody appears to play little role in protecting mice against infection with *E. multilocularis* cysts (Ali-Khan, 1974). In experiments where an effect of immune serum on transplanted metacestodes has been detected, the protection has been only partial and not comparable with the striking results obtained against egg-induced infections (Kowalski and Thorson, 1972a; Niederkorn, 1977; Siebert *et al.*, 1978a,b). Antibody and complement induce permeability defects in *T. taeniaeformis* (Murrell, 1971; Hustead and Williams, 1977) and *T. crassiceps* (Hustead and Williams, 1977) larvae, but these are reversed when complement levels are depleted. Hustead and Williams (1977) suggested that larvae may protect themselves *in vivo* by liberating anticomplementary substances to reduce complement levels in their vicinity.

Reactions observed around larvae *in situ* implicate inflammatory cells in the destruction of metacestodes, and this concept has been supported by some *in vitro* experiments. Eosinophils adhered to *T. taeniaeformis* larvae *in vitro* and incorporated tegumentary cytoplasm into large phagosomes, but release of eosinophil granules was not observed (Engelkirk *et al.*, 1981). In the same experiments, mast cells accumulated and underwent degranulation, but the significance of this was not clear. *T. crassiceps* larvae implanted into the peritoneal cavity of immune mice are eventually encapsulated and destroyed, and eosinophils, mast cells, macrophages and lymphocytes participate in this reaction (Siebert *et al.*, 1979). Macrophages have a protoscolicidal effect on *E. multilocularis in vitro* (Rau and Tanner, 1976; Baron and Tanner, 1977), and Ali-Khan and Siboo (1980) suggested that neutrophils could attack antibody-coated *E. multilocularis* cysts.

Immune responses to metacestodes are probably T-cell-dependent (Baron and Tanner, 1976; Mitchell *et al.*, 1977b; Anderson and Griffin, 1979a,b), but a conclusive role for true cell-mediated immunity has not been established. Fibrous encapsulation of metacestodes is also T-cell-dependent (Pollacco *et al.*, 1978), and this process is potentially beneficial to both the parasite and the host. The capsule may limit access of the host defences to the parasite, but cestodes contain substances that are highly toxic to host cells (Siebert *et al.*, 1979; Engelkirk and Williams, 1982), and the capsule may also serve to protect the host from these. When the parasite dies the capsule provides little barrier to the entry of host cells.

(b) Pseudophyllidea

There is a little information available concerning immunity to the plerocercoid stage of *Spirometra* spp. in laboratory mice. This metacestode has a wide host range including both cold and warm-blooded vertebrates. Infected mice develop specific cell-mediated and humoral responses against plerocercoids which seem to have minimal adverse effects on the parasite, although they may play a role in stopping migration and facilitating encapsulation (Bennett, 1971, 1978; Hensen, 1969; Opuni, 1973; Opuni and Muller, 1975). Stephanson (personal communication) found that the infectivity of plerocercoids for mice was not significantly affected by either immunosuppressive treatment of the host, e.g. thymectomy, irradiation, corticosteroids, or non-specific immunostimulation following inoculation with BCG, *Corynebacterium parvum* or complete Freund's adjuvant. After infection of mice, footpad responses were detected 3 days post-infection (dpi), eosinophilia 4 dpi and IgE levels, which started to rise by 7 dpi, rose to very high levels by 30 dpi. IgM and IgG were detected by 7 dpi. Massive degranulation of peritoneal mast cells had little effect on the parasite, but rather had adverse effects on the host.

(5) Evasion of the immune response

Taeniid cestodes rely for their transmission on prey–predator relationships between the intermediate and definitive hosts. To maximize the opportunity for completion of the life-cycle, two important features characterize the relationship between the larval stage and its host. First, a very high level of resistance to challenge infection prevents death of the intermediate host from overwhelming parasite burdens, and second, prolonged survival of larvae established from the initial infection increases the chance of them being eaten by the definitive host. This phenomenon of "concomitant immunity" has attracted much research in many host–parasite systems, schistosomiasis being a classic example (Smithers and Terry, 1969). Space does not permit a detailed analysis here, but Rickard and Williams (1982) present a more comprehensive review.

Comment was made earlier in this chapter that as the early developmental stages become resistant to the effects of antibody, striking changes take place in their surface morphology. A simple mechanism for escaping the immune response would be to alter the antigenicity of sites exposed to host attack, and further research is necessary to determine whether the morphological changes in the surface membrane also represent antigenic shifts. Evidence for antigens specific to the oncosphere stage has been presented (Craig and Rickard, 1981, 1982), and some other experiments provide additional support for this notion. Sheep with long-standing infections of *T. ovis* (Rickard *et al.*, 1976) and *T. hydatigena* (Gemmell and Johnstone, 1981) became heavily infected following oral challenge with eggs, but the newly established larvae were quickly destroyed. Thus, persistent infection with metacestode stages did not maintain high levels of immunity directed against invading oncospheres. Leid and Williams (1974a) found that strobilocercus antigens failed to absorb out all protective antibodies from the serum of rats infected with *T. taeniaeformis*. The fact that rats can be immunized against oral challenge with *T. taeniaeformis* eggs using strobilocercus antigens (Campbell, 1936; Kwa and Liew, 1977; Ayuya and Williams, 1979) appears contrary to the view that oncospheres possess unique antigenic determinants, but it has not been demonstrated whether the immunity that operates in this instance is directed against the oncosphere or later stages of development.

Another mechanism parasites may adopt to evade the immune response of the host is to mask their own antigens by adsorbing host protein onto the surface or by synthesizing host-like antigens. Hypotheses for antigenic convergence of host and parasite (Sprent, 1962) and molecular mimicry by the parasite (Damian, 1964) have been proposed. Larval cestodes adsorb host immunoglobulins, perhaps specific antibody, onto their surface, and these molecules may act as "blocking antibody" and prevent other host defence mechanisms from recognizing or attacking the parasite (Varela-Díaz *et al.*, 1972; Rickard, 1974; Willms and Arcos, 1977; Kwa and Liew, 1978). Recently, Willms *et al.* (1980) used *in vitro* translation of parasite-derived RNA to produce a protein which was precipitated by rabbit anti-pig IgG. Provided that the RNA extract used was not contaminated with material from adherent host cells, this experiment constitutes striking evidence for molecular mimicry by the cysticercus of *T. solium*.

Studies on the tegument of adult cestodes (Oaks and Lumsden, 1971; Lumsden, 1975) have shown that it is a very active structure with rapid turnover. Although there is no evidence yet available for antigenic modulation at the surface membrane of larval cestodes, such as has been described for schistosomes (Kemp *et al.*, 1980), it is a possibility worthy of investigation.

The fibrous capsule formed around metacestodes may act as a partial barrier to the immune defences of the host. As well, metacestodes in particular locations often seem to provoke minimal host responses, e.g. *T. solium* in the central nervous system (Marquez-Monter, 1971; Showramma and Reddy, 1963; Flisser *et al.*, 1980) and pulmonary cysts of *E. granulosus* in humans (Todorov *et al.*, 1979). In these situations parasite antigens may be sequestered from the host response.

There are several experiments describing suppression of host immune responses during larval cestode infections. Thus, Good and Miller (1976) showed that mice infected with *T. crassiceps* had impaired antibody responses to sheep red blood cells (SRBC), although Ali-Khan (1979) found that mice with 12-week infections of *E. multilocularis* had heightened antibody responses to the same antigen. Ali-Khan (1978) also described a reduction in the numbers of T-cells in the lymphoreticular tissues of mice with long-standing infections of *E. granulosus*. BALB/c mice infected with either protoscolices or cysts of *E. granulosus equinus* showed non-specific immunosuppression, and when their mesenteric lymph node cells (MLNC) were adoptively transferred to other mice, they caused marked suppression in the response of the recipients to SRBC (Allan *et al.*, 1981). These authors found that there was a significant decrease in the numbers of Thy-1 cells in the MLNC transplants, and suggested that chronic infection with *E. granulosus* had altered the composition of the T-cell population in favour of non-specific T-cell suppressor activity. Rickard and Outteridge (unpublished) found that factors in the serum of rabbits heavily infected with *T. pisiformis* inhibit lymphocyte reactivity in *in vitro* transformation assays, and this may explain the anergic state to intradermal inoculation of antigen that develops in such animals (Rickard and Outteridge, 1974). Polyclonal B-cell activation by extracts of *T. solium* (Sulivan-Lopez *et al.*, 1980) could also cause immunosuppression due to lymphocyte exhaustion.

It has been suggested that immunosuppression may occur during infection of rats and hampsters with *Spirometra* spp. (Phares *et al.*, 1976; Tachovsky *et al.*, 1973). Stephanson (personal communication) considers that glycoprotein secretory products from the tail region of *Spirometra plerocercoids* may play a role in causing immunosuppression in infected mice.

Live metacestodes in the tissue of their host do not excite a vigorous inflammatory response, but when they die the host reaction develops rapidly. Evidence is accumulating that cestode larvae may directly influence the host defences in a number of ways, e.g. by cytotoxic effects (Siebert *et al.*, 1979; Engelkirk and Williams, 1982), inhibiting host enzymes (Németh and Juhász, 1980), reducing responsiveness to mitogens (Annen *et al.*, 1980), and by releasing anticomplementary substances (Hammerberg *et al.*, 1976). This latter phenomenon has been studied in some detail (Hammerberg and Williams, 1978a,b), and the soluble complement activating agent of *T. taeniaeformis* has been found to be a highly sulphated acidic proteoglycan. Anticomplementary activity may be important in interfering with the effectiveness of complement-fixing anti-parasite antibodies.

(6) The "functional" antigens

Killed parasites or extracts of dead parasites have been used successfully in immunization experiments, especially against infections with *T. taeniaeformis*, *T. pisiformis* and *H. nana* in laboratory animals (see Gemmell and Soulsby, 1968; Clegg and Smith, 1978). In domesticated ruminants, the most consistent results have been obtained using antigens from oncospheres. Early work (Gemmell, 1964, 1965a,b, 1969; Wikerhauser *et al.*, 1971) strongly suggested that living oncospheres were much more effective than killed parasites in stimulating immunity, giving rise to the concept that so-called "metabolic" or "excretion-secretion" (ES) antigens released by living parasites were the important antigens. Experiments showed that when Millipore membrane diffusion chambers containing living oncospheres were implanted into the peritoneal cavities of rats and lambs, diffusible antigens released from the parasites stimulated solid immunity against a challenge infection with eggs (Rickard and Bell, 1971a). When methods were developed for *in vitro* cultivation of oncospheres (Heath and Smyth, 1970) the logical step was to determine whether antigens collected in the medium used to cultivate larvae contained "functional" antigens, and successful immunization using this method was reported for *T. ovis* in lambs (Rickard and Bell, 1971b). Since then, *in vitro* culture antigens have been used effectively to prevent infection with *T. hydatigena*, *T. saginata*, *T. taeniaeformis* and *T. pisiformis* (see Rickard and Williams, 1982). Antigens collected during *in vitro* incubation of metacestode stages have also been used to immunize rats against *T. taeniaeformis* (Kwa and Liew, 1977; Ayuya and Williams, 1979) and mice against *M. corti* (Kowalski and Thorson, 1972b).

It has been widely assumed that the *in vitro* culture antigens were ES antigens or metabolic products of the living parasite. However, there is little doubt that most preparations of this kind contain breakdown products released by degenerating parasites, or by sloughing of the tegument in the case of metacestode stages. Rajasekariah *et al.* (1980a,b) attempted to purify ES antigens from oncosphere and metacestode stages of *T. taeniaeformis* and found that high-speed centrifugation reduced the immunogenicity of the supernatant and concentrated protective activity in the pellet. This suggested strongly that the "functional" antigens were particulate or membrane-bound and not entirely soluble metabolic products. This does not necessarily exclude oncosphere secretory products (Rajasekariah *et al.*, 1980a,b) because the penetration glands contain membrane-bound inclusions which are shed in groups contained within an envelope of surface membrane (Lethbridge and Gijsbers, 1974; Furukawa *et al.*, 1977). However, the initial attack by antibody appears to be on the surface membrane (Furukawa *et al.*, 1981) indicating that important antigenic sites are located here.

Recent experiments with *T. taeniaeformis* in mice (Rajasekariah *et al.*, 1980a,b), *T. saginata* in cattle (Rickard and Brumley, 1981) and *T. ovis* in sheep (Osborn *et al.*, 1981) have shown that *in vitro* cultivation is not necessary to collect "functional" antigens, and that oncospheres or eggs disrupted by ultrasonic disintegration are highly effective immunogens. After sonication, the oncospheral antigens are still largely particulate and substantially removed by centrifugation at 100 000 *g* (Rajasekariah *et al.*, 1980b; Rickard and Brumley, 1981). Further experiments have shown that sodium deoxycholate can be used to solubilize the oncospheral antigens without affecting their immunogenicity (Rajasekariah *et al.*, 1982), and this should be of considerable assistance in further purification studies. It would be interesting to compare the properties of oncospheral antigens with the immunizing factor of molecular weight 140 000 purified from metacestodes of *T. taeniaeformis* by Kwa and Liew (1977).

C. Immunization against infection

The very high level of immunity developed by animals infected with larval cestodes makes these parasites attractive candidates for the development of practical vaccines. Early experiments have been extensively reviewed (Gemmell and Soulsby, 1968; Weinmann, 1970; Gemmell and Macnamara, 1972; Gemmell, 1976; Gemmell and Johnstone, 1977), and more recent findings have been discussed by Clegg and Smith (1978), Flisser *et al.* (1979), Williams (1979, 1983), Rickard and Howell (1982) and Rickard and Williams (1982).

Experiments have shown that young animals can be protected against infection by immunization of their mothers prior to parturition (Rickard *et al.*, 1977a,b; Lloyd, 1979), and that a single inoculation of *T. ovis* culture antigen emulsified in oil adjuvant maintains a significant level of immunity for at least 12 months in sheep (Rickard *et al.*, 1977b). However, there are several other important problems to be addressed before large scale immunization against larval cestodes could become a practical control measure.

Immunization of sheep and cattle against field infection with *T. ovis* and *T. saginata*, respectively, has generally not given levels of protection as high as that achieved in laboratory experiments (Rickard *et al.*, 1976, 1981a, 1982), and some workers have reported negative results using culture antigens to immunize calves against infection with *T. saginata* (Wikerhauser *et al.*, 1978; Mitchell and Armour, 1980). Difficulties in standardizing crude antigen preparations is one problem, but another important consideration is variations in innate resistance of the animals. For example, Rickard *et al.* (1982) showed that male cattle responded poorly to immunization against *T. saginata* infection by comparison with females. A major difficulty in assessing the value of vaccination is that no information is available concerning the reduction in infection required to achieve significant control of these parasites in an endemic situation.

Vaccination with non-living oncospheres appears to have little impact on the survival of established metacestodes, whereas immunization with live oncospheres does affect larval survival (Gemmell, 1970, 1972). An effect on established larvae is important for two practical reasons. First, any parasites which evade the initial lethal host response must be killed so that they are unavailable for transmission to the definitive host. Second, at the start of a control programme, there would be a large reservoir of mature larvae in the population which, unless destroyed, would significantly slow down initial progress towards eradication. Studies must be made of antigens from life-cycle stages other than oncospheres to determine whether a "cocktail" vaccine could be produced which would kill established metacestodes as well as prevent new infection.

If all of these problems are overcome, there still remains the major difficulty of an adequate supply of antigen at an economic cost. Parasites such as *T. solium* and *T. saginata* can only be obtained from man and eggs of *T. solium* and *E. granulosus* are dangerous to handle because of their infectivity to humans. One approach to overcome the difficulty of obtaining material from humans has been to use oncospheres from cestodes that are readily grown in dogs to stimulate cross-immunity in the host. Cross-immunity has been shown to occur between several taeniid cestodes of sheep, cattle and laboratory rodents (see Rickard and Williams, 1982). In field trials where cattle were immunized against *T. saginata* infection using antigens prepared from *T. hydatigena* oncospheres, the level of protection achieved was less than that attained using the homologous *T. saginata* antigen, and was probably not sufficient to be of practical value (Rickard *et al.*, 1981a, 1982). Perhaps other cross-reacting antigens would have achieved better results, and both *T. ovis* (Rickard and Adolph, 1976) and *T.*

taeniaeformis (Lloyd, 1979) have shown promise in immunizing cattle against *T. saginata*.

IV. CONCLUSIONS

Clearly, fundamental information concerning immune responses to larval and adult cestodes is lacking for those species of economic or public health significance. Results obtained in laboratory animals have yielded much important information, but studies are urgently needed to identify mechanisms that may be peculiar to particular host–parasite relationships. It is interesting to compare, for instance, the amount of fundamental data available on immunity to fascioliasis in ruminants with that available on cysticercosis (Rickard and Howell, 1982). Sheep develop little resistance to infection with *Fasciola hepatica* and immunization experiments have had little success. Thus, considerable research has been devoted to analysing the fundamental relationships between this parasite and its ruminant hosts in order to improve vaccination procedures. On the other hand, immunization has been extremely effective against cysticercosis in domesticated ruminants, so that while many successful vaccination trials have been carried out, little effort has been directed at obtaining fundamental information on the mechanisms of immunity in these animals.

A major obstacle to the development of practical vaccines against larval cestodes will be to obtain adequate supplies of antigen. Use of living animals will be unacceptable on a large scale, and conventional methods for *in vitro* culture of parasite stages are expensive and inadequate. Because of this, there is an urgent need to identify and purify the "functional" antigen(s). Monoclonal antibody methods have been applied to cestodes (Craig *et al.*, 1980, 1981) and this technology may be useful for identification of the important antigen(s) and the cells that produce them. Once these antigens are identified, methods for their mass production can be investigated, for example recombinant DNA methods or *in vitro* culture of cell lines if the appropriate cells can be identified and propagated. Sakamoto (1978) has achieved some success with culture of cells of *Echinococcus* spp. Howell (1981) reported experiments in which he fused rat fibroblast cells with cells from *F. hepatica* and the progeny produced an identifiable parasite protein. Research along these lines could be most rewarding because of the large numbers of mammalian cell lines which can be propagated *in vitro*.

Experiments on immunity to the adult stages of taeniid cestodes do not hold out much hope for the development of useful vaccines. Perhaps this could be anticipated because the immunological regulation of natural infection with these parasites operates through the intermediate host

(Gemmell, 1976). In contrast, where an effective immune response to the larval phase may not exist, e.g. *H. citelli* in camel crickets, resistance of the definitive host to infection with adult tapeworms may regulate the incidence of infection within the host population as well as restrict the size of worm burdens (Wassom *et al.*, 1973).

Little success has been achieved in stimulating an immune response to destroy established metacestodes. A better understanding of the mechanisms which allow these parasites to evade immune aggression may yield important clues for improving vaccination procedures. Oral administration of cestode antigens stimulates strong immunity against infection (Ayuya and Williams, 1979; Lloyd, 1979; Rickard *et al.*, 1981b), and because this route of immunization is potentially of great benefit, especially in man, more research should be directed toward understanding the events that take place in the gut during infection with both oncospheres and adult cestodes. Adjuvants are obviously going to play an important role in parenteral immunization, and considerable research is necessary to determine the most satisfactory material to use.

Variations in innate resistance will play a major role in determining the success of vaccination strategies. A better understanding of these phenomena may help to identify genetic markers which can be used to predict the susceptibility of individual animals, and these markers would be useful in breeding programmes to select for genetic resistance. Techniques such as embryo transfer and cloning may greatly accelerate selection methods.

Although there is still much research work to be done before vaccination against these parasites becomes feasible on a large scale, their importance to agriculture and human health, and the impressive results obtained thus far, give good reason for optimism.

V. ACKNOWLEDGEMENTS

I am grateful to Drs J. Andreassen, M. A. Gemmell, D. D. Heath, C. A. Hopkins, A. Ito, J. Stephanson, R. C. A. Thompson and J. F. Williams for providing me with unpublished information or manuscripts in press. Miss Jane Brumley's constructive criticism was of invaluable assistance and Mrs Barbara Chambers prepared the manuscript expertly. Financial support for work carried out in this laboratory was provided by: National Health and Medical Research Council of Australia; Australian Meat Research Committee; Ministry of Water Resources of Victoria; Melbourne and Metropolitan Board of Works.

VI. REFERENCES

Ali-Khan, Z. (1974). Host–parasite relationship in *Echinococcosis.* I. Parasite biomass and antibody response in three strains of inbred mice against graded doses of *Echinococcus multilocularis. J. Parasit.* **60**, 231–235.

Ali-Khan, Z. (1978). Pathological changes in the lymphoreticular tissues of Swiss mice infected with *Echinococcus granulosus* cysts. *Z. ParasitKde*, **58**, 47–54.

Ali-Khan, Z. (1979). Humoral response to sheep red blood cells in C57 L/J mice during early and chronic stages of infection with *Echinococcus multilocularis* cysts. *Z. ParasitKde*, **59**, 259–265.

Ali-Khan, Z. and Siboo, R. (1980). Pathogenesis and host response in subcutaneous alveolar hydatidosis. 1. Histogenesis of alveolar cyst and a qualitative analysis of the inflammatory infiltrates. *Z. ParasitKde*, **62**, 241–254.

Allan, D., Jenkins, P., Connor, R. J. and Dixon, J. B. (1981). A study of immunoregulation of BALB/c mice by *Echinococcus granulosus equinus* during a prolonged infection. *Parasit. Immun.* **3**, 137–142.

Anderson, M. J. D. and Griffin, J. F. T. (1979a). *Taenia crassiceps* in the rat. I. Differences in susceptibility to infection and development of immunocompetence in relation to age and host strain. *Int. J. Parasit.* **9**, 229–233.

Anderson, M. J. D. and Griffin, J. F. T. (1979b). *Taenia crassiceps* in the rat. II. Transfer of immunity and immunocompetence with lymph node cells. *Int. J. Parasit.* **9**, 235–239.

Andreassen, J. and Hopkins, C. A. (1980). Immunologically mediated rejection of *Hymenolepis diminuta* by its normal host, the rat. *J. Parasit.* **66**, 898–903.

Andreassen, J., Hinsbo, 'O. and Ruitenburg, E. J. (1978). *Hymenolepis diminuta* infections in congenitally athymic (nude) mice: worm kinetics and intestinal histopathology. *Immunology*, **34**, 105–113.

Andreassen, J., Hinsbo, O. and Vienberg, S. (1982). Responsiveness of congenitally thymus deficient nude mice to the intestinal cestode, *Hymenolepis diminuta. Int. J. Parasit.* **12**, 215–219.

Annen, J., Kohler, P., Eckert, J. and Speck, S. (1980). Cytotoxic properties of *Echinococcus granulosus* cyst fluid. *In* "The Host–Invader Interplay", (H. Van den Bossche, ed.), pp. 339–342. Elsevier/North-Holland Biomedical Press, Amsterdam.

Ansari, A. and Williams, J. F. (1976). The eosinophilic response of the rat to infection with *Taenia taeniaeformis. J. Parasit.* **62**, 728–736.

Ansari, A., Williams, J. F. and Musoke, A. J. (1976). Antibody mediated secondary eosinophilic response to *Taenia taeniaeformis* in the rat. *J. Parasit.* **62**, 737–740.

Ayuya, J. M. and Williams, J. F. (1979). Immunological response of the rat to infection with *Taenia taeniaeformis.* VII. Oral and parenteral immunization with parasite antigens. *Immunology*, **36**, 825–834.

Bailey, W. S. (1951). Host-tissue reactions to initial and superimposed infections with *Hymenolepis nana* var. *fraterna. J. Parasit.* **37**, 440–444.

Bannerjee, D. and Singh, K. S. (1969). Studies on *Cysticercus fasciolaris.* IV. Immunity to *Cysticercus fasciolaris* in rat. *Indian J. Anim. Sci.* **39**, 250–253.

Baron, R. W. and Tanner, C. E. (1976). The effect of immunosuppression on secondary *Echinococcus multilocularis* infections in mice. *Int. J. Parasit.* **6**, 31–42.

Baron, R. W. and Tanner, C. E. (1977). *Echinococcus multilocularis* in the mouse: the *in vitro* protoscolicidal activity of peritoneal macrophages. *Int. J. Parasit.* **7**, 489–495.

Befus, A. D. (1975). Secondary infection of *Hymenolepis diminuta* in mice: effects of varying worm burdens in primary and secondary infections. *Parasitology*, **71**, 61–75.

Befus, A. D. (1977). *Hymenolepis diminuta* and *H. microstoma*: Mouse immunoglobulins binding to the tegumental surface. *Expl. Parasit.* **41**, 242–251.

Befus, A. D. and Featherston, D. W. (1974). Delayed rejection of single *Hymenolepis diminuta* in primary infections of young mice. *Parasitology*, **69**, 77–85.

Befus, A. D. and Threadgold, L. T. (1975). Possible immunological damage to the tegument of *Hymenolepis diminuta* in mice and rats. *Parasitology*, **71**, 525–534.

Bennett, L. J. (1971). "The Immunological Responses Produced by Mice and Amphibians to Spargana", Ph.D. Thesis, University of Adelaide.

Bennett, L. J. (1978). The immunological responses of mice to Australian spargana. *J. Parasit.* **64**, 182–185.

Beveridge, I. and Rickard, M. D. (1975). The development of *Taenia pisiformis* in various definitive host species. *Int. J. Parasit.* **5**, 633–639.

Bland, P. W. (1976). Immunity to *Hymenolepis diminuta*: unresponsiveness of the athymic nude mouse to infection. *Parasitology*, **72**, 93–97.

Bout, D. T., Joseph, M., David, J. R. and Capron, A. (1981). *In vitro* killing of *S. mansoni* schistosomula by lymphokine-activated mouse macrophages. *J. Immun.* **127**, 1–5.

Butterworth, A. E., Vadis, M. A. and David, J. R. (1980). Mechanisms of eosinophil-mediated helminthotoxicity. *In* "The Eosinophil in Health and Disease", (A. A. F. Mahmoud and K. F. Austen, eds), pp. 253–273. Grune and Stratton, New York.

Campbell, D. H. (1936). Active immunization of albino rats with protein fractions from *Taenia taeniaeformis* and its larval form *Cysticercus fasciolaris*. *Am. J. Hyg.* **23**, 104–113.

Campbell, D. H. (1938a). The specific protective property of serum from rats infected with *Cysticercus crassicollis*. *J. Immun.* **35**, 195–204.

Campbell, D. H. (1938b). The specific absorbability of protective antibodies against *Cysticercus crassicollis* in rats and *C. pisiformis* in rabbits from infected and artificially immunized animals. *J. Immun.* **35**, 205–216.

Chandler, A. C. (1939). The effects of number and age of worms on development of primary and secondary infections with *Hymenolepis diminuta* in rats, and an investigation into the true nature of "premunition" in tapeworm infections. *Am. J. Hyg.* **29**, 105–114.

Chandler, A. C. (1940). Failure of artificial immunization to influence *Hymenolepis diminuta* infections in rats. *Am. J. Hyg.* **31**, 17–22.

Chappell, L. H. and Pike, A. W. (1976). Loss of *Hymenolepis diminuta* from the rat. *Int. J. Parasit.* **6**, 333–339.

Choromanski, L. (1978). The influence of cyclophosphamide on the development of *Hymenolepis diminuta* in mice. Short Communications IVth Int. Cong. Parasitol., Section E, pp. 32–33.

Christie, P. R. and Moqbel, R. (1980). Effect of high doses of gamma-irradiation on growth of *Hymenolepis diminuta. J. Helminth.* **54**, 267–269.

Clapham, P. A. (1940). Studies on *Coenurus glomeratus. J. Helminth.* **18**, 45–52.

Clegg, J. A. and Smith, M. A. (1978). Prospects for the development of dead vaccines against helminths. *Adv. Parasit.* **16**, 165–218.

Coleman, E. J. and De Salva, J. J. (1963). Mast cell responses to cestode infection. *Proc. Soc. exp. Biol. Med.* **112**, 432–434.

Coleman, R. M., Carty, J. M. and Graziadei, W. D. (1968). Immunogenicity and phylogenetic relationship of tapeworm antigens produced by *Hymenolepis nana* and *Hymenolepis diminuta. Immunology*, **15**, 297–304.

Cook, R. W. and Williams, J. F. (1981). Pathology of *Taenia taeniaeformis* infection in the rat: gastrointestinal changes. *J. comp. Path.* **91**, 205–217.

Craig, P. S. and Rickard, M. D. (1981). Anti-oncospheral antibodies in the serum of lambs experimentally infected with either *Taenia ovis* or *Taenia hydatigena. Z. ParasitKde*, **64**, 169–177.

Craig, P. S. and Rickard, M. D. (1982). Antibody responses of experimentally infected lambs to antigens collected during *in vitro* maintenance of the adult, metacestode or oncosphere stages of *Taenia hydatigena* and *Taenia ovis*, with further observations on anti-oncospheral antibodies. *Z. ParasitKde*, **67**, 197–209.

Craig, P. S., Mitchell, G. F., Cruise, K. M. and Rickard, M. D. (1980). Hybridoma antibody immunoassays for the detection of parasitic infection: attempts to produce an immunodiagnostic reagent for a larval taeniid cestode infection. *Aust. J. exp. Biol. med. Sci.* **58**, 339–350.

Craig, P. S., Hocking, R. E., Mitchell, G. F. and Rickard, M. D. (1981). Murine hybridoma-derived antibodies in the processing of antigens for the immunodiagnosis of hydatid (*Echinococcus granulosus*) infection in sheep. *Parasitology*, **83**, 303–317.

Damian, R. T. (1964). Molecular mimicry: Antigen sharing by parasite and host and its consequences. *Am. Na.* **98**, 129–149.

Dow, C. and Jarrett, W. F. H. (1960). Age, strain and sex differences in susceptibility to *Cysticercus fasciolaris* in the mouse. *Expl. Parasit.* **10**, 72–74.

Engelkirk, P. G. and Williams, J. F. (1982). *Taenia taeniaeformis* in the rat: ultrastructure of the host–parasite interface on days 1–7 post infection. *J. Parasit.* **68**, 620–633.

Engelkirk, P. G., Williams, J. F. and Signs, M. M. (1981). Interactions between *Taenia taeniaeformis* and host cells *in vitro*. I. Rapid adherence of peritoneal cells to strobilocerci. *Int. J. Parasit.* **11**, 463–474.

Farooqi, H. U. (1958). The occurrence of certain specialised glands in the rostellum of *Taenia solium* L. *Z. ParasitKde*, **18**, 308–311.

Flisser, A., Pérez-Montfort, R. and Larralde, C. (1979). The immunology of human and animal cysticercosis: a review. *Bull. Wld. Hlth. Org.* **57**, 839–856.

Flisser, A., Woodhouse, E. and Larralde, C. (1980). Human cysticercosis: antigens, antibodies and non-responders. *Clin. Expl. Immun.* **39**, 27–37.

Forsek, Z. and Rukavina, J. (1959). [Experimental immunization of dogs against *Echinococcus granulosus.* 1. Preliminary findings.] *Veterinaria, Saraj.* **8**, 479–482.

Friedberg, W., Neas, B. R., Faulkner, D. N. and Congdon, C. C. (1979).

Hymenolepis nana: Intestinal tissue phase in actively immunized mice. *J. Parasit.* **65**, 61–64.

Furukawa, T. (1974). Adherence reactions with mouse lymphoid cells against the oncosphere larvae of *Hymenolepis nana*. *Jap. J. Parasit.* **23**, 236–249.

Furukawa, T., Miyazato, T., Okamato, K. and Nakai, Y. (1977). The fine structure of the hatched oncospheres of *Hymenolepis nana*. *Jap. J. Parasit.* **26**, 49–62.

Furukawa, T., Niwa, A. and Miyazato, T. (1981). Ultrastructural aspects of immune damage to *Hymenolepis nana* oncospheres in mice. *Int. J. Parasit.* **11**, 287–300.

Geerts, S., Kumar, V. and Vercruysse, J. (1977). *In vivo* diagnosis of bovine cysticercosis. *Vet. Bull.* **47**, 653–665.

Gemmell, M. A. (1962). Natural and acquired immunity factors interfering with development during the rapid growth phase of *Echinococcus granulosus* in dogs. *Immunology*, **5**, 496–503.

Gemmell, M. A. (1964). Immunological responses of the mammalian host against tapeworm infections. 1. Species specificity of hexacanth embryos in protecting sheep against *Taenia hydatigena*. *Immunology*, **7**, 489–499.

Gemmell, M. A. (1965a). Immunological responses of the mammalian host against tapeworm infections III. Species specificity of hexacanth embryos in protecting sheep against *Taenia ovis*. *Immunology*, **8**, 281–290.

Gemmell, M. A. (1965b). Immunological responses of the mammalian host against tapeworm infections. II. Species specificity of hexacanth embryos in protecting rabbits against *Taenia pisiformis*. *Immunology*, **8**, 270–280.

Gemmell, M. A. (1969). Immunological responses of the mammalian host against tapeworm infections XI. Antigen sharing among *Taenia pisiformis*, *T. hydatigena* and *T. ovis*. *Expl. Parasit.* **26**, 67–72.

Gemmell, M. A. (1970). Hydatidosis and cysticercosis. 3. Induced resistance to the larval phase. *Aust. Vet. J.* **46**, 366–369.

Gemmell, M. A. (1972). Some problems of inducing resistance to *Taenia hydatigena* under conditions of a strong infection pressure. *Aust. Vet. J.* **48**, 29–31.

Gemmell, M. A. (1976). Immunological responses and regulation of the cestode zoonoses. *In* "Immunology of Human Parasitic Infections", (S. Cohen and E. Sadun, eds), pp. 333–358. Blackwell Scientific Publications, Oxford.

Gemmell, M. A. (1978). Perspectives on options for hydatidosis and cysticercosis control. *Vet. Med. Rev.* **1**, 3–48.

Gemmell, M. A. and Johnstone, P. D. (1977). Experimental epidemiology of hydatidosis and cysticercosis. *Adv. Parasit.* **15**, 311–369.

Gemmell, M. A. and Johnstone, P. D. (1981). Factors regulating tapeworm populations: estimations of the duration of acquired immunity by sheep to *Taenia hydatigena*. *Res. vet. Sci.* **30**, 53–56.

Gemmell, M. A. and Macnamara, F. N. (1972). Immune responses to tissue parasites II. Cestodes. *In* "Immunity to Animal Parasites", (E. J. L. Soulsby, ed.), pp. 235–272. Academic Press, London and New York.

Gemmell, M. A. and Soulsby, E. J. L. (1968). The development of acquired immunity to tapeworms and progress towards active immunization, with special reference to *Echinococcus* spp. *Bull. Wld. Hlth. Org.* **39**, 45–55.

Good, A. H. and Miller, K. L. (1976). Depression of the immune response to sheep erythrocytes in mice infected with *Taenia crassiceps* larvae. *Infec. Immun.* **14**, 449–456.

Gray, J. S. (1972). Studies on the course of infection with the poultry cestode *Raillietina cesticillus* (Molin, 1858) in the definitive host. *Parasitology*, **65**, 243–250.

Gray, J. S. (1973). Studies on host resistance to secondary infections of *Raillietina cesticillus* Molin, 1858 in the fowl. *Parasitology*, **67**, 375–382.

Gray, J. S. (1976). The cellular response of the fowl small intestine to primary and secondary infections of the cestode *Raillietina cesticillus* (Molin). *Parasitology*, **73**, 189–204.

Hamilton, D. R. (1974). Immunosuppressive effects of predator induced stress in mice with acquired immunity to *Hymenolepis nana*. *J. psychosom. Res.* **18**, 143–153.

Hammerberg, B. and Williams, J. F. (1978a). Interaction between *Taenia taeniaeformis* and the complement system. *J. Immun.* **120**, 1033–1037.

Hammerberg, B. and Williams, J. F. (1978b). Physico-chemical characterization of complement-interacting factors from *Taenia taeniaeformis*. *J. Immun.* **120**, 1039–1045.

Hammerberg, B., Musoke, A. J., Hustead, S. T. and Williams, J. F. (1976). Anticomplementary substances associated with taeniid metacestodes. *In* "Pathophysiology of Parasitic Infection", (E. J. L. Soulsby, ed.), pp. 233–240. Academic Press, New York and London.

Hammerberg, B., Musoke, A. J., Williams, J. F. and Leid, R. W. (1977). Uptake of colostral immunoglobulins by the suckling rat. *Lab. Anim. Sci.* **27**, 50–53.

Harris, W. G. and Turton, J. A. (1973). Antibody response to tapeworm (*Hymenolepis diminuta*) in rat. *Nature, Lond.* **246**, 521–522.

Hearin, J. T. (1941). Studies on the acquired immunity to the dwarf tapeworm *Hymenolepis nana* var. *fraterna* in the mouse host. *Am. J. Hyg.* **33**, 71–87.

Heath, D. D. (1970). "The Developmental Biology of Larval Cyclophyllidean Cestodes in Mammals", Ph.D. Thesis, Australian National University.

Heath, D. D. (1971). The migration of oncospheres of *Taenia pisiformis*, *T. serialis* and *Echinococcus granulosus* within the intermediate host. *Int. J. Parasit.* **1**, 145–152.

Heath, D. D. (1973a). Resistance to *Taenia pisiformis* larvae in rabbits. II. Temporal relationships and the development phase affected. *Int. J. Parasit.* **3**, 491–498.

Heath, D. D. (1973b). An improved technique for the *in vitro* culture of taeniid larvae. *Int. J. Parasit.* **3**, 481–484.

Heath, D. D. and Lawrence, S. B. (1976). *Echinococcus granulosus*: development *in vitro* from oncosphere to immature hydatid cyst. *Parasitology*, **73**, 417–423.

Heath, D. D. and Lawrence, S. B. (1981). *Echinococcus granulosus* cysts: early development *in vitro* in the presence of serum from infected sheep. *Int. J. Parasit.* **11**, 261–266.

Heath, D. D. and Pavloff, P. (1975). The fate of *Taenia taeniaeformis* oncospheres in normal and passively protected rats. *Int. J. Parasit.* **5**, 83–88.

Heath, D. D. and Smyth, J. D. (1970). *In vitro* cultivation of *Echinococcus granulosus*, *Taenia hydatigena*, *T. ovis*, *T. pisiformis* and *T. serialis* from oncosphere to cystic larva. *Parasitology*, **61**, 329–343.

Heath, D. D., Parmeter, S. N. and Osborn, P. J. (1980). An attempt to immunise dogs against *Taenia hydatigena*. *Res. vet. Sci.* **29**, 388–389.

Hensen, H. L. (1969). "Immunological Studies on the Genus *Spirometra* (Cestoda: Pseudophyllidea)", Ph.D. Thesis, Louisiana State University.

Herd, R. P. (1976). The cestocidal effect of complement in normal and immune sera *in vitro*. *Parasitology*, **72**, 325–334.

Herd, R. P. (1977). Resistance of dogs to *Echinococcus granulosus*. *Int. J. Parasit.* **7**, 135–138.

Herd, R. P., Chappel, R. J. and Biddell, D. (1975). Immunization of dogs against *Echinococcus granulosus* using worm secretory antigens. *Int. J. Parasit.* **5**, 395–399.

Hesselberg, C. A. and Andreassen, J. (1975). Some influences of population density on *Hymenolepis diminuta* in rats. *Parasitology*, **71**, 517–523.

Heyneman, D. (1962). Studies on helminth immunity: I. Comparison between lumenal and tissue phases of infection in the white mouse by *Hymenolepis nana* (Cestoda: Hymenolepididae). *Am. J. trop. Med. Hyg.* **11**, 46–63.

Heyneman, D. (1963). Host–parasite resistance patterns—Some implications from experimental studies with helminths. *Ann. N.Y. Acad. Sci.* **113**, 114–129.

Heyneman, D. and Welsh, J. F. (1959). Action of homologous antiserum *in vitro* against life-cycle stages of *Hymenolepis nana*, the dwarf mouse tapeworm. *Expl. Parasit.* **8**, 119–128.

Hinsbo, O., Andreassen, J. and Hesselberg, C. A. (1975). Immunity to *Hymenolepis diminuta* in the rat. *Norw. J. Zool.* **23**, 197.

Hinsbo, O., Andreassen, J. and Ruitenberg, J. (1976). Immunity to *Hymenolepis diminuta* in the rat: effects of ATS treatment. *Parasitology*, **73**, XXX.

Hopkins, C. A. (1980). Immunity and *Hymenolepis diminuta*. *In* "Biology of the Tapeworm *Hymenolepis diminuta*", (H. Arai, ed.), pp. 551–614. Academic Press, London and New York.

Hopkins, C. A. (1982). Immunological memory in mice to adult *Hymenolepis diminuta* (Cestoda). *J. Parasit.* **68**, 32–38.

Hopkins, C. A. and Barr, I. F. (1982). The source of antigen in an adult tapeworm. *Int. J. Parasit.* **12**, 327–333.

Hopkins, C. A. and Stallard, H. E. (1974). Immunity to intestinal tapeworms: the rejection of *Hymenolepis citelli* by mice. *Parasitology*, **69**, 63–76.

Hopkins, C. A. and Stallard, H. E. (1976). The effect of cortisone on the survival of *Hymenolepis diminuta* in mice. *Rice Univ. Studies*, **62**, 145–159.

Hopkins, C. A. and Zajac, A. (1976). Transplantation of *Hymenolepis diminuta* into naive, immune and irradiated mice. *Parasitology*, **73**, 73–81.

Hopkins, C. A., Subramanian, G. and Stallard, H. (1972a). The development of *Hymenolepis diminuta* in primary and secondary infections in mice. *Parasitology*, **64**, 401–412.

Hopkins, C. A., Subramanian, G. and Stallard, H. (1972b). The effect of immunosuppressants on the development of *Hymenolepis diminuta* in mice. *Parasitology*, **65**, 111–120.

Hopkins, C. A., Andreassen, J. and Barr, I. F. (1980). Duration of immunological memory evoked by adult tapeworms. *Parasitology*, **81**, xl–xli.

Howard, R. J. (1976). The growth of secondary infections of *Hymenolepis microstoma* in mice: the effect of various primary infection regimes. *Parasitology*, **72**, 317–323.

Howell, M. J. (1981). An approach to the production of helminth antigens *in vitro*: The formation of hybrid cells between *Fasciola hepatica* and a rat fibroblast cell line. *Int. J. Parasit.* **11**, 235–242.

Hunninen, A. V. (1935). Studies on the life history and host–parasite relations of *Hymenolepis fraterna* (*H. nana* var. *fraterna* Stiles) in white mice. *Am. J. Hyg.* **22**, 414–443.

Hustead, S. T. and Williams, J. F. (1977). Permeability studies on taeniid metacestodes: II. Antibody-mediated effects on membrane permeability in larvae of *Taenia taeniaeformis* and *Taenia crassiceps*. *J. Parasit.* **63**, 322–326.

Incani, R. N. and McLaren, D. J. (1981). Neutrophil-mediated cytotoxicity to schistosomula of *Schistosoma mansoni in vitro*: studies on the kinetics of complement and/or antibody-dependent adherence and killing. *Parasite Immun.* **3**, 107–126.

Ingram, D. G. and Smith, A. N. (1965). Immunological responses of young animals. I. Review of the literature. *Can. vet. J.* **61**, 194–203.

Isaak, D. D., Jacobson, R. H. and Reed, N. D. (1975). Thymus dependence of tapeworm (*Hymenolepis diminuta*) elimination from mice. *Infec. Immun.* **12**, 1478–1479.

Isaak, D. D., Jacobson, R. H. and Reed, N. D. (1977). The course of *Hymenolepis nana* infections in thymus-deficient mice. *Int. Archs. Allergy appl. Immun.* **55**, 504–513.

Ito, A. (1977). The mode of passive protection against *Hymenolepis nana* induced by serum transfer. *Int. J. Parasit.* **7**, 67–71.

Ito, A. (1978). *Hymenolepis nana*: Protective immunity against mouse-derived cysticercoids induced by initial inoculation with eggs. *Expl. Parasit.* **46**, 12–19.

Ito, A. (1980). *Hymenolepis nana*: Survival in the immunized mouse. *Expl. Parasit.* **49**, 248–257.

Ito, A. (1982). *Hymenolepis nana*: Immunogenicity of a lumen phase of the direct cycle and failure of autoinfection in BALB/c mice. *Expl. Parasit.* **54**, 113–120.

Kagan, I. G. (1976). Serodiagnosis of hydatid disease. *In* "Immunology of Parasitic Infections", (S. Cohen and E. Sadun, eds), pp. 130–142. Blackwell Scientific Publications, Oxford.

Kassis, A. I. and Tanner, C. E. (1976). The role of complement in hydatid disease: *in vitro* studies. *Int. J. Parasit.* **6**, 25–35.

Kemp, W. M., Brown, P. R., Merritt, S. C. and Miller, R. E. (1980). Tegument associated antigen modulation by adult male *Schistosoma mansoni*. *J. Immun.* **124**, 806–811.

Kondo, K., Yoshimura, H., Ohnishi, Y., Nishida, K. and Kamimura, K. (1977). Immunological studies on diphyllobothriasis I. Immunoglobulin and precipitation tests using ouchterlony and immunoelectrophoresis in the patients. *Jap. J. Parasit.* **26**, 265–270.

Kowalski, J. C. and Thorson, R. E. (1972a). Protective immunity against tetrathyridia of *Mesocestoides corti* by passive transfer of serum in mice. *J. Parasit.* **58**, 244–246.

Kowalski, J. C. and Thorson, R. E. (1972b). Immunization of laboratory mice against tetrathyridia of *Mesocestoides corti* (Cestoda) using a secretory and excretory antigen and a soluble somatic antigen. *J. Parasit.* **58**, 732–734.

Kwa, B. H. and Liew, F. Y. (1977). Immunity in taeniasis-cysticercosis. I.

Vaccination against *Taenia taeniaeformis* in rats using purified antigen. *J. exp. Med.* **146**, 118–131.

Kwa, B. H. and Liew, F. Y. (1978). Studies on the mechanism of long-term survival of *Taenia taeniaeformis* in rats. *J. Helminth.* **52**, 1–6.

Larsh, J. E. (1951). Host–parasite relationships in cestode infections, with emphasis on host resistance. *J. Parasit.* **37**, 343–352.

Lawrence, S. B., Heath, D. D., Parmeter, S. N. and Osborn, P. J. (1980). Development of early larval stages of *Taenia ovis in vitro* using a cell monolayer. *Parasitology*, **81**, 35–40.

Leid, R. W. and Williams, J. F. (1974a). The immunological response of the rat to infection with *Taenia taeniaeformis*. I. Immunoglobulin classes involved in passive transfer of resistance. *Immunology*, **27**, 195–208.

Leid, R. W. and Williams, J. F. (1974b). The immunological response of the rat to infection with *Taenia taeniaeformis*. II. Characterization of reaginic antibody and an allergen associated with the larval stage. *Immunology*, **27**, 209–226.

Leid, R. W. and Williams, J. F. (1975). Reaginic antibody response in rabbits infected with *Taenia pisiformis*. *Int. J. Parasit.* **5**, 203–208.

Leid, R. W. and Williams, J. F. (1979). Helminth parasites and the host inflammatory system. *Chem. Zool.* **11**, 229–271.

Lethbridge, R. C. and Gijsbers, M. F. (1974). Penetration gland secretion by hexacanths of *Hymenolepis diminuta*. *Parasitology*, **68**, 303–311.

Lindsay, M. (1981). "The Localisation and Identification of Antibody-forming Cells in the Tissues of Rats Infected with *Taenia taeniaeformis*", M.S. Thesis, Michigan State University.

Lloyd, S. (1979). Homologous and heterologous immunization against the metacestodes of *Taenia saginata* and *Taenia taeniaeformis* in cattle and mice. *Z. ParasitKde*, **60**, 87–96.

Lloyd, S. and Soulsby, E. J. L. (1976). Passive transfer of immunity to neonatal calves against the metacestodes of *Taenia saginata*. *Vet. Parasit.* **2**, 355–362.

Lloyd, S. and Soulsby, E. J. L. (1978). The role of IgA immunoglobulins in the passive transfer of protection to *Taenia taeniaeformis* in the mouse. *Immunology*, **34**, 939–945.

Lumsden, R. D. (1975). Surface ultrastructure and cytochemistry of parasitic helminths. *Expl. Parasit.* **37**, 267–339.

Machado, A. J., Gazzinelli, G., Pellegrino, J. and Días da Silva, W. (1975). *Schistosoma mansoni*: The role of complement C3-activating system in the cercaricidal action of normal serum. *Expl. Parasit.* **38**, 20–29.

Machnicka, B. and Choromanski, L. (1978). Immunosuppressive effect of *Trypanosoma cruzi* infection on the development of *Hymenolepis diminuta* in CFW mice. *Short Communications IVth Int. Cong. Parasitol., Warsaw.* Section E, p. 32.

Machnicka-Roguska, B. and Zwierz, C. (1970). Intradermal test with antigenic fractions in *Taenia saginata* infection. *Acta parasit. pol.* **17**, 293–299.

Machnicka, B. and Zwierz, C. (1974). The immunologic reactivity of the sera of people infected with *Taenia saginata* to *Cysticercus bovis* antigens. *Bull. Acad. pol. Sci. Sér. Sci. tech.* **22**, 259–261.

Marquez-Monter, H. (1971). Cysticercosis. *In* "Pathology of Protozoal and

Helminthic Diseases", (R. A. Marcial-Rojas, ed.), pp. 592–617. Williams and Wilkins, Baltimore, USA.

Matossian, R. M., Rickard, M. D. and Smyth, J. D. (1977). Hydatidosis: A global problem of increasing importance. *Bull. Wld. Hlth. Org.* **55**, 499–501.

Matov, K. and Vasilev, I. D. (1955). [Active immunization of dogs against intestinal *Echinococcus*.] *Izv. tsent. khelmint. Lab. sof.* **1**, 111–125.

Miller, H. M. (1932). Superinfection of cats with *Taenia taeniaeformis*. *J. Prev. Med.* **6**, 17–29.

Mitchell, G. B. B. and Armour, J. (1980). Failure to protect calves against *Taenia saginata* using antigens prepared from *in vitro* cultivation of the larval stage. *Res. vet. Sci.* **29**, 373–377.

Mitchell, G. F., Goding, J. W. and Rickard, M. D. (1977a). Studies on the immune response to larval cestodes in mice. I. Increased susceptibility of certain mouse strains and hypothymic mice to *Taenia taeniaeformis* and analysis of passive transfer of resistance with serum. *Aust. J. exp. Biol. med. Sci.* **55**, 165–186.

Mitchell, G. F., Marchalonis, J. J., Smith, P. M., Nicholas, W. L. and Warner, N. L. (1977b). Studies on immune responses to larval cestodes in mice: Immunoglobulins associated with the larvae of *Mesocestoides corti*. *Aust. J. exp. Biol. med. Sci.* **55**, 187–211.

Mitchell, G. F., Rajasekariah, G. R. and Rickard, M. D. (1980). A mechanism to account for mouse strain variation in resistance to the larval cestode, *Taenia taeniaeformis*. *Immunology*, **39**, 481–489.

Miyazato, T., Furukawa, T. and Inoue, T. (1979). Intestinal pathology associated with primary and secondary infections of *Hymenolepis nana* in mice. *Jap. J. Parasit.* **28**, 185–195.

Moss, G. D. (1971). The nature of the immune response of the mouse to the bile duct cestode, *Hymenolepis microstoma*. *Parasitology*, **62**, 285–294.

Moss, G. D. (1972). The effect of cortisone acetate treatment on the growth of *Hymenolepis microstoma* in mice. *Parasitology*, **64**, 311–320.

Movsesijan, M. and Mladenović, Ž. (1970). [Active immunization of dogs against *Echinococcus granulosus*]. *Vet. Glasn.* **24**, 189–193.

Murrell, K. D. (1971). The effect of antibody on the permeability control of larval *Taenia taeniaeformis*. *J. Parasit.* **57**, 875–880.

Musoke, A. J. and Williams, J. F. (1975a). The immunological response of the rat to infection with *Taenia taeniaeformis*. V. Sequence of appearance of protective immunoglobulins and the mechanism of action of 7Sγ2a antibodies. *Immunology*, **29**, 855–866.

Musoke, A. J. and Williams, J. F. (1975b). Immunoglobulins associated with passive transfer of resistance to *Taenia taeniaeformis* in the mouse. *Immunology*, **28**, 97–102.

Musoke, A. J. and Williams, J. F. (1976). Immunological response of the rat to infection with *Taenia taeniaeformis*: protective antibody response to implanted parasites. *Int. J. Parasit.* **6**, 265–269.

Musoke, A. J., Williams, J. F., Leid, R. W. and Williams, C. S. F. (1975). The immunological response of the rat to infection with *Taenia taeniaeformis*. IV. Immunoglobulins involved in passive transfer of resistance from mother to offspring. *Immunology*, **29**, 845–853.

Musoke, A. J., Williams, J. F. and Leid, R. W. (1978). Immunological response of the rat to infection with *Taenia taeniaeformis*: VI. The role of immediate hypersensitivity in resistance to reinfection. *Immunology*, **34**, 565–570.

Németh, I. (1970). Immunological study of rabbit cysticercosis. II. Transfer of immunity to *Cysticercus pisiformis* (Bloch, 1780) with parenterally administered immune serum or lymphoid cells. *Acta vet. hung.* **20**, 69–79.

Németh, I. and Juhász, S. (1980). A trypsin and chymotrypsin inhibitor from the metacestodes of *Taenia pisiformis*. *Parasitology*, **80**, 433–446.

Niederkorn, J. Y. (1977). Immunization of rats against *Mesocestoides corti* (Cestoda) by a subcutaneous vaccination of living tetrathyridia and by passive transfer with serum. *Proc. Ark. Acad. Sci.* **31**, 79–80.

Nieland, M. L. (1968). Electron microscopic observations on the egg of *Taenia taeniaeformis*. *J. Parasit.* **54**, 957–969.

Oaks, J. A. and Lumsden, R. D. (1971). Cytological studies on the absorptive surfaces of cestodes. V. Incorporation of carbohydrate-containing macromolecules into tegument membranes. *J. Parasit.* **57**, 1256–1268.

Öhman-James, C. (1973). Cytology and cytochemistry of the scolex gland cells in *Diphyllobothrium ditremum* (Creplin, 1825). *Z. ParasitKde*, **42**, 77–86.

Okamoto, K. (1968). Effect of neonatal thymectomy on acquired resistance to *Hymenolepis nana* in mice. *Jap. J. Parasit.* **17**, 53–59.

Okamoto, K. and Koizumi, M. (1972). *Hymenolepis nana*: effect of antithymocyte serum on acquired immunity in mice. *Expl. Parasit.* **32**, 56–61.

Olivier, L. (1962). Studies on natural resistance to *Taenia taeniaeformis* in mice. II. The effect of cortisone. *J. Parasit.* **48**, 758–762.

Opuni, E. K. (1973). "Laboratory Studies on *Spirometra theileri* (Baer, 1925) n. comb. (Cestoda: Pseudophyllidea) from East Africa", Ph.D. Thesis, University of London.

Opuni, E. K. and Muller, R. L. (1975). Studies on *Spirometra theileri* (Baer, 1925) n. comb. 3. Acquired immunity to reinfection with plerocercoids in mice. *J. Helminth.* **49**, 199–204.

Osborn, P. J., Heath, D. D. and Parmeter, S. N. (1981). Immunization of lambs against infection with *Taenia ovis* using an extract of *Taenia ovis* eggs. *Res. vet. Sci.* **31**, 90–92.

Pappas, P. W. and Read, C. P. (1975). Membrane transport in helminth parasites: A review. *Expl. Parasit.* **37**, 469–530.

Phares, C. K., Hofert, J. F. and Pettinger, C. L. (1976). Growth stimulation of lymphatic tissue by plerocercoid larvae of the tapeworm, *Spirometra mansonoides*. *Gen. comp. Endocr.* **28**, 103–106.

Pollacco, S., Nicholas, W. L., Mitchell, G. F. and Stewart, A. C. (1978). T-cell dependent collagenous encapsulating response in the mouse liver to *Mesocestoides corti*. *Int. J. Parasit.* **8**, 457–467.

Rajasekariah, G. R., Mitchell, G. F. and Rickard, M. D. (1980a). *Taenia taeniaeformis* in mice: protective immunisation with oncospheres and their products. *Int. J. Parasit.* **10**, 155–160.

Rajasekariah, G. R., Rickard, M. D. and Mitchell, G. F. (1980b). Immunization of mice against infection with *Taenia taeniaeformis* using various antigens prepared from eggs, oncospheres, developing larvae and strobilocerci. *Int. J. Parasit.* **10**, 315–324.

Rajasekariah, G. R., Rickard, M. D., Mitchell, G. F. and Anders, R. F. (1982). Immunization of mice against *Taenia taeniaeformis* using solubilized oncospheral antigens. *Int. J. Parasit.* **12**, 111–116.

Ramazanov, V. T. (1971). [Immunity of sheep and dogs to experimental infections.] *Mater. Nauchn. Konf. Vses. Obshch. Gelmintol.* **23**, 219–225.

Rau, M. E. and Tanner, C. E. (1976). *Echinococcus multilocularis* in the cotton rat: the *in vitro* protoscolicidal activity of peritoneal cells. *Int. J. Parasit.* **6**, 195–198.

Rickard, M. D. (1974). Hypothesis for the long-term survival of *Taenia pisiformis* cysticerci in rabbits. *Z. ParasitKde*, **44**, 203–209.

Rickard, M. D. (1979). The immunological diagnosis of hydatid disease. *Aust. vet. J.* **55**, 99–104.

Rickard, M. D. and Adolph, A. J. (1976). Vaccination of calves against *Taenia saginata* infection using a "parasite-free" vaccine. *Vet. Parasit.* **1**, 389–392.

Rickard, M. D. and Arundel, J. H. (1974). Passive protection of lambs against infection with *Taenia ovis* via colostrum. *Aust. vet. J.* **50**, 22–24.

Rickard, M. D. and Bell, K. J. (1971a). Immunity produced against *Taenia ovis* and *T. taeniaeformis* infection in lambs and rats following *in vivo* growth of their larvae in filtration membrane diffusion chambers. *J. Parasit.* **57**, 571–575.

Rickard, M. D. and Bell, K. J. (1971b). Successful vaccination of lambs against infection with *Taenia ovis* using antigens produced during *in vitro* cultivation of the larval stages. *Res. vet. Sci.* **12**, 401–402.

Rickard, M. D. and Brumley, J. L. (1981). Immunisation of calves against *Taenia saginata* infection using antigens collected by *in vitro* incubation of *T. saginata* oncospheres or ultrasonic disintegration of *T. saginata* and *T. hydatigena* oncospheres. *Res. vet. Sci.* **30**, 99–103.

Rickard, M. D. and Howell, M. J. (1982). Comparative aspects of immunity in fascioliasis and cysticercosis in domesticated animals. *In* "Biology and Control of Endoparasites", (L. E. A. Symons, A. D. Donald, and J. K. Dineen, eds) pp. 343–373. Academic Press, Australia.

Rickard, M. D. and Outteridge, P. M. (1974). Antibody and cell-mediated immunity in rabbits infected with the larval stages of *Taenia pisiformis*. *Z. ParasitKde*, **44**, 187–201.

Rickard, M. D. and Williams, J. F. (1982). Hydatidosis/cysticercosis: Immune mechanisms and immunization against infection. *Adv. Parasit.* **21**, 229–296.

Rickard, M. D., Parmeter, S. N. and Gemmell, M. A. (1975). The effect of development of *Taenia hydatigena* larvae in the peritoneal cavity of dogs on resistance to a challenge infection with *Echinococcus granulosus*. *Int. J. Parasit.* **5**, 281–283.

Rickard, M. D., White, J. B. and Boddington, E. B. (1976). Vaccination of lambs against infection with *Taenia ovis*. *Aust. vet. J.* **52**, 209–214.

Rickard, M. D., Adolph, A. J. and Arundel, J. H. (1977a). Vaccination of calves against *Taenia saginata* infection using antigens collected during *in vitro* cultivation of larvae: passive protection via colostrum from vaccinated cows and vaccination of calves protected by maternal antibody. *Res. vet. Sci.* **23**, 365–367.

Rickard, M. D., Boddington, E. B. and McQuade, N. (1977b). Vaccination of lambs against *Taenia ovis* infection using antigens collected during *in vitro* cultivation of larvae: passive protection via colostrum from vaccinated ewes and

the duration of immunity from a single vaccination. *Res. vet. Sci.* **23**, 368–371.
Rickard, M. D., Mackinlay, L. M., Kane, G. J., Matossian, R. M. and Smyth, J. D. (1977c). Studies on the mechanism of lysis of *Echinococcus granulosus* protoscoleces incubated in normal serum. *J. Helminth.* **51**, 221–228.
Rickard, M. D., Coman, B. J. and Cannon, R. M. (1977d). Age resistance and acquired immunity to *Taenia pisiformis* infection in dogs. *Vet. Parasit.* **3**, 1–9.
Rickard, M. D., Arundel, J. H. and Adolph, A. J. (1981a). A preliminary field trial to evaluate the use of immunisation for the control of naturally acquired *Taenia saginata* infection in cattle. *Res. vet. Sci.* **30**, 104–108.
Rickard, M. D., Rajasekariah, G. R. and Mitchell, G. F. (1981b). Immunisation of mice against *Taenia taeniaeformis* using antigens prepared from *T. pisiformis* and *T. hydatigena* eggs or oncospheres. *Z. ParasitKde*, **66**, 49–56.
Rickard, M. D., Brumley, J. L. and Anderson, G. A. (1982). A field trial to evaluate the use of antigens from *Taenia hydatigena* oncospheres to prevent infection with *T. saginata* in cattle grazing on sewage-irrigated pasture. *Res. vet. Sci.* **32**, 189–193.
Sakamoto, T. (1978). Development of echinococcal tissue cultured *in vitro* and *in vivo*. *Mem. Fac. Agric. Kagomisha Univ.* **14**, 109–115.
Seddon, H. R. (1931). The development in sheep of immunity to *Moniezia expansa*. *Ann. trop. Med. Parasit.* **25**, 431–435.
Shield, J. M. (1969). *Dipylidium caninum*, *Echinococcus granulosus* and *Hydatigera taeniaeformis*: Histochemical identification of cholinesterases. *Expl. Parasit.* **25**, 217–231.
Showramma, A. and Reddy, D. B. (1963). Silent cysticercosis of the brain. An analysis of five cases with special reference to histopathology. *Indian J. Path. Bact.* **6**, 142–147.
Siebert, A. E., Jr, Good, A. H. and Simmons, J. E. (1978a). Kinetics of primary and secondary infections with *Taenia crassiceps* metacestodes (Zeder, 1800) Rudolphi, 1810 (Cestoda: Cyclophyllidea). *Int. J. Parasit.* **8**, 39–43.
Siebert, A. E., Jr, Good, A. H. and Simmons, J. E. (1978b). Ultrastructural aspects of early immune damage to *Taenia crassiceps* metacestodes. *Int. J. Parasit.* **8**, 45–53.
Siebert, A. E., Jr, Good, A. H. and Simmons, J. E. (1979). Ultrastructural aspect of the host cellular immune response to *Taenia crassiceps* metacestodes. *Int. J. Parasit.* **9**, 323–331.
Silverman, P. H. (1955). A technique for studying the *in vitro* effect of serum on activated taeniid hexacanth embryos. *Nature, Lond.* **176** 598–599.
Singh, B. B. and Rao, B. V. (1967). On the development of *Cysticercus fasciolaris* in albino rat liver and its reaction on the host tissue. *Ceylon vet. J.* **15**, 121–129.
Šlais, J. (1970). "The Morphology and Pathogenicity of the Bladder Worms, *Cysticercus cellulosae* and *Cysticercus bovis*", W. Junk, N.V., The Hague, Netherlands.
Slusarski, W. and Zapart, W. (1971). Diagnostic value of intradermal test with acid-soluble protein fractions in *Taenia* infections in man. *Acta parasit. pol.* **19**, 445–455.
Smithers, S. R. and Terry, R. J. (1969). Immunology of schistosomiasis. *Adv. Parasit.* **7**, 41–93.

Smyth, J. D. (1964). Observations on the scolex of *Echinococcus granulosus*, with special reference to the occurrence and cytochemistry of secretory cells in the rostellum. *Parasitology*, **54**, 515–526.

Smyth, J. D. (1969). "The Physiology of Cestodes", Oliver and Boyd, Edinburgh.

Smyth, J. D. and Haselwood, G. A. D. (1963). The biochemistry of bile as a factor in determining host specificity in intestinal parasites, with particular reference to *Echinococcus granulosus. Ann. N.Y. Acad. Sci.* **113**, 234–260.

Smyth, J. D. and Heath, D. D. (1970). Pathogenesis of larval cestodes in mammals. *Helm. Abstr.* **39**, 1–23.

Smyth, J. D., Gemmell, M. and Smyth, M. M. (1970). Establishment of *Echinococcus granulosus* in the intestine of normal and vaccinated dogs. *In* "H. D. Srivastava Commemoration Volume", (K. S. Singh and B. K. Tandan, eds), pp. 167–178. Division of Parasitology, Indian Veterinary Research Institute, Izatnagar, Uttar Pradesh.

Solomon, G. B. (1969). Host hormones and parasitic infection. *In* "International Review of Tropical Medicine, Vol. 3", pp. 101–158. Academic Press, New York and London.

Sprent, J. F. A. (1962). Parasitism, immunity and evolution. *In* "The Evolution of Living Organisms", (J. N. Leeper, ed.), pp. 149–165. Melbourne University Press.

Stoll, N. R. (1937). Tapeworm studies. V. Absence of *M. expansa* from the sheep intestine early after infection. *Am. J. Hyg.* **26**, 148–161.

Sulivan-Lopez, J., Sealey, M., Ramos, C., Melendro, E. J., Willms, K. and Ortiz-Ortiz, L. (1980). B lymphocyte stimulation by parasitic organisms. *In* "Molecules, Cells and Parasites in Immunology", (A. Flisser, K. Willms and C. Larralde, eds), pp. 113–124. Academic Press, New York and London.

Tachovsky, T. G., Hare, J. D., Ritterson, A. L. and Mueller, J. F. (1973). Enhanced growth of a polyoma transformed tumour cell in *Spirometra mansonoides*—infected hampsters. *Proc. Soc. exp. Biol. Med.* **143**, 780–782.

Tan, B. D. and Jones, A. W. (1968). Resistance of mice to reinfection with the bile-duct cestode, *Hymenolepis microstoma. Expl. Parasit.* **22**, 250–255.

Thompson, R. C. A., Dunsmore, J. D. and Hayton, A. R. (1979). *Echinococcus granulosus*: Secretory activity of the rostellum of the adult cestode *in situ* in the dog. *Expl. Parasit.* **48**, 144–163.

Thompson, R. C. A., Penhale, W. J., White, T. R. and Pass, D. A. (1982). BCG-induced inhibition and destruction of *Taenia taeniaeformis* in mice. *Parasite Immun.* **4**, 93–99.

Todorov, T., Dakov, I., Kosturkova, M., Tenev, S. and Dimitrov, A. (1979). Immunoreactivity of pulmonary echinococcosis. 1. A comparative study of immunodiagnostic tests. *Bull. Wld. Hlth. Org.* **57**, 735–740.

Turner, H. M. and McKeever, S. (1976). The refractory responses of the White Swiss strain of *Mus musculus* to infection with *Taenia taeniaeformis. Int. J. Parasit.* **6**, 483–487.

Turner, E. L., Berberian, D. A. and Dennis, E. W. (1933). Successful artificial immunization of dogs against *Taenia echinococcus. Proc. Soc. exp. Biol. Med.* **30**, 618–619.

Turner, E. L., Berberian, D. A. and Dennis, E. W. (1936). The production of artificial immunity in dogs against *Echinococcus granulosus*. *J. Parasit.* **22**, 14–28.

Varela-Díaz, V. M., Gemmell, M. A. and Williams, J. F. (1972). Immunological responses of the mammalian host against tapeworm infections. XII. Observations on antigenic sharing between *Taenia hydatigena* and *Taenia ovis*. *Expl. Parasit.* **32**, 96–101.

Varute, A. T. (1971). Mast cells in cyst-wall of hydatid cyst of *Taenia taeniaeformis* (Batsch). *Indian J. exp. Biol.* **9**, 200–203.

Verheyen, A., Vanparijs, O., Borgers, M. and Thienpont, D. (1978). Scanning electron microscopic observations of *Cysticercus fasciolaris* (= *Taenia taeniaeformis*) after treatment of mice with mebendazole. *J. Parasit.* **64**, 411–425.

Verster, A. (1974). The golden hampster as a definitive host of *Taenia solium* and *Taenia saginata*. *Onderstepoort J. vet. Res.* **41**, 23–28.

Vukovic, V. (1949). [Infection and superinfection of the dog with *Taenia hydatigena*.] *Arch. Sci. Biol., Belgrade*, **1**, 258–261.

Wakelin, D. (1976). Genetic control of susceptibility and resistance. *Adv. Parasit.* **16**, 217–308.

Wassom, D. L., Guss, V. M. and Grundmann, A. W. (1973). Host resistance in a natural host–parasite system. Resistance to *Hymenolepis citelli* by *Peromyscus maniculatus*. *J. Parasit.* **59**, 117–121.

Weinmann, C. J. (1966). Immunity mechanisms in cestode infections. *In* "Biology of Parasites", (E. J. L. Soulsby, ed.), pp. 301–320. Academic Press, New York and London.

Weinmann, C. J. (1970). Cestodes and Acanthocephala. *In* "Immunity to Parasitic Animals", (G. J. Jackson, R. Herman, I. Singer, eds), Vol. 2, pp. 1021–1059. Appleton-Century-Crofts, New York.

Weinmann, C. J. and Rothman, A. H. (1967). Effects of stress upon acquired immunity to the dwarf tapeworm, *Hymenolepis nana*. *Expl. Parasit.* **21**, 61–67.

Wikerhauser, T., Žuković, M. and Džakula, N. (1971). *Taenia saginata* and *T. hydatigena*: Intramuscular vaccination of calves with oncospheres. *Expl. Parasit.* **30**, 36–40.

Wikerhauser, T., Brglez, J., Džakula, N., Asaj, R. and Matíc-Piantanida, D. (1978). Experimental immunization of calves against cysticercosis, using a vaccine prepared from metabolic products of activated *Taenia saginata* oncospheres. *Acta Parasit. Jugoslav.* **9**, 51–56.

Williams, J. F. (1971). Immunological responses of dogs to tapeworm infection. *In* "21st Gaines Veterinary Symposium, Ames, Iowa", pp. 35–40.

Williams, J. F. (1979). Recent advances in the immunology of cestode infections. *J. Parasit.* **65**, 337–349.

Williams, J. F. (1983). Cestode Infections. *In* "Immunology of Parasitic Infections", (S. Cohen and K. Warren, eds), in press. Blackwell Scientific Publications, Oxford.

Williams, J. F. and Pérez-Esandi, M. V. (1971). Reaginic antibodies in dogs infected with *Echinococcus granulosus*. *Immunology*, **20**, 451–455.

Williams, J. F., Shearer, A. M. and Ravitch, M. M. (1981). Differences in susceptibility of rat strains to experimental infection with *Taenia taeniaeformis*. *J. Parasit.* **67**, 540–547.

Willms, K. and Arcos, L. (1977). *Taenia solium*: host serum proteins on the cysticercus surface identified by an ultrastructural immunoenzyme technique. *Expl. Parasit.* **43**, 396–406.

Willms, K., Merchant, M. T., Arcos, L., Sealey, M., Díaz, S. and de Leon, L. D. (1980). Immunopathology of cysticercosis. *In* "Molecules, Cells, and Parasites in Immunology", (A. Flisser, K. Willms and C. Larralde, eds), pp. 145–162. Academic Press, New York and London.

Yamashita, J., Ohbayashi, M., Sakamoto, T., Suzuki, K. and Okugi, M. (1963). Studies on *Echinococcosis* XIV. Further observations on the difference of susceptibility to *Echinococcus multilocularis* among uniform strains of the mouse. *Jap. J. vet. Res.* **11**, 50–54.

Chapter 15

Chemotherapy and the Effects of Chemotherapeutic Agents

G. C. Coles

I. INTRODUCTION

The pathological effects of most cestode infections have encouraged the search for cestodicides for many years. Initially drugs were based on herbal remedies, later on compounds extracted from plants and on synthetic chemicals. Since it is relatively easy to remove some tapeworms from the host intestine, particularly using the laboratory model of *Hymenolepis nana* in the mouse, there are many reports of chemicals possessing cestodicidal activity. Details of cestodicides and compounds showing some activity against cestodes are contained in reviews by Standen (1963), Keeling (1968), Arundel (1972), de Carneri and Vita (1973), Robertson (1977), Sharma *et al.* (1980) and Van den Bossche (1980a) but most of the chemicals listed are no longer of practical significance.

Although modern cestodicides are adequate for the control of intestinal dwelling cestodes, there are no really satisfactory drugs for killing the larval stages in the intermediate host. Compounds are also required that could be used under practical conditions for the destruction of cestode eggs.

BIOLOGY OF THE EUCESTODA Vol. 2
ISBN 0-12-062102-9

II. SCREENING FOR CESTODICIDES

Development of a modern drug is a very expensive process requiring extensive toxicological research which includes long-term carcinogenicity testing as well as detailed confirmation of activity. If the drug is given to a ruminant, residue data and the withdrawal period (time required between dosing and killing for human consumption) have to be determined. Costs could easily exceed $20 million and it is doubtful therefore whether the market size for cestodicides, and particularly the market share likely to be gained by a novel drug, warrants the development of a drug only showing activity against cestodes. Activity against cestodes is therefore likely to be an added bonus to a molecule already showing other anthelmintic properties, and this is true of all recent commercial developments of cestodicides.

The object of screening chemicals is to determine as rapidly as possible, and with the minimum number of tests, whether a compound has the potential for commercial development rather than to determine its full spectrum of activity. Certain criteria are therefore decisive, and these are displayed in Fig. 1. Having established activity in rodent models, either activity against *Moniezia* species in ruminants is an added bonus for a potential anthelmintic, or activity is required against tapeworms in dogs and cats. Since good cestodicides now exist, a new drug should be technically as good as the best and therefore be effective against immature *Echinococcus granulosus* in dogs. Alternatively a drug should have activity against larval cestodes, and if it is for the ruminant market it should be cost effective and not have residue problems. Only when the key tests have indicated a compound is of considerable commercial interest, and initial toxicological tests are satisfactory, can further evaluation to establish the full spectrum of cestodicidal activity be justified.

A. Chemotherapy of *Hymenolepis* species

The simplest test available is to determine whether *H. nana* survives when incubated *in vitro* at 37°C with a chemical. The advantages of such a test are cost, time and sparing of animals, but these are in practice outweighed by the relatively large numbers of false positives detected, which include biocides, the failure to detect certain types of drugs (e.g. a diisothiocyanate, Katiyar *et al.*, 1967), the lack of information of pharmacokinetics in the host and the potential metabolism of an inactive chemical by the host to give an active metabolite. Sen and Hawking (1960) described a simple peptone broth for maintenance of *H. nana* which permitted the detection of some of what are now old cestodicides. This test was used by Blakemore *et al.* (1964)

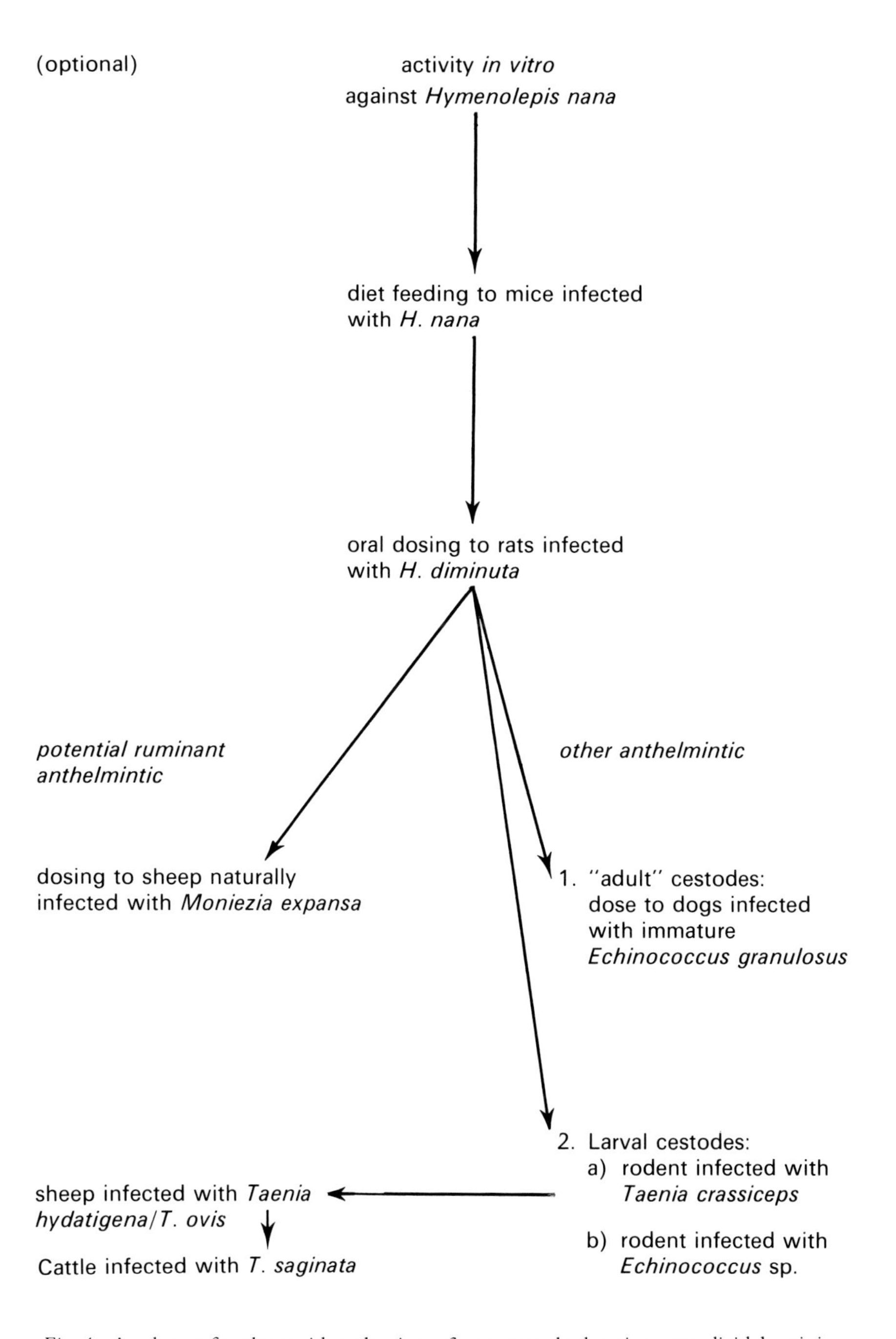

Fig. 1. A scheme for the rapid evaluation of compounds showing cestodicidal activity.

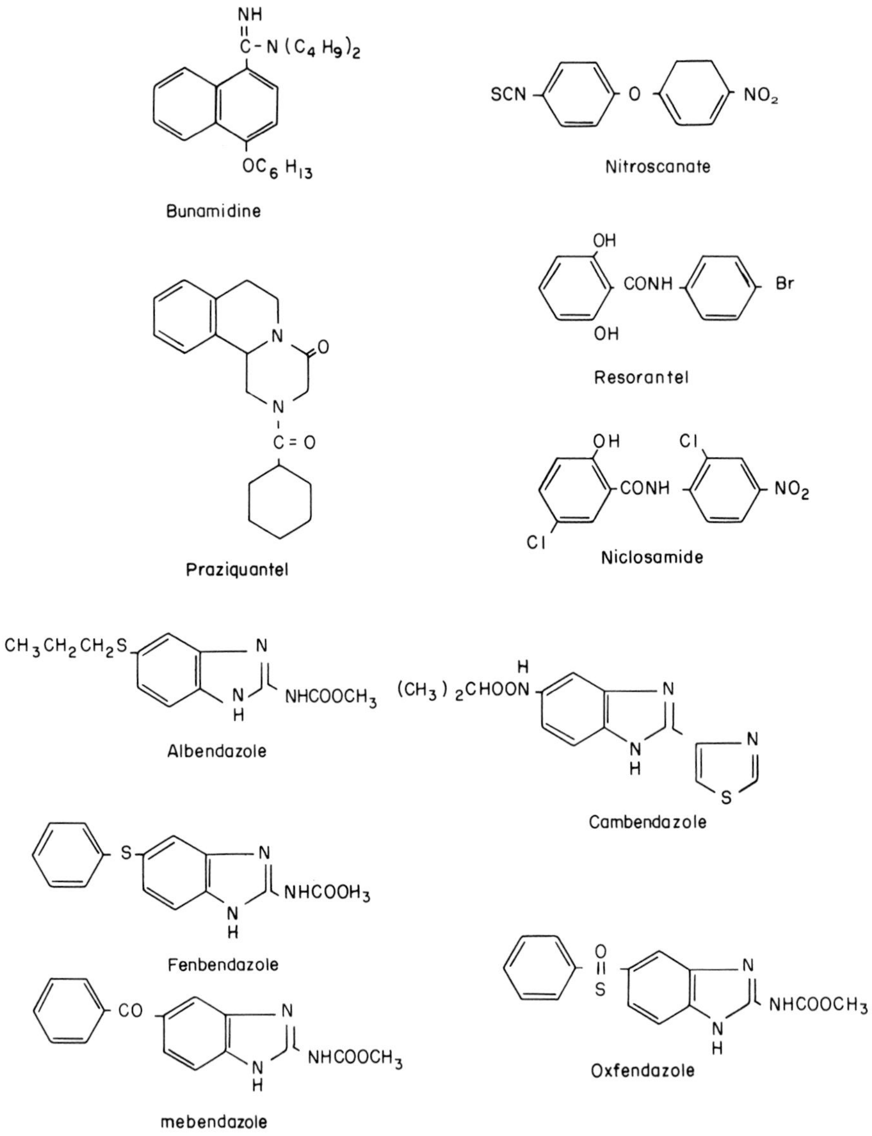

Bunamidine

Nitroscanate

Resorantel

Praziquantel

Niclosamide

Albendazole

Cambendazole

Fenbendazole

Oxfendazole

mebendazole

Fig. 2. Structures of the more commonly used cestodicides.

to examine the anthelmintic constituents of ferns. The method also allows the detection of some modern cestodicides. Minimum active doses in μg/ml after 24 h incubation of adult worms were: bunamidine 1, resorantel 1, niclosamide 0.05, nitroscanate 2.5 and praziquantel 0.1. However, a number of benzimidazoles active *in vivo* were not active *in vitro* in this test (Coles, unpublished).

Removal of *H. nana* from infected mice is now viewed as a standard screen test (Theodorides, 1976). Young mice previously infected with 200–400 eggs of *H. nana* are usually dosed with the test compound by intraperitoneal injection, gavage or in-diet medication. Injection guarantees that the drug is in the animal but insoluble drugs may never reach the worms. Gavage dosing of nocturnal feeders during the daytime may reduce the length of exposure of worms to the drugs, especially if, unlike *H. nana*, the worms are in the upper part of the small intestine, but in-diet medication means that the exact dosage per animal is not known. Perhaps the most dramatic difference between gavage and dietary dosing is the much larger amount of fenbendazole required to remove the nematode *Nematospiroides dubius* by gavage than by dietary feeding (Kirsch, 1975; Coles and McNeillie, 1977). Parbendazole in-diet removed *H. nana*, but not when administered three times by gavage (Brody and Elward, 1971). These authors administered drugs for 18 days in the diet, but five days is sufficient to detect all modern cestodicides and all benzimidazoles examined with the exception of thiabendazole (Coles and McNeillie, 1977). The activity of a range of drugs against *H. nana* in mice is summarized in Table I. However although *H. nana* is a good screening model, it will not necessarily detect low cestodicidal activity. For example, amidantel is active at 250 mg/kg × 1 against *H. diminuta* in rats but not against *H. nana* in mice (Wollweber *et al.*, 1979).

A number of drugs have been used in the diet to remove *H. nana* or *H. diminuta* in colonies of rodents including 0.3% thiabendazole for 14 days to mice (Taffs, 1975, 1976), 0.1% niclosamide for two or three weeks to rats (Hughes *et al.*, 1973), 0.33% niclosamide for seven days to hamsters (Ronald and Wagner, 1975) and 0.0125% uredofos for six days to mice (Tetzlaff and Weir, 1978).

Although *H. nana* has been used in rats to examine potential cestodicides (e.g. Singh *et al.*, 1976), the more usual infection for screening chemicals is *H. diminuta*. As more work has been performed with diet feeding of drugs to mice and oral dosing of drugs to rats, direct comparisons of the two screens cannot always be made. Available data (Table II) however, suggest that *H. diminuta* is more sensitive to cestodicides than *H. nana*. Multiple dosing with cestodicides may give higher cure rates than single dosing depending on the pharmacokinetics of the drug, e.g. with the benzimidazoles. The most relevant feature is possibly the length of exposure of the

Table I. The activity of some drugs against *Hymenolepis nana* in the mouse.

(a) *in diet administration of drug*

Drug	Concentration in diet (%)	Length of feeding (days)	Removal of worms %	Reference
bunamidine	0.1	18	99	Brody and Elward (1971)
	0.05	5	100	Coles and McNeillie (1977)
niclosamide	0.1	18	93	Brody and Elward (1971)
	0.1	5	97	Coles and McNeillie (1977)
nitroscanate	0.1	5	100	Coles and McNeillie (1977)
albendazole	0.05	5	100	Coles and McNeillie (1977)
cambendazole	0.05	5	100	Coles and McNeillie (1977)
fenbendazole	0.05	5	99	Coles and McNeillie (1977)
fenbendazole	<.03	5	>90	Düwel (1978)
mebendazole	0.025	5	100	Coles and McNeillie (1977)
oxfendazole	0.025	5	100	Coles and McNeillie (1977)
oxibendazole	0.05	5	100	Coles and McNeillie (1977)
	0.05	7	100	Theodorides (1976)
parbendazole	0.1	18	99	Brody and Elward (1971)
	0.1	5	100	Coles and McNeillie (1977)
praziquantel	0.025	5	98	Coles and McNeillie (1977)
streptothricin	0.2	3	100	Brown *et al.* (1977)

(b) *oral dosing*

Drug	Dose (mg/kg)	Number of days dosed	Removal of worms %	Reference
axenomycin D	20	1	97	Della Bruna *et al.* (1973)
emetine	48	1	100	Seth and Lovekar (1972)
emetine	20	3	100	Seth and Lovekar (1972)
fenbendazole	50	5	>90	Düwel (1978)
furodazole	25	3 (twice daily)	100	Alaimo *et al.* (1978)
niclosamide	200	3	100	Brody and Elward (1971)
compound 77-6[a]	100	1	82 (cure rate)	Gupta *et al.* (1980)
nitroscanate	100	1	"nearly active"	Middleton *et al.* (1979)
amoscanate	50	1	100	Sen and Deb (1981)
praziquantel	25	1	100[b]	Thomas and Gonnert (1977) Thomas and Andrews (1977)
praziquantel	35	1	100	Gupta *et al.* (1980)

[a] compound 77-6 is a niclosamide analogue, 3,5-dibromo-2′-chloro salicylanilide-4′-isothiocyanate.

[b] less with very young worms (Thomas and Gonnert, 1978a).

Table II. The activity of some drugs dosed orally against *Hymenolepis diminuta* in the rat.

Drug	Dosed once		Dosed 3 consecutive days		Reference
	Dose (mg/kg)	Removal of worms %	Dose (mg/kg)	Removal of worms %	
albendazole	100	100	50	100	Coles (unpublished)
cambendazole	50	97	12.5	98	Coles (unpublished)
fenbendazole	50	>90			Düwel (1978)
fenbendazole	50	78	6	100	Coles (unpublished)
oxfendazole	25	84	12.6	100	Coles (unpublished)
oxibendazole	400	83	100	100	Coles (unpublished)
parbendazole	100	74	100	100	Coles (unpublished)
niclosamide	50	100	25	90	Coles (unpublished)
compound 77-6[a]	50	100			Gupta *et al.* (1980)
resorantel	24	100			Düwel and Kirsch (1971)
oxyclozanide	50	"usually 100"			Hopkins *et al.* (1973)
nitroscantate	50	100	50	100	Coles (unpublished)
praziquantel	5	100			Thomas and Gonnert (1977)
praziquantel	2.5	100			Thomas and Andrews (1977) Coles (unpublished)

[a] compound 77-6 is a niclosamide analogue, 3,5-dibromo-2′-chloro-salicylandide-4′-isothiocyanate.

cestode to the drug, permitting its concentration by the worm. While *M. expansa* may be inherently more sensitive to benzimidazoles than *H. diminuta*, the gradual release of drug from the rumen of the host may provide a continuous exposure of the cestode to the drug in a way not possible in the rat.

Tests of potential drugs are not, of course, confined to *H. nana* and *H. diminuta*. The action of praziquantel against *Diphyllobothrium latum*, a cestode of man, was first demonstrated in a *D. latum* infection in golden hamsters (Bylund *et al.*, 1977) where a single dose of 50 mg/kg removed all adult worms. Similarly, praziquantel removed *H. microstoma*, a tapeworm of the mouse bile duct, at 5 mg/kg (Thomas and Gönnert, 1977) thus predicting its successful use against *Stilesia hepatica*.

B. Chemotherapy of larval cestodes

Larval cestodes of economic importance (see review by Arundel, 1972) can be controlled by interruption of the cestode life-cycle by chemotherapy of the adult worms, but mathematical models of the infection in New Zealand suggest therapy of larval and adult stages is the optimum method of control (Harris *et al.*, 1980). It is neither possible to prevent access of all people infected with *T. saginata* to pasture land, nor to ensure sludge used for fertilization of pasture is free from *T. saginata* eggs, so cost-effective drugs would be of use in treating cyst-infected cattle. Similarly, unless dogs are regularly treated with drugs, sheep will develop tapeworm cysts and people will continue to become infected with *E. granulosus* (or very rarely *E. multilocularis*), and not all infections in man are amenable to surgery. There is thus a real need for drugs effective against larval cestodes, although production of a drug specifically for larval tapeworms would not be economic. Recently, therefore, drugs have been examined against larval cestodes in rodents.

A range of chemicals claimed to have some activity against *E. granulosus* in man were shown to lack efficacy against *E. granulosus* in mice and birds (Kammerer and Péréz-Esandi, 1975). However, since 1974, there have been a number of papers demonstrating activity of some benzimidazoles and praziquantel against several larval cestodes (Table III). Praziquantel only showed activity against certain larval cestodes after repeated dosing and then was better against older rather than younger cestodes (Thomas and Gönnert, 1978a). Of a range of benzimidazoles tested against *T. crassiceps* as a single dose, only mebendazole showed high activity, but mebendazole, flubendazole (p-fluoromebendazole) and fenbendazole showed excellent activity against several cestodes if administered, preferably by diet, over a prolonged period. The results of greatest significance are with *E. granulosus* because of infections in man, but whether data can be extrapolated from mouse to man is doubtful. Heath *et al.* (1975) reported that mebendazole is most effective when cysts are free of the fibrous capsule but, despite a high efficacy, tissue regrowth occurred after medication was terminated. It was also reported that two strains of *E. multilocularis* differed in their response to mebendazole (Hinz, 1978) suggesting that the strain of worm could be of importance in therapy in man. Kammerer and Judge (1976) used 10-month-old *E. granulosus* in mice suggesting it was more analogous to the human situation. Mebendazole reduced cysts in the mice, but some mice suddenly died indicating possible toxaemia due to rupture of the cysts. Although highly encouraging, data with mebendazole and *E. granulosus* are not unequivocal, and this has subsequently been confirmed by use in man.

The reduced activity of mebendazole against larval stages of *H. nana* in

Table III. Chemotherapy of larval cestodes in small mammals.

Species	Drug	Dose level and route	Removal of worms %	References
Echinococcus granulosus	1 flubendazole	0.1% in diet for 21 days	100 (no change in body weight over 140 days)	Thienpont *et al.* (1978)
	1 mebendazole	50 mg/kg × 14 oral	100	Heath and Chevis (1974)
	1 mebendazole	0.1% in diet for 14 days	very little germinal tissue left	Heath *et al.* (1975)
	1 mebendazole	50 mg/kg × 10 oral	85 reduction in cysts	Kammerer and Judge (1976)
E. multilocularis	4 cambendazole	0.1% in diet for 63 days	98	Campbell and Blair (1974b)
	4 fenbendazole	0.05% in diet for 28 days	100	Düwel (1978)
	4 fenbendazole	0.05% in diet for 80 days	>90	Eckert *et al.* (1978)
	1 fenbendazole	0.05% in diet for 102 days	95	Hinz (1978)
	1 mebendazole	150 mg/kg × 3 i.p.	100	Campbell *et al.* (1975)
	4 mebendazole	0.05% in diet for 120 days	96	Eckert *et al.* (1977)
	4 mebendazole	0.5% in diet for 80 days	>90	Eckert *et al.* (1978)
	4 mebendazole	0.05% in diet for 300 days	>99	Eckert and Burkhardt (1980)
	1 praziquantel	250 mg/kg × 10 oral and s.c.	87	Thomas and Gönnert (1978a)
	4 thiabendazole	0.5% in diet for 63 days	99	Campbell *et al.* (1975)
Hymenolepis nana	1 mebendazole	0.025% in diet for 6 days	not effective	Novak and Evans (1981)
(24 hours-old)	1 praziquantel	500 mg/kg	25	Thomas and Gönnert (1978a)
(72 hours-old)	1 praziquantel	100 mg/kg	100	Thomas and Gönnert (1978a)
Mesocestoides corti	1 fenbendazole	0.03% in diet for 40 days	100	Düwell (1978)
	1 mebendazole	50 mg/kg × 14	100	Heath and Chevis (1974)
	1 mebendazole	100 mg/kg × 1 s.c.	100	Heath *et al.* (1975)
	1 mebendazole	0.1% in diet for 5 days	100	Bennet *et al.* (1978)

Table III contd.

	1 praziquantel	1000 mg/kg × 1 s.c.	not effective	Novak (1977)
	1 praziquantel	0.5% in diet for 20 days	95	Novak (1977)
Taenia crassiceps	1 cambendazole	0.1% in diet for 42 days	96	Campbell and Blair (1974b)
	1 mebendazole	25 mg/kg × 1 i.p.	94	Campbell *et al.* (1975)
	2 mebendazole	500 mg/kg × 1 oral	100	Heath and Lawrence (1979)
	2 mebendazole	100 mg/kg once every 2 weeks (3 ×)	100	Heath and Lawrence (1979)
	1 praziquantel	0.5% in diet for 20 days	39	Novak (1977)
	1 thiabendazole	0.5% in diet for 42 days	56	Campbell and Blair (1974b)
T. pisiformis	3 closantel	20 mg/kg intramuscular	killed liver but not intraperitoneal cysts	Chevis *et al.* (1980)
	3 mebendazole	50 mg/kg × 14 oral	100	Heath and Chevis (1974)
	3 mebendazole	0.1% in diet for 14 days	100	Heath *et al.* (1975)
	3 mebendazole	25 mg/kg × 5 oral	100	Hörchner *et al.* (1976)
	3 praziquantel	50 mg/kg × 5 oral	93	Hörchner *et al.* (1976)
	3 praziquantel	25 mg/kg × 5 s.c.	100	Thomas and Gönnert (1978a)
T. taeniaeformis	1 cambendazole	0.1% in diet for 64 days	100	Campbell and Blair (1974a)
	1 flubendazole	0.025% in diet for 21 days	100	Thienpont *et al.* (1978)
	1 mebendazole	0.025% in diet for 21 days	100	Thienpont *et al.* (1974)
	1 mebendazole	62.5 mg/kg × 10 oral	very low	Hörchner *et al.* (1976)
	1 praziquantel	125 mg/kg × 10 s.c.	100	Hörchner *et al.* (1976)
(28–32 days-old)	1 praziquantel	500 mg/kg × 5 oral every other day	0	Thomas and Gönnert (1978a)
(49–53 days-old)	1 praziquantel	100 mg/kg every other day × 5 oral	100	Thomas and Gönnert (1978a)
	1 thiabendazole	0.5% in diet for 64 days	100	Campbell and Blair (1974a)

1 in mice, 2 in rats, 3 in rabbits, 4 in jirds

the mouse (Novak and Evans, 1981) has also been reported for larval stages in the intermediate host *Tenebrio molitor*. Mebendazole (1%) in flour reduced the numbers of *H. diminuta* cysts in beetles but did not kill them all (Evans and Novak, 1976), and the effects of 10% thiabendazole, cambendazole and albendazole were reversible (Evans *et al.*, 1980). *H. nana* cysts were less affected by mebendazole than *H. diminuta* cysts (Evans *et al.*, 1979) and *H. microstoma* cysts were not significantly affected by any of the four drugs (Evans *et al.*, 1979, 1980). As both the pharmacokinetics of the drug within the host and the host immunological response to drug-induced damage in the cyst will differ widely between insect and mammal, the response of the parasite within the insect to chemotherapy is most unlikely to have any predictive or useful applications.

Prior to the *in vivo* demonstration of activity of certain anthelmintics against larval cestodes in rodents, some evaluation of compounds against larval cestodes had been undertaken *in vitro* (reviewed by Sakamoto, 1973). Halogenated salicylanilides and bisphenols were the most active compounds against both *E. multilocularis* and *E. granulosus* (Sakamoto, 1973; Sakamoto and Gemmell, 1975). Some of these compounds were evaluated *in vivo*, but with little success (Sakamoto, 1979). However, the economics of the development of larval cestodicides and the uncertainty of the usefulness of the *in vitro* test system have limited their application.

C. Chemical killing of cestode eggs

The cestode life-cycle can be interrupted by chemotherapy of the adult worm in the host intestine or bile duct, and in some instances by killing the larval stage in the body tissue. An alternative possibility is to chemically kill the cestode eggs after they have passed out of the host. While this is unlikely ever to assume much importance, there are specific cases where it might be of value.

Contamination of pasture with *T. saginata* eggs is unlikely to be controlled with specific cestode ovicides, but treatment of pasture with lime nitrogen reduced the survival of the eggs to three days (Jelenova, 1981) which could be of value if contamination with human faeces is suspected. However, since contamination will more usually come from use of sewage effluent for irrigation or sewage sludge for fertilization, additional use of fertilizer may be inappropriate. Of greater importance may be chemical killing of *E. granulosus* eggs. When infected dogs are treated with praziquantel, large numbers of worms containing viable eggs are rapidly excreted and the resultant faeces are extremely dangerous. A chemical that would effectively disinfect the areas where the dogs are retained after dosing would be of value. Ferrous iodide and potassium permanganate were the most effective inorganic compounds for killing *T. hydatigena* eggs (Parnall, 1965), although sodium hypochlorite will destroy taeniid eggs (Laws, 1967; Crewe and Owen, 1979). The most effective chemical against *E. granulosus* eggs at room temperature was glutaraldehyde (5%), but 70% ethanol, 0.015% ammonium dodecylsulphate and Lugol's iodine significantly reduced infectivity after 1 h (Pérez-Esandi *et al.*, 1975). Two hours exposure at 37°C to a saturated solution of bunamidine killed *E. granulosus* eggs and greatly reduced the infectivity of *T. taeniaeformis* eggs (Williams *et al.*, 1973), but neither praziquantel nor bunamidine killed *E. granulosus* eggs within the proglottides (Thakur *et al.*, 1979). Further work is thus required before chemical killing of cestode eggs becomes a reliable practical proposition.

III. THE THERAPY OF IMPORTANT CESTODES

As already stressed the object of screening large numbers of chemicals in rodents is to identify the very few chemicals which have significant activity against cestodes. These are then chemically modified to optimize the structure–activity relationship of the chemical series, or if this is not appropriate (e.g. where the molecule already has other interesting biological activity) to decide if tests in non-rodents are warranted. Tests in humans are not likely unless the chemical shows great promise in canine or ruminant cestode infections.

The principal targets for which compounds found active in rodent cestode infections are (1) adult cestodes of sheep and cattle, primarily *Moniezia* sp., (2) adult cestodes of dogs, especially *E. granulosus* but also *Taenia hydatigena* and *T. ovis*, (3) adult cestodes of humans, (4) larval cestodes of ruminants, especially *T. saginata* (*Cysticercus bovis*), and (5) larval cestodes of humans (i.e. *E. granulosus*). Cost-effective drugs exist against all but the larval cestodes (groups 4 and 5). The therapies available for these categories of cestode infections are described below.

A. Adult cestodes of sheep and cattle

The economic importance of *Moniezia* infections in lambs has not been adequately evaluated, largely due to the great difficulty of producing infected oribatid mites in the laboratory for controlled infections in lambs. While Kates and Goldberg (1951) found no retardation of growth in lambs infected with *M. expansa*, Hansen *et al.* (1950) and Stampa (1967) did. With moderate infection rates and abundant nutrition there may be little effect on the growth of young animals, but many farmers do not like to see tapeworm segments in animal faeces and medicate for the condition. In addition, *Moniezia* sp. may cause scouring and thus increase fly "breech strike" (McBeath *et al.*, 1977).

Earlier therapies included lead arsenate, copper aceto-arsenite and tricholorophen, but these have been displaced by bunamidine, salicylanilides and more recently benzimidazoles. Bunamidine, niclosamide and resorantel are useful therapies for *Moniezia* sp. (Table IV) but, as a single modern benzimidazole can cure both nematode and *Moniezia* sp. at the same time, this would now seem preferable. Four compounds which are widely sold are effective: albendazole; cambendazole; fenbendazole and oxfendazole. Although mebendazole is a good cestodicide, it is not widely used in sheep and cattle. Somewhat more variable results have been found with fenbendazole than with albendazole, cambendazole or oxfendazole (Table IV), but this may only represent wider testing under differing conditions. Given the probable insignificance of small numbers of cestodes surviving after treatment, the modern benzimidazoles would seem to be the drugs of choice for treatment of *Moniezia* infections. Although benzimidazole-resistant nematodes represent a growing problem, the length of the *Moniezia* life-cycle and the relatively low selection pressure suggest that emergence of benzimidazole-resistant *Moniezia* will be unlikely. Praziquantel was also effective against *M. expansa* (complete removal at 2.5 mg/kg Thomas and Gönnert, 1978b), but it is too expensive for regular use in ruminants.

Niclosamide at 100 mg/kg controlled *Avitellina* sp., but not *Stilesia hepatica* at 150 mg/kg (Graber, 1969). However, the fasciolicide, oxyclozanide, possibly because of its secretion into the bile, killed *S. hepatica* at 17–21 mg/kg (Harrow, 1969) as did praziquantel at 15 mg/kg. However, the high cost of praziquantel makes it uneconomical for this use (Verster and Marincowitz, 1980). Both oxyclozanide at 45 mg/kg (Colombo *et al.*, 1968) and albendazole at 7.5 mg/kg (Craig and Shepherd, 1980) killed *Thysanosoma actinoides*, but oxyclozanide was not effective against *Moniezia* sp. (Guralp and Oguz, 1971) even though it caused release of segments (Walley, 1966).

Tapeworms in horses are not usually clinically significant. The commonest is *Anoplocephala perfoliata.* Kelly and Bain (1975) reported 20 mg/kg of mebendazole to be 99% effective, Slocombe (1979) found it to have no effect up to 35.2 mg/kg, although he reported pyrantel to be completely effective at 19.8 mg/kg. This result is surprising as mebendazole has widespread anticestodal activity, but pyrantel is devoid of activity in laboratory tests (Coles and McNeillie, 1977).

Table IV. Activity of cestodicides against *Moniezia* species in cattle and sheep.

Drug	Dose (mg/kg)	Removal of worms %	Reference
albendazole	2.5	100	Van Schalkwyk *et al.* (1979)
	3.8	100	Led *et al.* (1979)
	5	97.8	Ciordia *et al.* (1978)
	10	100	Theodorides *et al.* (1976)
	10	100	Ciordia *et al.* (1978)
bunamidine	50	100	Czipri *et al.* (1968), Guralp and Oguz (1971)
cambendazole	25–30	99–100	Gibbs and Gupta (1972), Horak *et al.* (1972), Campbell and Butler (1973), Foix (1979)
fenbendazole	3.5	81	Kennedy and Todd (1975)
	5	53	Kennedy and Todd (1975)
	5	100	McBeath *et al.* (1977)
	5	99 (egg count only)	Bezubik *et al.* (1978)
	7.5	94	Kennedy and Todd (1975)
	7.5	98	Townsend *et al.* (1977)
	10	100	Townsend *et al.* (1977)
	10	>95	Düwel and Tiefenbach (1978)
	10	90 & 67 (cure rates)	Malan (1980)
	15	100	Corba *et al.* (1979)
niclosamide	75	100	Hall (1966); Graber (1969)
oxfendazole	5	100	Reuss (1979), Corba *et al.* (1980)
resorantel	62.5	>95	Düwel (1970)

B. Cestodes of dogs and cats

The most important cestode of dogs is *E. granulosus* because of the serious nature of the infection in man caused by accidential ingestion of eggs. Infection of ruminants with *E. granulosus* causes loss of carcass value as do cysts of *T. hydatigena* and *T. ovis*. The majority of trials have therefore concentrated on these first two species. As immature *E. granulosus* are more difficult to remove than mature ones, drug efficacy is usually now evaluated against 21–28-day-old worms.

The activity of a series of experimental and commercial anthelmintics against *E. granulosus* is displayed in Table V. Although some of the treatments give 100% cure rates against immature worms, most of these require high or multiple doses. With multiple dosing, palatability can be a problem (e.g. with bunamidine), and several drugs (e.g. bunamidine, niclosamide and nitroscanate) can cause diarrhoea and vomiting although with micronization, the efficacy of nitroscanate can be improved without increasing toxicity (Gemmell and Oudemans, 1975a; Gemmell *et al.*, 1977b, 1979b; Richards and Somerville, 1980). Micronization is crucial for the activity of mebendazole since if coarse particles were used, no activity was found (Gemmell *et al.*, 1977c, 1978a). Available data suggest that mebendazole is the best of the benzimidazoles followed by oxfendazole. Gemmell *et al.* (1977c) rated a series of benzimidazoles against *T. hydatigena*. The efficacy was fenbendazole > parbendazole > cambendazole > thiabendazole, though none was highly effective. Mebendazole and nitroscanate have an advantage over bunamidine as both have activity against nematodes as well as cestodes, but the drug of choice for *E. granulosus* is praziquantel. It is active orally and by intramuscular injection (Gemmell *et al.*, 1980), is almost devoid of side effects and, as a result, is being widely used in *Echinococcus* eradication schemes. It is also effective against *E. multilocularis* (Rommel *et al.*, 1976).

A representative survey of efficacy data of cestodicides against cestodes of dogs and cats (Table VI) indicate that control of several cestode infections is possible with a variety of drugs. As complete removal of tapeworms is desirable and possible, doses not completely effective have been omitted from the table. However, unless species of tapeworms are identified, complete removal of worms cannot be ensured with most of the drugs, with the noted exception of praziquantel. This is much more active and has a broader spectrum than other cestodicides and is thus the drug of choice for all cestode infections of dogs and cats.

Table V. Activity of experimental and commercial cestodicides against *Echinoccocus granulosus* in the dog.

Drug	Dose (mg/kg)	Stage	Removal of worms %	Reference
arecoline	4	immature	98	Trejos *et al.* (1975)
bithional*	50 × 2	adult	100	Gemmell *et al.* (1975b)
bunamidine	25	adult	100	Andersen *et al.* (1975)
	50	immature	up to 98.8	Andersen *et al.* (1975)
	50 × 1	immature	78	Gemmell and Shearer (1968)
	50 × 2	immature	98	Gemmell and Shearer (1968)
	25 × 2	immature	73.5	Shearer and Gemmell (1969)
	32 × 3	immature	100	Gemmell and Oudemans (1974)
diuredosan*	50 × 3	immature	>99	Gemmell *et al.* (1978c)
fospirate*	40 × 3	immature	100	Gemmell and Oudemans (1975a)
	80 × 2	immature	90 (cure rate)	Schantz and Prezioso (1976)
mebendazole	20 × 2	immature	100	Gemmell *et al.* (1975a)
	160 × 1	immature	100	Gemmell *et al.* (1975a)
niclosamide	300	immature	inactive	Gemmell *et al.* (1977a)
nitroscanate	400 × 3	immature	92.6	Schantz *et al.* (1976)
	250 × 2	immature	100	Gemmell and Oudemans (1975b)
oxfendazole	20 × 2	immature	>99	Gemmell *et al.* (1979a)
praziquantel	5	immature	>99	Gemmell *et al.* (1977d)
	5	immature	100	Thakur *et al.* (1978)
	5	immature	100	Thomas and Gönnert (1978b)
	5	adult	100	Andersen *et al.* (1978)
	5	immature	100	Andersen *et al.* (1979)
	2.5	immature	100	Gemmell *et al.* (1980)
SQ 27104*	100	immature	>99	Gemmell *et al.* (1978b)

* bithional = 2,2′-sulphinyl bis (4,6-dichloro-phenol).

diuredosan = Diethyl [0-[3-(ptolylsulphonyl) ureido]phenyl]carbamoyl phosphoramidate.

fospirate = 3,5,6-tri-chlor-2 pyridyl phosphate.

SQ 27104 = streptothricin antibiotic complex.

Table VI. The activity of commercial and experimental cestodicides against cestodes of dogs and cats (excluding *E. granulosus*).

Species	Drug	Dose (mg/kg)	Removal of worms %	Reference
Diplydium caninum	niclosamide	100	100	Poole *et al.* (1971)
	niclosamide	157	100	Cruthers *et al.* (1979)
	praziquantel	2.5	100	Thomas and Gönnert (1978b)
	praziquantel	4.4–6.3*	100	Kruckenberg *et al.* (1981)
	uredofos	50	100	Roberson (1976), Roberson and Ager (1976)
Diphyllobothrium sp.	axenomycin D	5	100	Della Bruna *et al.* (1973)
Mesocestoides corti	albendazole	100	100	Todd (1978)
	bunamidine	24	100 (1 dog)	Todd *et al.* (1978)
	niclosamide	160–470	variable	Todd *et al.* (1978)
	praziquantel	5	100	Thomas and Gönnert (1978b)
	uredofos	28	>99 (1 dog)	Todd *et al.* (1978)
Taenia crassiceps	bunamidine	52	100	Alaimo *et al.* (1978)
	furodazole	100	100	Alaimo *et al.* (1978)
	niclosamide	153	100	Alaimo *et al.* (1978)
T. hydatigena	bithional	50 × 2	100	Gemmell *et al.* (1975b)
	bunamidine	32 × 3	<100	Shearer and Gemmell (1969) Gemmell and Oudemans (1974)
	diuredosan	25	100	Gemmell *et al.* (1978c)
	fenbendazole	20 × 4	74	Gemmell *et al.* (1977c)
	fenbendazole	100	100	Gautam *et al.* (1979)
	fospirate	10	100	Gemmell and Oudemans (1975a)
	mebendazole	20	100	Gemmell *et al.* (1975a)
	niclosamide	32	100	Gemmell *et al.* (1977a)
	nitroscanate	64	100	Gemmell and Oudemans (1975b)

Table VI contd.

	nitroscanate	64	>97	Gemmell *et al.* (1979b)
	oxfendazole	80	100	Gemmell *et al.* (1979a)
	praziquantel	2.5	100	Rommell *et al.* (1976)
	praziquantel	5	<100	Thomas and Gönnert (1978b)
	praziquantel	1.25	100	Gemmell *et al.* (1977d)
(2-day-old)	praziquantel	5	100	Baldock *et al.* (1977)
(14-day-old)	praziquantel	1	100	Baldock *et al.* (1977)
	SQ 21704	25	100	Gemmell *et al.* (1978b)
T. ovis	niclosamide	50	100	Gemmell *et al.* (1977a)
	praziquantel	1.25	100	Gemmell *et al.* (1977d)
T. pisiformis	axenomycin D	10	100	Della Bruna *et al.* (1973)
	bunamidine	25	99	Hatton (1965)
	furodazole	50	100	Alaimo *et al.* (1978)
	niclosamide	110	100	Poole *et al.* (1971)
	niclosamide	157	67	Cruthers *et al.* (1979)
	praziquantel	0.5	100	Thomas and Gönnert (1978b)
	SQ 21704	25	100	Szanto *et al.* (1979)
T. taeniaeformis	bunamidine	25	variable	Hatton (1965)
	fenbendazole	50 × 3	100	Roberson and Burke (1980)
	niclosamide	100	100	Wescott (1967)
	praziquantel	2.5	100	Rommel *et al.* (1976)
	praziquantel	0.5	100	Thomas and Gönnert (1978b)
	praziquantel	3.3–8.4*	100	Kruckenberg *et al.* (1981)
	SQ 21704	45	100	Szanto *et al.* (1979)

* Smaller animals received higher mg/kg doses.

Table VII. Activity of cestodicides against tapeworms in man.

Worm species	Drug	Dose	Patients cured %	Number of patients	Reference
Diphyllobothrium latum	niclosamide	26–44 mg/kg	100	4	Perera *et al.* (1970)
	paromomycin	4 gr/patient	100	5	Tanowitz and Wittner (1973)
	paromomycin	4 gr/adults 75 mg/kg children	100	4	Wittner and Tanowitz (1971)
	praziquantel	25 mg/kg	100	33	Apajalahti (1977)
D. pacificum	praziquantel	10 mg/kg	100	25	Espejo (1977)
Diplydium caninum	niclosamide	1–2 gr/patient	100	13	Jones (1979)
Hymenolepis diminuta	niclosamide	1·2 gr/patient	89	19	Jones (1979)
H. nana	niclosamide	26–44 mg/kg	100	7	Perera *et al.* (1970)
	niclosamide	100 mg/kg × 1	20	10	Most *et al.* (1971)
	niclosamide	40–50 mg/kg × 5	80	10	Most *et al.* (1971)
	niclosamide	60–80 mg/kg × 5	90	52	Most *et al.* (1971)
	niclosamide	60 mg/kg × 1 + 15 × 6	88	68	Khalil (1969)
	paromomycin	45 mg/kg × 5	100	10	Wittner and Tanowitz (1971)
	praziquantel	15 mg/kg	80	10	Baranski (1977)
	praziquantel	20 mg/kg	100	6	Baranski (1977)
	praziquantel	13.5–16	81	26	Canzonieri *et al.* (1977)
	praziquantel	25 mg/kg	93	15	Canzonieri *et al.* (1977)
	praziquantel	15 mg/kg	100	25	Espejo (1977)
	praziquantel	15 mg/kg	100	29	Rim *et al.* (1978)
	praziquantel	25 mg/kg	97	31	Rim *et al.* (1978)
H. nana	praziquantel	10 mg/kg	77	22	Schenone *et al.* (1977)
	praziquantel	15 mg/kg	95	19	Schenone *et al.* (1977)
	praziquantel	25 mg/kg	100	30	Schenone *et al.* (1977)
	praziquantel	25 mg/kg	98.5	65	Schenone (1980)
	praziquantel	15 mg/kg	93.8	65	Schenone (1980)
	praziquantel	10 mg/kg	76	25	Schenone (1980)
Mesocestoides sp.	niclosamide	1 gr (child)	100	1	Gleason *et al.* (1973)
Taenia sp.	paromomycin	75 mg/kg	92	15	Botero (1970)

Table VII contd.

T. saginata	mebendazole	100 mg/patient twice daily for 3 days	95	47	Arambulo *et al.* (1978)
	mebendazole	300 mg/patient twice daily for 3 days	0	5	Vakil *et al.* (1975)
	niclosamide		95.5	45	Ahkami and Hadjian (1969)
	niclosamide	60 mg/kg	86.4	13	Khalil (1969)
	paromomycin	adults 4 g children 75 mg/kg	100	3	Wittner and Tanowitz (1971)
	praziquantel	10 mg/kg	100	8	Baranski (1977)
	praziquantel	9–10 mg/kg	100	39	Canzonieri *et al.* (1977)
	praziquantel	10 mg/kg	100	25	Espejo (1977)
	praziquantel	10 mg/kg	100	50	Paz (1977)
T. solium	mebendazole	300 mg/patient twice daily for 3 days	100	10	Chavarría *et al.* (1977)
		200 mg/patient twice daily for 3 days	73	25	Chavarría *et al.* (1977)
T. solium	niclosamide	26–44 mg/kg	100	1	Perera *et al.* (1970)
	paromomycin	adults 4 g children 75 mg/kg	100	3	Wittner and Tanowitz (1971)
	praziquantel	25 mg/kg	100	25	Espejo (1977)

C. Adult cestodes of man

Probably the two most common cestodes of the intestinal tract of man are *T. saginata* and *H. nana. T. solium* occurs where pigs and people are in close proximity and is dangerous because of possible self-infection with cysticerci. Diagnosis and therapy have been discussed by Gönnert (1974). It is clear from Table VII, which summarizes a number of representative trials, that these tapeworms can be removed with more than one drug. *H. nana* is more refractory to therapy than the other cestodes, and niclosamide does not usually give a complete cure of this species. The choice of drug must therefore rest primarily on freedom from side-effects and the convenience of administration. In rural third world areas, it can be difficult to ensure drugs are taken regularly over several days and the ideal therapy is a single dose. Although mebendazole is a valuable drug because of its activity against parasitic nematodes, because it is almost devoid of side-effects, the regime of two doses per day for three days is a disadvantage. Similarly, even the four capsules of paromomycin every 15 min for four doses for adults as used by Wittner and Tanowitz (1971) is a disadvantage and, when used for five days for *H. nana*, over half the patients complained of diarrhoea. Niclosamide controls many cestode infections effectively and is an inexpensive drug with no side-effects (Most *et al.*, 1971), but therapy for *H. nana* lasted for five days. The disadvantages with other therapies are overcome by the single oral dose used for praziquantel. From extensive clinical trials, the recommended doses are: taeniasis, 5–10 mg/kg; hymenolepiasis, 15 mg/kg; diphyllobothriasis (pacificum) 10 mg/kg; and diphyllobothriasis (latum), 25 mg/kg (Groll, 1977, 1980).

D. Larval cestodes of sheep and cattle

The activities of drugs against larval cestodes in rodents transfer to larger animals. For example, Barsanti *et al.* (1979) reported that 50 mg/kg/day of mebendazole for 107 days led to 16 months freedom from clinical symptoms in a dog infected with *Mesocestoides* sp. Drugs showing activity in rodents also show activity in pigs and ruminants (Table VIII). However, prolonged treatment or high doses are usually required, and this will not be economical except for very valuable breeding animals. Deep freeze storage is a practical alternative for killing cysts in meat animals.

The most economically important cyst is that of *T. saginata*, and good control has been obtained with praziquantel at high doses (Table VIII), except for the young cysts (Gallie and Sewell, 1978). Similarly, *T. hydatigena* is controlled with praziquantel and mebendazole, but the effectiveness of the therapy may depend on the amount of cyst material in the animal (Heath and Lawrence, 1978). The location of the parasite can be important as well. *T. multiceps* and *T. solium* in the sheep brain appear resistant to therapy (Verster *et al.*, 1978; Telléz-Giron *et al.*, 1981). A really effective single treatment drug that is cost effective thus still awaits discovery.

E. *E. granulosus* therapy in man

Although the usual way of treating *E. granulosus* cysts in man is by surgery, not all cysts can be removed and there is also a danger from rupture of the cyst leading to anaphylactic shock or reinfection. Sterilization of the cyst prior to removal has therefore become usual, and cetrimide (0.5% or less) is the preferred chemical having replaced 20% sodium chloride, 2%

formalin and 1% $AgNO_3$ (Eslami *et al.*, 1978). Use of two 5-minute treatments with 0.5% cetrimide is adequate and has reduced the cases returning after surgery from 10% to zero (Frayha *et al.*, 1981). However, chemical treatment as well as or instead of surgery is required, and mebendazole has been fairly widely evaluated (data reviewed by Amman *et al.*, 1979; Beard *et al.*, 1978; Bekhti *et al.*, 1980; Hinz, 1980; Nolla Panades *et al.*, 1980 and Schantz *et al.*, 1982).

There has been considerable debate over the effectiveness of mebendazole treatment of hydatidosis and the significance of fever that is sometimes associated with the therapy (Bryceson, 1980; Larkworthy and Atiyeh, 1980; MacNair, 1980; Miskovitz and Javitt, 1980; Osborne, 1980a,b). Some therapy does not appear to have been effective (Quilici *et al.*, 1979), and there is little surgical proof of efficacy following treatment. There is nevertheless considerable clinical evidence for stabilization of the condition of many patients or their improvement.

A range of dose levels and lengths of treatment have been tried (tabulated in Bekhti *et al.*, 1980; Schantz *et al.*, 1982). The doses used have sometimes been dependent on a lack of toleration of the therapy by the patient. Collapse of the cyst wall as revealed by tomography (Bryceson, 1980), apparent cure of a ruptured cyst (Mulhall, 1980), improved radiological presentation in a young patient (Loughran and McCarey, 1980), striking decrease in abdominal girth (associated with fever) (Murray-Lyon and Renolds, 1979), decrease in size and number of cysts (Kammerer, 1978), disappearance of allergic manifestations associated with disseminated hydatidosis and round lesions reduced (Werczberger *et al.*, 1979), improved pathological findings (Kern *et al.*, 1979), and complete regression of intrahepatic cysts as shown by ultrasonic echotomography (Bekhti *et al.*, 1977) indicate the types of beneficial changes observed after mebendazole therapy of hydatidosis due to *E. granulosus*. With *E. multilocularis* infection the disease was arrested (Wilson *et al.*, 1978).

In the most detailed study reported, Bekhti *et al.* (1980) found that doses of 16–35 mg/kg/day for 21–81 days in eight cases, and 50 mg/kg/day for 3–11½ months in eight cases, successfully treated single or multiple primary cysts and prevented secondary hydatidosis. Efficacy was monitored both by physical and immunological methods, and the importance of determining parasite-specific IgE antibodies as used in their earlier study (Bekhti *et al.*, 1977), was stressed. Compared with their trials, the results obtained were better, probably due to the high dose-levels and prolonged therapy. Similar encouraging results with high doses over long periods have also been reported from South Africa (Kayser, 1980). The necessity of high doses for prolonged periods illustrates both the need for more active drugs and the problems associated with the insolubility of mebendazole. Systemic availability was improved by concomitant food intake, but the

Table VIII. The efficacy of cestodicides against larval cestodes in pigs, sheep and cattle.

Parasite		Drug	Dose	Removal of worms %	Reference
Echinococcus	3	mebendazole	25 mg/kg × 10	5/7 cured	Pawlowski *et al.* (1976a)
granulosus	1	mebendazole	50 mg/kg × 14	significantly retarded growth	Heath and Lawrence (1978)
	1	mebendazole	50 mg/kg × 90	little live tissue left	Gemmell *et al.* (1981)
	1	praziquantel	50 mg/kg	no effect	Heath and Lawrence (1978)
Taenia hydatigena	1	mebendazole	25 mg/kg × 5	100	Oguz (1976)
	3	mebendazole	25 mg/kg × 5	99	Hörchner *et al.* (1976)
	1	mebendazole	50 mg/kg × 14	almost all killed	Heath and Lawrence (1978)
		mebendazole	150 mg/kg × 30	100	Gemmel *et al.* (1981)
	3	praziquantel	50 mg/kg × 5	100	Hörchner *et al.* (1976)
	1	praziquantel	75 mg/kg × 3	100	Oguz (1976)
	1	praziquantel	50 mg/kg × 1	all killed if less than 100 cysts	Heath and Lawrence (1978)
	1	praziquantel	50 mg/kg × 1	100	Thomas and Gönnert (1978a)
	1	praziquantel	100 mg/kg × 1	>99	Bankov and Gradinarski (1980)
T. multiceps	1	mebendazole	100 mg/kg × 14	not effective	Verster *et al.* (1978)
T. ovis	1	mebendazole	50 mg/kg × 14	almost all killed	Heath and Lawrence (1978)
		praziquantel	50 mg/kg × 1	100	Heath and Lawrence (1978)
T. saginata	2	albendazole	50 mg/kg	86	Lloyd *et al.* (1978)
	2	albendazole	45 mg/kg	not effective	Craig and Ronald (1978)
	2	albendazole	50 mg/kg	not effective	Stevenson *et al.* (1981)
	2	mebendazole	40 mg/kg (i.p.)	not effective	Pawlowski *et al.* (1976b)
	2	mebendazole	100 mg/kg	not effective	Gallie and Sewell (1977)

Table VIII contd.

	2	praziquantel	50 mg/kg × 4	100 against 12 week old cysts not effective against 4 week old cysts	Gallie and Sewell (1978)
	2	praziquantel	100 mg/kg	100	Pawlowski *et al.* (1978)
	2	praziquantel	10 mg/kg × 10	100	Pawlowski *et al.* (1978)
	2	praziquantel	50 mg/kg	100	Thomas and Gönnert (1978a)
	2	praziquantel	10 mg/kg × 4	100	Thomas and Gönnert (1978a)
	2	praziquantel	100 mg/kg	100	Walther and Koske (1979) Walther and Grossklaus (1979)
T. solium	3	fenbendazole	5 mg/kg × 7	100	Araki and Cho (1980)
	3	flubendazole	8.3–28.5 mg/kg × 10	4/6 cured	Telléz-Giron *et al.* (1981)
	3	flubendazole	40 mg/kg × 10	4/5 cured	Telléz-Giron *et al.* (1981)

1 sheep, 2 cattle, 3 pigs.

large variations found between patients suggest that individual plasma levels need monitoring during therapy (Münst *et al.*, 1980). Similar wide values for plasma mebendazole have also been found, with ranges between 4 and 575 ng/ml following doses in the range 16–60 mg/kg (Witassek *et al.*, 1980). In two patients given 1g mebendazole orally three hours prior to surgery, cyst fluid contained 300 and 320 ng/ml (Brandimarte *et al.*, 1980) so mebendazole clearly enters the cysts, but what concentration is required for rapid killing of the cysts is not known. While not ideal, mebendazole represents a valuable aid for control of hydatid cysts that are not amenable to surgery.

IV. MECHANISMS OF ACTION OF CESTODICIDES

Research into the mode of action of cestodicides and their uptake by the parasite is not essential for practical chemotherapy, but can provide useful information. For example, knowledge of the pharmacology of the drug in the cestode might possibly lead to improved formulations with increased efficacy and reduced toxicity to the host. In addition, investigations into the mode of action of cestodicides should yield new information on the biochemistry of the worms and may reveal important host–parasite metabolic differences. There has, however, been little success in designing drugs to exploit host–parasite metabolic differences. While it may be safe to try to exploit host–parasite differences revealed by drug studies on cestodes, because the chances of cestode drug resistance occurring is low, this is not the case with nematodes where resistance is a growing problem.

Of the five major types of drugs currently used against cestodes, four (benzimidazoles, nitroscanate, praziquantel and salicylanilides) affect other helminths in addition to cestodes. Much of the work on these drugs has not been with cestodes, but it seems reasonable to assume that the basic mechanisms of action are similar in different helminths. Most work on mechanism of actions has been on benzimidazoles, praziquantel and salicylanilides, and discussion will be largely confined to these three drug types. Brief reference will be made to bunamidine, but there is insufficient information available to warrant discussion of other drugs listed in the tables. Previous reviews on anthelmintic mechanisms relevant to their actions on cestodes have been published by Coles (1977a,b), Rew (1978) and Van den Bossche (1976, 1978, 1980b, 1981).

A. Benzimidazoles

Benzimidazoles have a very broad spectrum of activity. In addition to killing some larval and adult cestodes, they are active against nematodes, are ovicidal against developing helminths' eggs (Coles and Simpkin, 1977b; Coles and Briscoe, 1978), some are active against liver fluke and thiabendazole is widely used as a fungicide. It is possible that different benzimidazoles work in a variety of ways on different parasites. However it is more likely that the drugs work in the same basic way but differ in their pharmacokinetics in both the host and parasite which would explain differences in efficacy and in the biochemical effects observed on the parasite.

Historically, the first proposed mode of action of benzimidazoles on helminths was the inhibition of fumarate reductase from *Haemonchus contortus* by thiabendazole (Prichard, 1970, 1973; Malkin and Camacho, 1972). The enzyme from benzimidazole-resistant *H. contortus* was resistant to thiabendazole and cambendazole (Malkin and Camacho, 1972; Prichard, 1973; Romanowski *et al.*, 1975), but mebendazole did not inhibit fumarate reductase, even though mebendazole and thiabendazole are side resistant in nematodes (Colglazier *et al.*, 1975; Round *et al.*, 1974). This difference between thiabendazole and mebendazole at the enzyme level could be explained if benzimidazole resistance was expressed as reduced drug uptake in resistant worms. If, however, uptake was similar in normal and resistant worms, then inhibition of fumarate reductase could not be the mode of benzimidazole action. This latter interpretation was supported by the failure to find fumarate reductase in eggs of *Nematodirus spathiger* and adult *Nippostrongylus brasiliensis* even though thiabendazole was highly ovicidal (Coles, 1977a). Studies with thiabendazole on the mitochondrial electron transport system of *Ascaris suum* showed that while it could act at more than one mitochondrial site this may not necessarily be the primary site of action (Köhler and Bachman, 1978).

Mebendazole reduced glycogen levels in a range of parasitic helminths, including the cestode *H. nana*, because of reduced glucose uptake (Van den Bossche, 1972a). In *A. suum*, the highest concentration of mebendazole was found in intestinal tissue, the site of glucose uptake (Van den Bossche and De Nollin, 1973). Decreased glucose uptake and a fall in amounts of glycogen also occurred in *M. expansa* treated *in vitro* with mebendazole and cambendazole. ATP levels were reduced (Rahman and Bryant, 1977), and this was also found within 3 h in worms taken from sheep dosed with mebendazole. The most dramatic fall was, however, in free glucose in the scolex (Rahman *et al.*, 1977). Detailed histological examination of tissues in nematodes and cestodes exposed to mebendazole showed that the first detectable change in morphology was the disappearance of cytoplasmic microtubules and subsequent block in the transport of secretory granules leading to ultimate degeneration of the cells (intestinal in nematodes, tegumentary in cestodes) (Borgers and de Nollin, 1975; Borgers *et al.*, 1975a,b,c; Verheyen *et al.*, 1976; Atkinson *et al.*, 1980). As viewed under the scanning electron microscope, mebendazole treatment of cysticerci of *T. taeniaeformis* resulted in gradual disappearance of microtriches, degeneration of the tegument and attachment of host cells (Verheyen *et al.*, 1978).

Biochemical research with other cell systems has confirmed that benzimidazoles bind to tubulin. In the fungus *Aspergillus nidulans*, methyl benzimidazole carbamate and thiabendazole bound to unpolymerized tubulin preventing mitosis and there was reduced binding to tubulin in a drug-resistant strain (Davidse, 1975; Davidse and Flach, 1977, 1978) which showed up by electrophoresis as the presence of altered β tubulins (Sheir-Neiss *et al.*, 1978). Parbendazole inhibited microtubule assembly *in vitro* and *in vivo* in the myxamoebae of *Physarum polycephalum* (Quinlan *et al.*, 1981).

In neoplastic mammalian cells, de Brabander *et al.* (1975) showed that nocodazole, a cytotoxic benzimidazole, interfered with microtubules, and specifically bound to rat brain tubulin (Hoebeke *et al.*, 1976). In tissue culture cells, parbendazole inhibited microtubule assembly and caused depolymerization of cytoplasmic microtubules (Havercroft *et al.*, 1981).

The affinity of a range of benzimidazoles for mammalian tubulin has been reported (Friedman and Platzer, 1978; Ireland *et al.*, 1979; Laclette *et al.*, 1980). Binding is competitive with colchicine.

A similar situation occurs within tubulin from *A. suum*, but the affinity of mebendazole for tubulin from *A. suum* embryos was reported to be much higher (Friedman and Platzer, 1980a,b) than that from the intestines of adult worms (Köhler and Bachman, 1980). While Friedman and Platzer (1980a,b) suggested that the difference in properties between embryonic *A. suum* tubulin and mammalian brain tubulin could explain the safety of benzimidazoles, Köhler and Bachman (1980) found only a twofold difference in affinity for mebendazole and suggested that a possible host–parasite difference in pharmacokinetics might explain the safety of benzimidazole anthelmintics. The affinity of tubulin from *A. suum* eggs for colchicine has been reported to change during embryogenesis (Friedman *et al.*, 1980), but this is unlikely to account for the discrepancies between the two sets of results.

Although the basic biochemical mechanism of action of benzimidazoles on helminths now seems clear, there is little information about the resultant biochemical effects on cestodes. How does cestode tubulin isolated from *H. diminuta* (Watts, 1981) differ from mammalian tubulin in affinity for benzimidazoles? Are benzimidazoles concentrated by adult cestodes as they appear to be in nematodes (Coles, 1977a, Köhler and Bachman, 1980) even though entry into the cyst (*E. granulosus*) *in vitro* (Reisin *et al.*, 1977) and in most, but possibly not all cysts, treated *in vivo* (Kammerer and Miller, 1981) is by diffusion? Do different benzimidazoles accumulate in the same parts of adult cestodes and if so, where? The first effects of fenbendazole on *H. diminuta in vivo* suggested a neurotoxic effect (Duwel and Schleich, 1978). Answers to these and other questions are required for a full understanding of the action of benzimidazoles on cestodes.

B. Bunamidine

In vitro studies with *H. nana* and *H. diminuta* suggest that the prime effect of bunamidine is on the cestode tegument, and it is relevant to note that bunamidine is highly irritant to mammalian mucous membranes. Ultrastructural studies revealed vacuolization followed by sloughing off of outer layers of *H. nana* maintained in 10^{-4}M bunamidine (Hart *et al.*, 1977), and structural alterations indicative of cellular degeneration in the cells of the tegument and nephridial system in *H. diminuta* maintained in 4.8×10^{-6} M bunamidine (the 24 h LC_{50} is 6.4×10^{-6} M) (Chatfield and Yeary, 1979). The higher concentrations used on *H. nana* resulted in reduced glucose uptake and release of surface phosphatase in the medium as might be expected from tegumentary damage. No similar studies were undertaken with *H. diminuta*, but fumarate reductase was inhibited by bunamidine. In view of the many anthelmintics that have been reported to inhibit this enzyme, no significance can be attributed to this observation without detailed biochemical analyses of worms treated both *in vitro* and *in vivo*. The molecular basis on which bunamidine affects the cestode tegument still requires determining.

C. Praziquantel

Although praziquantel is effective against cestodes, the majority of research has been on its schistosomicidal effect (Andrews, 1981). Praziquantel very rapidly caused contraction of *Schistosoma mansoni* (Andrews, 1978; Pax *et al.*, 1978; Coles, 1979; Chevasse *et al.*, 1979) followed within a short time by disruption of the tegument. This also occurred with the tegument of *Dicrocoelium dendriticum* but not *Fasciola hepatica* (Becker *et al.*, 1980a). Similarly, in cestodes there was a rapid contraction and paralysis which occurred at concentrations as low as 0.01 mg/ml (Andrews and Thomas, 1979), but it could be reversed in drug-free medium (Thomas and Andrews, 1977). Contraction *in vivo* occurred within 1 min of dosing (Thomas and Gönnert, 1977). This was followed in the growth zone of the neck region of *H. nana*, *H. diminuta* and *H. microstoma*, but not other parts of the body, by vacuolization and disruption of the tegument. The strobilocerci of *T. taeniaeformis* developed identical lesions, but the wall of the bladder remained unaffected, as did invaginated protoscoleces and the cyst wall of *E. multilocularis* (Becker *et al.*, 1980b; Becker *et al.*, 1981). Praziquantel also damaged the tegument of *D. latum* (Bylund *et al.*, 1977). Effects on the proper functioning of sucker movement in cestodes and on egg production by *S. mansoni* were caused by concentrations as low as 0.001 mg/ml (Andrews, 1978; Andrews and Thomas, 1979). This is far below the serum concentration of 0.2 mg/ml found 1–2 h after an oral dose of 20 mg/kg (Leopold *et al.*, 1978).

Experiments with ^{14}C praziquantel demonstrated rapid uptake in trematodes and cestodes with an even distribution of drug and with tissue: medium ratios up to 3.5. However, uptake into intact cysts (*T. taeniaeformis*) was much slower, with the greatest concentrations of drug in the cyst wall. Loss of drug from *S. mansoni* to medium was rapid with an estimated drug half-life of 7.5 min. There was no evidence of significant metabolism of praziquantel by *S. mansoni* (Andrews *et al.*, 1980).

The uptake of praziquantel by *H. diminuta in vitro* resulted, within 15 min, in reduced glucose uptake, increased lactate secretion, increased secretion of total acidic metabolites in the absence of glucose and an efflux of glucose and α-amino nitrogen into the medium (Thomas and Andrews, 1977; Andrews and Thomas, 1979). Other reported effects of praziquantel include inhibition of certain NADH-oxidizing enzymes in *A. suum* (Köhler and Bachman, 1978), a decrease in ATPase in whole *S. mansoni* (Nechay *et al.*, 1980), a depolarization of the tegumental membrane (Fetterer *et al.*, 1980a) and a displacement of ouabain from its binding site on whole worms (Fetterer *et al.*, 1980b). However praziquantel did not inhibit protein synthesis in an *in vitro* assay (Lukacs *et al.*, 1980).

All these effects are probably secondary, the prime effect of praziquantel being to permit rapid non-selective movement of calcium ions (Pax *et al.*, 1978; Coles, 1979; Fetterer *et al.*, 1980c). Depletion of calcium or alteration of Ca^{++}/Mg^{++} ratios affected the length of contraction, and fluoxetine delayed the onset of the praziquantel-induced contraction (Pax *et al.*, 1979). Praziquantel is not an ionophore (Pax *et al.*, 1978; Coles, 1979), and the detailed biochemical mechanisms by which praziquantel permits calcium influx into cells are not known. However, it is obvious that sustained contraction with impairment of sucker function in cestodes in the mammalian gut or bile duct will result in their rapid ejection from the host. How far the damage to the cestode tegument is related to calcium ion movement caused by praziquantel is not certain, but it seems likely that the two will be connected. The therapeutic effect of praziquantel on cysts will presumably result from the tegumentary damage described by Becker *et al.* (1981), followed by host-cell invasion. Melhorn *et al.* (1981) stated that 17 h after treatment, *S. mansoni* are fixed to the walls of blood vessels with fibroblasts and the interior of the worms are being destroyed by host cells. Evidence for an indirect effect of praziquantel on cestode cysts comes from the observation of Novak (1977) who reported that cysts of *T. crassiceps* "appeared dead" when removed from animals fed praziquantel, but resumed movement after 6 h *in vitro*. There is obviously much more research required before the detailed mechanism of action of praziquantel on tapeworms is fully understood.

D. Salicylanilides

Salicylanilides are broad spectrum anthelmintics used to control cestodes, trematodes and nematodes (adult *H. contortus*). Less substituted molecules (niclosamide and resorantel) control cestodes, while more substituted (more lipophilic) molecules control trematodes and *H. contortus*, presumably due to differential distribution within the host and different uptake by the parasites. Salicylanilides are uncouplers of oxidative phosphorylation in mammalian mitochondria (van Miert and Groeneveld, 1969; Corbett and Goose, 1971; Van den Bossche *et al.*, 1979), and niclosamide-stimulated ATPase in mitochondria from rats (Putter, 1970). There seems no reason to doubt that salicylanilides also interfere with energy metabolism in helminths, although exactly similar effects do not always occur presumably due to differing distribution of the compounds within the parasite.

The first effect of salicylanilides observed on *F. hepatica* was the emptying of the gut contents which suggested that oxyclozanide may have differentially affected the nervous system (Coles, 1975), a possibility endorsed by the findings of Edwards *et al.* (1981) that at low doses, flukes were affected without the expected lowering of ATP levels. In rats infected with *H. diminuta* there was a rapid posterior migration of the worms after dosing with resorantel (Coles, 1977a). Interestingly, in *F. hepatica* from sheep treated with closantel, the posterior parts of the worm are affected more than the anterior (Verheyen *et al.*, 1980) and, as already mentioned, oxyclozanide causes loss of segments of *M. expansa* without removal of the head (Walley, 1966). This could either be explained by the existence of a metabolic gradient along cestodes, as described for *H. diminuta* under aerobic conditions by Coles and Simpkin (1977a), or it could be due to differential absorption of drug, or both.

Histological examination of flukes treated with closantel revealed that the first effect is on the mitochondrial structure (Verheyen *et al.*, 1980) and this is reflected in biochemical measurements which showed reduced malate-induced phosphorylation, reduced ATP and increased succinate (Van den Bossche *et al.*, 1979; Kane *et al.*, 1980). Similar reduced levels of ATP were found after *in vitro* and *in vivo* treatment of *F. hepatica* with rafoxanide (Cornish and Bryant, 1976; Cornish *et al.*, 1977; Prichard, 1978). Reduced levels of ATP have also been reported following treatment of *H. diminuta* with resorantel (Schacht *et al.*, 1971). Uncoupling was demonstrated in isolated *H. diminuta* mitochondria by measurement of oxygen uptake (Yorke and Turton, 1974). Low drug-doses stimulated oxygen uptake while higher doses inhibited it, but the potency of the compounds was not directly related to their cestodicial activity, indicating the role of drug distribution and uptake in determining activity.

Other experiments with nematodes and cestodes have demonstrated further the effects of salicylanilides on parasite mitochondria. They reduced the esterification of labelled phosphate by mitochondria of *H. diminuta* and *A. suum* (Schiebel *et al.*, 1968; Saz and Lescure, 1968; Saz, 1972; Saz *et al.*, 1972; Van den Bossche, 1972b) and inhibited succinate dehydrogenase (Metzger and Düwel, 1973). They also stimulated ATPase in *A. suum* mitochondria (Van den Bossche, 1972b). The relative importance of the different possible actions has not been determined in cestodes, but all would have the same effect of reducing available ATP.

Although almost all available information is on the effect of drugs on helminths, two studies have included the effect of helminths on drugs (Douch and Gahagan, 1977; Douch, 1979). *M. expansa* did not hydrolyse niclosamide but did reduce the nitro group to the amine. Rosorantel was not hydrolysed or hydroxylated but clioxanide was deacetylated by a cytosolic enzyme. There is clearly need for further work of this type.

V. CONCLUSION

The withdrawal of several pharmaceutical companies from anthelmintic research combined with the small market size for cestodicides make it unlikely that many new drugs will be discovered. Hopefully, however, the length of the cestode life-cycle means that drug resistance will not become a problem. If this view is correct, then new drugs for adult cestodes are not required. For full control of *Taenia* sp. and *E. granulosus*, both adult and larval stages need treating with drugs or controlling by vaccination. An economical therapy for larval cestodes in sheep and cattle or an effective vaccine is thus still required and improved therapy for man would be an advantage.

Although there are now good indications how benzimidazoles, praziquantel and salicylanilides act on cestodes at the molecular level, there are many unanswered questions, particularly on drug distribution within the cestode, which will hopefully be resolved in the not too distant future.

VI. REFERENCES

Ahkami, S. and Hadjian, A. (1969). The appearance of the scolex of *Taenia saginata* in the stool after the eradication of the parasite by niclosamide. *Z. Tropenmed. Parasit.* **20**, 341–345.

Alaimo, R. J., Spencer, C. F., Sheffer, J. B., Storin, R. J., Hatton, C. J. and Kohls, R. E. (1978). Imidazo[4,5-f]quinolines. 4. Synthesis and anthelmintic activity of a series of imidazo[4,5-f]quinolin-9-ols. *J. Med. Chem.* **21**, 298–300.

Ammann, R., Akovbiantz, A., Eckert, J. and Largiader, F. (1979). Therapie der Echinokokkose. *Dt. Med. Wochenschr.* **104**, 1429–1431.

Andersen, F. L., Loveless, R. M. and Jensen, L. A. (1975). Efficacy of bunamidine hydrochloride against immature and mature stages of *Echinococcus granulosus*. *Am. J. Vet. Res.* **36**, 673–675.

Andersen, F. L., Conder, G. A. and Marsland, W. P. (1978). Efficacy of injectable and tablet formulations of praziquantel against mature *Echinococcus granulosus*. *Am. J. Vet. Res.* **39**, 1861–1862.

Andersen, F. L., Conder, G. A. and Marsland, W. P. (1979). Efficacy of injectable and tablet formulations of praziquantel against immature *Echinococcus granulosus*. *Am. J. Vet. Res.* **40**, 700–701.

Andrews, P. (1978). Praziquantel—a novel schistosomicide. *Parasitology*, **75**, xvii–xviii.

Andrews, P. (1981). A summary of the efficacy of praziquantel against schistosomes in animal experiments and notes on its mode of action. *Arzneim-Forsch.* **31**, 538–541.

Andrews, P. and Thomas, H. (1979). The effect of praziquantel on *Hymenolepis diminuta in vitro. Tropenmed. Parasit.* **30**, 391–400.

Andrews, P., Thomas, H. and Weber, H. (1980). The *in vitro* uptake of ^{14}C-praziquantel by cestodes, trematodes, and a nematode. *J. Parasit.* **66**, 920–925.

Apajalahti, J. (1977). Tratamiento de infecciones por *Diphyllobothrium latum* con una dosis oral única de praziquantel. *Bol. Chil. Parasit.* **32**, 43.

Araki, T. and Cho, K.-M. (1980). Chemotherapy of cysticercosis cellulosae. 10th Int. Cong. Trop. Med. Malaria, Manila, p. 272.

Arambulo, P. V., Cabrera, B. D. and Cabrera, M. G. (1978). Use of mebendazole in the treatment of *Taenia saginata* taeniasis in an endemic area in the Philippines. *Acta Trop.* **35**, 281–286.

Arundel, J. H. (1972). Cysticercosis of sheep and cattle. *Aust. Meat. Res. Comm. Rev.* **4**, 1–21.

Atkinson, C. A., Newsam, R. J. and Gull, K. (1980). Influence of the anti-microtubule agent, mebendazole, on the secretory activity of intestinal cells of *Ascaridia galli. Protoplasma*, **105**, 69–76.

Baldock, F. C., Flucke, W. J. and Hopkins, T. J. (1977). Efficiency of praziquantel, a new cesticide, against *Taenia hydatigena* in the dog. *Res. Vet. Sci.* **23**, 237–238.

Bankov, D. and Gradinarski, I. (1980). Treatment of *Cysticercus tenuicollis* with Droncit. *Vet. Sbirka*, **78**, 18–22.

Baranski, M. C. (1977). Tratamiento de teniasis e himenolepiasis humanas con praziquantel (Embay 8440). *Bol. Chil. Parasit.* **32**, 37–39.

Barsanti, J. A., Jones, B. D., Bailey, W. S. and Knipling, G. D. (1979). Diagnosis and treatment of peritonitis caused by a larval cestode, *Mesocestoides* spp., in a dog. *Cornell Vet.* **69**, 45–53.

Beard, T. C., Rickard, M. D. and Goodman, H. T. (1978). Medical treatment for hydatids. *Med. J. Aust.* **64**, 633–635.

Becker, B., Mehlhorn, H., Andrews, P. and Thomas, H. (1980a). Scanning and transmission electron microscope studies on the efficacy of praziquantel on *Hymenolepis nana* (Cestoda) *in vitro. Z. ParasitKde*, **61**, 121–133.

Becker, B., Mehlhorn, H., Andrews, P., Thomas, H. and Eckert, J. (1980b). Light and electron microscopic studies on the effect of praziquantel on *Schistosoma mansoni*, *Dicrocoelium dendriticum*, and *Fasciola hepatica* (Trematoda) *in vitro. Z. ParasitKde*, **63**, 113–128.

Becker, B., Mehlhorn, H., Andrews, P. and Thomas, H. (1981). Ultrastructural investigations on the effect of praziquantel on the tegument of five species of cestodes. *Z. ParasitKde*, **64**, 257–269.

Bekhti, A., Schaaps, J.-P., Capron, M., Dessaint, J. P., Santoro, F. and Capron, A. (1977). Treatment of hepatic hydatid disease with mebendazole: preliminary results in four cases. *Br. Med. J.* **2**, 1047–1051.

Bekhti, A., Nizet, M., Capron, M., Dessaint, J. P., Santoro, F. and Capron, A. (1980). Chemotherapy of human hydatid disease with mebendazole. Follow-up of 16 cases. *Acta Gastro-Enterol. Belg.* **43**, 48–65.

Bennet, E.-M., Behm, C. and Bryant, C. (1978). Effects of mebendazole and levamisole on tetrathyridia of *Mesocestoides corti* in the mouse. *Int. J. Parasit.* **8**, 463–466.

Bezubik, B., Borowik, M. M. and Brzozowska, W. (1978). The effect of Panacur on helminth parasites in naturally infected lambs. *Acta parasit. pol.* **26**, 75–82.

Blakemore, R. C., Bowden, K., Broadbent, J. L. and Drysdale, A. C. (1964). Anthelmintic constituents of ferns. *J. Pharm. Pharmacol.* **16**, 464–471.

Borgers, M. and De Nollin, S. (1975). The ultrastructural changes in *Ascaris suum* intestine after mebendazole treatment *in vivo*. *J. Parasit.* **61**, 110–122.

Borgers, M., De Nollin, S., de Brabander, M. and Thienpont, D. (1975a). Influence of the anthelmintic mebandazole on microtubules and intracellular organelle movement in nematode intestinal cells. *Am. J. Vet. Res.* **36**, 1153–1166.

Borgers, M., De Nollin, S., Verheyen, A., de Brabander, M. and Thienpont, D. (1975b). Effects of new anthelmintics on the microtubular system of parasites. *In* "Microtubules and Microtubule Inhibitors", (M. Borgers and M. de Brabander, eds), pp. 497–508. North Holland Publishing Company, Amsterdam.

Borgers, M., De Nollin, S., Verheyen, A., Vanparijs, D. and Thienpont, D. (1975c). Morphological changes in cysticerci of *Taenia taeniaeformis* after mebendazole treatment. *J. Parasit.* **61**, 830–843.

Botero, R. D. (1970). Paromomycin as effective treatment of *Taenia* infections. *Am. J. Trop. Med. Hyg.* **19**, 234–237.

Brabander, M. de, Van de Veire, R., Aerts, F., Geuens, G., Borgers, M., Desplenter, L. and De Cree, J. (1975). Oncodazole (R 17934): A new anticancer drug interfering with microtubules. Effects on neoplastic cells cultured *in vitro* and *in vivo*. *In* "Microtubules and Microtubule Inhibitors", (M. Borgers and M. de Brabander, eds), pp. 509–521. North Holland Publishing Company, Amsterdam.

Brandimarte, C., Attili, A. F., Cantafora, A., Martin, G. L. de, Giunchi, G. (1980). Serum and hydatid cystic fluid levels of mebendazole in patients with liver hydatidosis. *Ital. J. Gastroenterol.* **12**, 212–213.

Brody, G. and Elward, T. E. (1971). Comparative study of 29 known anthelmintics under standardized drug-diet and gavage medication regimens against four helminth species in mice. *J. Parasit.* **57**, 1068–1077.

Brown, W. E., Szanto, J., Meyers, E., Kawamura, T. and Arima, K. (1977). Taeniacidal activity of streptothricin antibiotic complex S15–1 (SQ 21,704). *J. Antibiot.* **30**, 886–889.

Bryceson, A. (1980). Mebendazole and hydatid disease. *Br. Med. J.* **280**, 796.

Bylund, G., Bång, B. and Wikgren, K. (1977). Tests with a new compound (Praziquantel) against *Diphyllobothrium latum*. *J. Helminth.* **51**, 115–119.

Campbell, W. C. and Blair, L. S. (1974a). Prevention and cure of hepatic cysticercosis in mice. *J. Parasit.* **60**, 1049–1052.

Campbell, W. C. and Blair, L. S. (1974b). Treatment of the cystic stage of *Taenia crassiceps* and *Echinococcus multilocularis* in laboratory animals. *J. Parasit.* **60**, 1053–1054.

Campbell, W. C. and Butler, R. W. (1973). Efficacy of cambendazole against tapeworm and roundworm infections in lambs. *Aust. Vet. J.* **49**, 517–519.

Campbell, W. C., McCracken, R. D. and Blair, L. S. (1975). Effect of parenterally injected benzimidazole compounds on *Echinococcus multilocularis* and *Taenia crassiceps* metacestodes in laboratory animals. *J. Parasit.* **61**, 844–852.

Canzonieri, C. J., Rodriguez, R. R., Castillo, H. E., Balella, C. I. de and Lucena, M.

(1977). Ensayos terapéuticos con praziquantel en infecciones por *Taenia saginata* e *Hymenolepis nana*. *Bol. Chil. Parasit.* **32**, 41–42.

Carneri, I. de and Vita, G. (1973). Drugs used in cestode diseases. *In* "Chemotherapy of Helminthiasis", (R. Cavier and F. Hawking, eds), Vol. 1, pp. 145–213. Pergamon Press, Oxford.

Chatfield, R. C. and Yeary, R. A. (1979). The effects of bunamidine HCl on *Hymenolepis diminuta in vitro*. *Vet. Parasit.* **5**, 177–193.

Chavarria, A. P., Villarejos, V. M. and Zeledón, R. (1977). Mebendazole in the treatment of *Taeniasis solium* and *Taeniasis saginata*. *Am. J. Trop. Med. Hyg.* **26**, 118–120.

Chavasse, C. J., Brown, M. C. and Bell, D. R. (1979). *Schistosoma mansoni*: Activity responses *in vitro* to praziquantel. *Z. ParasitKde*, **58**, 169–174.

Chevis, R. A. F., Kelly, J. D. and Griffin, D. L. (1980). The lethal effect of closantel on the metacestodes of *Taenia pisiformis* in rabbits. *Vet. Parasit.* **7**, 333–337.

Ciordia, H., McCampbell, H. C. and Stuedema, J. A. (1978). Cestodicidal activity of albendazole in calves. *Am. J. Vet. Res.* **39**, 517–518.

Coles, G. C. (1975). Fluke biochemistry—*Fasciola* and *Schistosoma*. *Helminth. Abstr.* **44**, 147–162.

Coles, G. C. (1977a). The biochemical mode of action of some modern anthelmintics. *Pestic. Sci.* **8**, 536–543.

Coles, G. C. (1977b). The mechanism of action of some veterinary anthelmintics. *In* "Perspectives in the Control of Parasitic Disease in Animals in Europe", (D. W. Jolly and J. M. Somerville, eds), pp. 53–63. Association of Veterinarians in Industry, London.

Coles, G. C. (1979). The effect of praziquantel on *Schistosoma mansoni*. *J. Helminth.* **53**, 31–33.

Coles, G. C. and Briscoe, M. G. (1978). Benzimidazoles and fluke eggs. *Vet. Rec.* **103**, 360–361.

Coles, G. C. and McNeillie, R. M. (1977). The response of nematodes *in vivo* and *in vitro* to some anthelmintics. *J. Helminth.* **51**, 323–326.

Coles, G. C. and Simpkin, K. G. (1977a). Metabolic gradient in *Hymenolepis diminuta* under aerobic conditions. *Int. J. Parasit.* **7**, 127–128.

Coles, G. C. and Simpkin, K. G. (1977b). Resistance of nematode eggs to the ovicidal activity of benzimidazoles. *Res. Vet. Sci.* **22**, 386–387.

Colglazier, M. L., Kates, K. C. and Enzie, F. D. (1975). Cross resistance to other anthelmintics in an experimentally produced cambendazole-resistant strain of *Haemonchus contortus* in lambs. *J. Parasit.* **61**, 778–779.

Colombo, E. G., Zerega, R. E., Rambeaud, J. and Zingoni, H. (1968). Acción antiparasitaria de oxyclozanida (3,3′, 5,5′, 5-pentacloro-2,2′-dihydroxybenzanilida), Parte I. *Gac. Vet.* **30**, 3–6.

Čorba, J., Legeny, J., Štoffa, P., Andraško, H., Kilik, J., Štafura, J., Baláž, M. and Popovič, Š. (1980). Klinické hodnotenie efektívnosti oxfendazolu (Systamex) pri najdôkežitejších helmintózach oviec, hovädzieho dobytka a ošípaných. *Veterinářstvi*, **30**, 121–122.

Čorba, J., Lietava, P., Düwell, D. and Reisenleiter, R. (1979). Efficacy of fenbendazole against the most important trematodes and cestodes of ruminants. *Br. Vet. J.* **135**, 318–323.

Corbett, J. R. and Goose, J. (1971). A possible biochemical mode of action of the fasciolicides nitroxynil, hexachlorophene, and oxyclozanide. *Pestic. Sci.* **2**, 119–121.

Cornish, R. A., Behm, C. A., Butler, R. W. and Bryant, C. (1977). The *in vivo* effects of rafoxanide on the energy metabolism of *Fasciola hepatica. Int. J. Parasit.* **7**, 217–220.

Cornish, R. A. and Bryant, C. (1976). Changes in energy metabolism due to anthelmintics in *Fasciola hepatica* maintained *in vitro. Int. J. Parasit.* **6**, 393–398.

Craig, T. M. and Ronald, N. C. (1978). Preliminary studies on the effect of albendazole on cysticerci of *Taenia saginata. Southwest. Vet.* **31**, 121–124.

Craig, T. M. and Shepherd, E. (1980). Efficacy of albendazole and levamisole in sheep against *Thysanosoma actinioides* and *Haemonchus contortus* from the Edwards Plateau, Texas. *Am. J. Vet. Res.* **41**, 425–426.

Crewe, W. and Owen, R. R. (1979). Action of certain chlorine-based disinfectants on *Taenia* eggs. *Trans. R. Soc. Trop. Med. Hyg.* **73**, 324.

Cruthers, L. R., Linkenheimer, W. H. and Maplesden, D. C. (1979). Taeniacidal efficacy of SQ 21,704 in dogs by various types of oral administration and in comparison with niclosamide and bunamidine hydrochloride. *Am. J. Vet. Res.* **40**, 676–678.

Czipri, D. A., Nunns, V. J. and Shearer, G. C. (1968). Bunamidine hydroxynaphthoate-activity against *Moniezia expansa* in sheep. *Vet. Rec.* **82**, 505–506 and 521.

Davidse, L. C. (1975). Antimitotic activity of methyl benzimidazol-2-ylcarbamate in fungi and its binding to cellular protein. *In* "Microtubules and Microtubule Inhibitors", (M. Borgers and M. de Brabander, eds), pp. 483–495. North Holland Publishing Company, Amsterdam.

Davidse, L. C. and Flach, W. (1977). Differential binding of methyl benzimidazol-2-YL carbamate to fungal tubulin as a mechanism of resistance to this antimitotic agent in mutant strains of *Aspergillus nidulans. J. Cell Biol.* **72**, 174–193.

Davidse, L. C. and Flach, W. (1978). Interaction of thiabendazole with fungal tubulin. *Biochim. Biophys. Acta*, **543**, 82–90.

Della Bruna, C., Ricciardi, M. L. and Sanfilippo, A. (1973). Axenomycins, new cestocidal antibiotics. *Antimicrob. Agents Chemother.* **3**, 708–710.

Douch, P. G. C. (1979). The metabolism of the anthelmintics clioxanide and resorantel and related compounds *in vitro* by *Moniezia expansa*, *Ascaris suum* and mouse- and sheep-liver enzymes. *Xenobiotica*, **9**, 263–268.

Douch, P. G. C. and Gahagan, H. M. (1977). The metabolism of niclosamide and related compounds by *Moniezia expansa*, *Ascaris lumbricoides* var *suum*, and mouse- and sheep-liver enzymes. *Xenobiotica*, **7**, 301–307.

Düwel, D. (1970). Ein neues Zestizid: Terenol[R]—seine Wirkung gegen Bandwürmer von Laboratoriums—und Haustieren. *Dt. Tieraerztl. Wochenschr.* **77**, 97–101.

Düwel, D. (1978). Activity of fenbendazole on metacestodes of different tapeworms in small and domestic animals. *Curr. Chemother.*, 142–144.

Düwel, D. and Kirsch, R. (1971). *In vivo*-und *in vitro*-Untersuchungen mit Terenol[R], einem neuen Anthelminthikum. *Zentralb. Veterinaermed. Reihe B.* **18**, 465–472.

Düwel, D. and Schleich, H. (1978). *In vivo*-Untersuchungen zur Wirkungsweise von Fenbendazol. *Zentralb. Veterinaermed. Reihe B.* **25**, 800–805.

Düwel, D. and Tiefenbach, B. (1978). Die Behandlung des Bandwurmbefalls bei Schafen mit Pancur. *Z. alle Geb. Veterinamed.* **33**, 1–6.

Eckert, J. and Burkhardt, B. (1980). Chemotherapy of experimental echinococcosis. *Acta Trop.* **37**, 297–300.

Eckert, J., Annen, J. and Barandun, G. (1977). Untersuchungen zur Chemotherapie der Echinokokkose. *Tropenmed. Parasit.* **28**, 274–275.

Eckert, J., Barandun, G. and Pohlenz, J. (1978). Chemotherapie der larvalen Echinokokkose bei Labortieren. *Schweiz. Med. Wschr.* **108**, 1104–1112.

Edwards, S. R., Campbell, A. J., Sheers, M., Moore, R. J. and Montague, P. E. (1981). Studies of the effect of diamphenethide and oxyclozanide on the metabolism of *Fasciola hepatica. Mol. Biochem. Parasit.* **2**, 323–338.

Eslami, A., Ahrari, H. and Saadatzadeh, H. (1978). Scolicidal effects of Cetrimide (R) on hydatid cyst (*Echinococcus granulosus*). *Trans. R. Soc. Trop. Med. Hyg.* **72**, 307–308.

Espejo, H. (1977). Tratamiento de infecciones por *Hymenolepis nana, Taenia saginata, Taenia solium* y *Diphyllobothrium pacificum* con praziquantel (Embay 8440). *Bol. Chil. Parasit.* **32**, 39–40.

Evans, W. S. and Novak, M. (1976). The effect of mebendazole on the development of *Hymenolepis diminuta* in *Tribolium confusum. Can. J. Zool.* **54**, 1079–1083.

Evans, W. S., Gray, B. and Novak, M. (1979). Effect of mebendazole on the larval development of three hymenolepidid cestodes. *J. Parasit.* **65**, 31–34.

Evans, W. S., Hardy, M. and Novak, M. (1980). A comparison of the effect of albendazole, cambendazole, and thiabendazole on the larval development of three hymenolepidid cestodes. *J. Parasit.* **66**, 935–940.

Fetterer, R. H., Pax, R. A. and Bennett, J. L. (1980a). *Schistosoma mansoni*: Characterization of the electrical potential from the tegument of adult males. *Expl. Parasit.* **49**, 353–365.

Fetterer, R. H., VandeWaa, J. A. and Bennett, J. L. (1980b). Characterization and localization of ouabain receptors in *Schistosoma mansoni. Mol. Biochem. Parasit.* **1**, 209–219.

Fetterer, R. H., Pax, R. A., Thompson, D., Bricker, C. and Bennett, J. L. (1980c). Praziquantel: Mode of its antischistosomal action. *In* "The Hose Invader Interplay", (H. Van den Bossche, ed.), pp. 695–698. Elsevier/North Holland Biomedical Press, Amsterdam.

Foix, J. (1979). Efficacité du cambendazole contre les *Moniezia* et les strongyles gastro-intestinaux de l'agneau. *Rev. Méd. Vét.* **130**, 1511–1522.

Fräyha, G. J., Bikhazi, K. J. and Kachachi, T. A. (1981). Treatment of hydatid cysts (*Echinococcus granulosus*) by Cetrimide[R]. *Trans. R. Soc. Trop. Med. Hyg.* **75**, 447–450.

Friedman, P. A. and Platzer, E. G. (1978). Interaction of anthelmintic benzimidazoles and benzimidazole derivatives with bovine brain tubulin. *Biochim. Biophys. Acta*, **544**, 605–614.

Friedman, P. A. and Platzer, E. G. (1980a). Interaction of anthelmintic benzimidazoles with *Ascaris suum* embryonic tubulin. *Biochem. Biophys. Acta*, **630**, 271–278.

Friedman, P. A. and Platzer, E. G. (1960b). The molecular mechanism of benzimidazoles in embryos of *Ascaris suum*. *In* "The Host Invader Interplay", (H. Van den Bossche, ed.), pp. 595–604. Elsevier/North Holland Biomedical Press, Amsterdam.

Friedman, P. A., Platzer, E. G. and Carroll, E. J. Jr (1980). Tubulin characterization during embryogenesis of *Ascaris suum*. *Dev. Biol.* **76**, 47–57.

Gallie, G. J. and Sewell, M. M. H. (1977). The effect of mebendazole on cysticerci of *Taenia saginata* in calves. *Trop. Anim. Health Prod.* **9**, 24.

Gallie, G. J. and Sewell, M. M. H. (1978). The efficacy of praziquantel against cysticerci of *Taenia saginata* in calves. *Trop. Anim. Health Prod.* **10**, 36–38.

Gautam, D. P., Malhotra, D. V., Banerjee, D. P. and Ram, S. T. (1979). Efficacy of fenbendazole in experimental *Taenia hydatigena* infection in dogs. *Indian J. Parasit.* **3**, 207–208.

Gemmell, M. A. and Oudemans, G. (1974). Treatment of *Echinococcus granulosus* and *Taenia hydatigena* in dogs with bunamide hydroxynaphthoate in a prepared food. *Res. Vet. Sci.* **16**, 85–88.

Gemmell, M. A. and Oudemans, G. (1975a). Effect of fospirate on *Echinococcus granulosus* and *Taenia hydatigena* infections in dogs. *Res. Vet. Sci.* **19**, 216–217.

Gemmell, M. A. and Oudemans, G. (1975b). Effect of nitroscanate on *Echinococcus granulosus* and *Taenia hydatigena* infections in dogs. *Res. Vet. Sci.* **19**, 217–219.

Gemmell, M. A. and Shearer, G. C. (1968). Bunamidine hydrochloride—its efficiency against *Echinococcus granulosus*. *Vet. Rec.* **82**, 252–256.

Gemmell, M. A., Johnstone, P. D. and Oudemans, G. (1975a). Effect of mebendazole on *Echinococcus granulosus* and *Taenia hydatigena* infections in dogs. *Res. Vet. Sci.* **19**, 229–230.

Gemmell, M. A., Oudemans, G. and Sakamoto, T. (1975b). The effect of biothional sulphoxide on *Echinococcus granulosus* and *Taenia hydatigena* infections in dogs. *Res. Vet. Sci.* **18**, 109–110.

Gemmell, M. A., Johnstone, P. D. and Oudemans, G. (1977a). The effect of niclosamide on *Echinococcus granulosus*, *Taenia hydatigena* and *Taenia ovis* infections in dogs. *Res. Vet. Sci.* **22**, 389–391.

Gemmell, M. A., Johnstone, P. D. and Oudemans, G. (1977b). The effect of micronised nitroscanate on *Echinococcus granulosus* and *Taenia hydatigena* infections in dogs. *Res. Vet. Sci.* **22**, 391–392.

Gemmell, M. A., Johnstone, P. D. and Oudemans, G. (1977c). The lethal effect of some benzimidazoles on *Taenia hydatigena* in dogs. *Res. Vet. Sci.* **23**, 115–116.

Gemmell, M. A., Johnstone, P. D. and Oudemans, G. (1977d). The effect of praziquantel on *Echinococcus granulosus*, *Taenia hydatigena* and *Taenia ovis* infections in dogs. *Res. Vet. Sci.* **23**, 121–123.

Gemmell, M. A., Johnstone, P. D. and Oudemans, G. (1978a). The effect of mebendazole in food on *Echinococcus granulosus* and *Taenia hydatigena* infections in dogs. *Res. Vet. Sci.* **25**, 107–108.

Gemmell, M. A., Johnstone, P. D. and Oudemans, G. (1978b). The effect of an antibiotic of the streptothricin family against *Echinococcus granulosus* and *Taenia hydatigena* infections in dogs. *Res. Vet. Sci.* **25**, 109–110.

Gemmell, M. A., Johnstone, P. D. and Oudemans, G. (1978c). The effect of diuredosan on *Echinococcus granulosus* and *Taenia hydatigena* infections in dogs. *Res. Vet. Sci.* **25**, 111–112.

Gemmell, M. A., Johnstone, P. D. and Oudemans, G. (1979a). The effect of oxfendazole on *Echinococcus granulosus* and *Taenia hydatigena* infections in dogs. *Res. Vet. Sci.* **26**, 389–390.

Gemmell, M. A., Johnstone, P. D. and Oudemans, G. (1979b). The effect of nitroscanate tablets on *Echinococcus granulosus* and *Taenia hydatigena* infections in dogs. *Res. Vet. Sci.* **27**, 255–257.

Gemmell, M. A., Johnstone, P. D. and Oudemans, G. (1980). The effect of route of administration on the efficacy of praziquantel against *Echinococcus granulosus* infections in dogs. *Res. Vet. Sci.* **29**, 131–132.

Gemmell, M. A., Johnstone, P. D. and Oudemans, G. (1981). Effect of mebendazole against *Echinococcus granulosus* and *Taenia hydatigena* cysts in naturally infected sheep and relevance to larval tapeworm infections in man. *Z. ParasitKde*, **64**, 135–147.

Gibbs, H. C. and Gupta, R. P. (1972). The anthelmintic activity of cambendazole in calves and lambs. *Can. J. Comp. Med.* **36**, 108–115.

Gleason, N. N., Kornblum, R. and Walzer, P. (1973). *Mesocestoides* (Cestoda) in a child in New Jersey treated with niclosamide (Yomesan^R). *Am. J. Trop. Med. Hyg.* **22**, 757–760.

Gönnert, R. (1974). Die Bandwurm-Infektionen des Menschen und ihre Behandlung. *Münch. Med. Wochenschr.* **116**, 1531–1538.

Gräber, M. (1969). A propos du pouvoir anthelminthique de N-(2′-chloro-4′-nitrophenyl)-5 chlorosalicylamide chez le mouton. *Rev. Elev. Med. Vet. Pays Trop.* **22**, 217–228.

Groll, E. (1977). Panorama general del tratamiento de las infecciones humanas por cestodes con praziquantel (Embay 8440). *Bol. Chil. Parasit.* **32**, 27–31.

Groll, E. (1980). Praziquantel for cestode infections in man. *Acta Trop.* **37**, 293–296.

Gupta, S., Katiyar, J. C., Sen, A. B., Dubey, S. K., Singh, H., Sharma, S. and Iyer, R. N. (1980). Anticestode activity of 3,5-dibromo-2′-chlorosalicylanilide-4′-isothiocyanate—a preliminary report. *J. Helminth.* **54**, 271–273.

Güralp, N. and Oğuz, T. (1971). Cihanbeyli ilçesinde kuzularda görülen *Moniezia* karsi degiŝik antelmentiklerle yapilan sağitma deneyleri ve alican sonuclar. *Vet. Fak. Derg. Ankara Univ.* **18**, 65–74.

Hall, C. A. (1966). Mansonil^R a new cestocide for sheep. *Vet. Med. Rev., Leverkusen*, **1**, 59–66.

Hansen, M. F., Kelley, G. W. and Todd, A. C. (1950). Observations on the effects of a pure infection of *Moniezia expansa* on lambs. *Trans. Am. Microsc. Soc.* **69**, 148–155.

Harris, R. E., Revfeim, K. J. A. and Heath, D. D. (1980). Simulating strategies for control of *Echinococcus granulosus, Taenia hydatigena* and *Taenia ovis*. *J. Hyg.* **84**, 389–404.

Harrow, W. T. (1969). *Stilesia* infestation in sheep. *Vet. Rec.* **34**, 564.

Hart, R. J., Turner, R. and Wilson, R. G. (1977). A biochemical and ultrastructural study of the mode of action of bunamidine against *Hymenolepis nana. Int. J. Parasit.* **7**, 129–134.

Hatton, C. J. (1965). A new taeniacide, bunamidine hydrochloride: its efficiency against *Taenia pisiformis* and *Diphylidium caninum* in the dog and *Hydatigena taeniaeformis* in the cat. *Vet. Rec.* **77**, 408–411.

Havercroft, J. C., Quinlan, R. A. and Gull, K. (1981). Binding of parbendazole to tubulin and its influence on microtubules in tissue-culture cells as revealed by immunofluorescence microscopy. *J. Cell. Sci.* **49**, 195–204.

Heath, D. D. and Chevis, R. A. F. (1974). Mebendazole and hydatid cysts. *Lancet*, **2**, 218–219.

Heath, D. D. and Lawrence, S. B. (1978). The effect of mebendazole and praziquantel on the cysts of *Echinococcus granulosus*, *Taenia hydatigena* and *T. ovis* in sheep. *N.Z. Vet. J.* **26**, 11–15.

Heath, D. D. and Lawrence, S. B. (1979). A single oral treatment with mebendazole for the control of *Taenia crassiceps* larval infections in rats. *Int. J. Parasit.* **9**, 73–76.

Heath, D. D., Christie, M. J. and Chevis, R. A. F. (1975). The lethal effect of mebendazole on secondary *Echinococcus granulosus*, cysticerci of *Taenia pisiformis* and tetrathyridia of *Mesocestoides corti*. *Parasitology*, **70**, 273–285.

Hinz, E. (1978). Zur Chemotherapie der experimentellen larvalen Echinococcose mit Fenbendazol. 1. Der Einfluß von Fenbendazol auf Befallsstärke und Protoscolex-Bildung bei *Echinococcus multilocularis*. *Zentralbl. Bakteriol. Parasitenkd. Infectionskr. Hyg. Abt.* **240**, 542–548.

Hinz, E. (1980). Zum heutigen Stand der Chemotherapie der Echinokokkose. *Ther. Ggw.* **119**, 867–883.

Hoebeke, J., Van Nijen, G. and Brabander, M. de (1976). Interaction of oncodazole (R 17934), a new antitumoral drug, with rat brain tubulin. *Biochem. Biophys. Res. Comm.* **69**, 319–324.

Hopkins, C. A., Grant, P. M. and Stallard, H. (1973). The effect of oxyclozanide on *Hymenolepis microstoma* and *H. diminuta*. *Parasitology*, **66**, 355–365.

Horak, I. G., Snijders, A. J. and Pienaar, I. (1972). The efficacy of cambendazole against cestode and nematode infestations in sheep and cattle. *J. S. Afr. Vet. Assoc.* **43**, 101–106.

Hörchner, F., Langnes, A. and Oğuz, T. (1976). Die Wirkung von Mebendazole und Praziquantel auf larvale Taenienstadien bei Maus, Kaninchen und Schwein. *Tropenmed. Parasit.* **28**, 44–50.

Hughes, H. C. Jr, Barthel, C. H. and Lang, C. M. (1974). Niclosamide as a treatment for *Hymenolepis nana* and *Hymenolepis diminuta* in rats. *Lab. Anim. Sci.* **23**, 72–73.

Ireland, C. M., Gull, K., Gutteridge, W. E. and Pogson, C. I. (1979). The interaction of benzimidazole carbamates with mammalian microtubule protein. *Biochem. Pharm.* **28**, 2680–2682.

Jelenová, I. (1981). Effects of fertilizers on eggs of *Taenia saginata* Goeze, 1782 under field conditions. *Folia Parasitol.* **28**, 285–287.

Jones, W. E. (1979). Niclosamide as a treatment for *Hymenolepis diminuta* and *Dipylidium caninum* infection in man. *Am. J. Trop. Med. Hyg.* **28**, 300–302.

Kammerer, W. S. (1978). Human pulmonary hydatid disease. *Echinococcus granulosus* treated with mebendazole. *Clin. Res.* **26**, 627A.

Kammerer, W. S. and Judge, D. M. (1976). Chemotherapy of hydatid disease

(*Echinococcus granulosus*) in mice with mebendazole and biothional. *Am. J. Trop. Med. Hyg.* **25**, 714–717.

Kammerer, W. S. and Miller, K. L. (1981). *Echinococcus granulosus*: permeability of hydatid cysts to mebendazole in mice. *Int. J. Parasit.* **11**, 183–185.

Kammerer, W. S. and Pérez-Esandi, M. V. (1975). Chemotherapy of experimental *Echinococcus granulosus* infection—trials in CF1 mice and birds (*Meriones unguiculatus*). *Am. J. Trop. Med. Hyg.* **24**, 90–95.

Kane, H. J., Behm, C. A. and Bryant, C. (1980). Metabolic studies on the new fasciolocidal drug, closantel. *Mol. Biochem. Parasitol.* **1**, 347–355.

Kates, K. C. and Goldberg, A. (1951). The pathogenicity of the common sheep tapeworm, *Moniezia expansa. Proc. Helminth. Soc., Wash.* **18**, 87–101.

Katiyar, J. C., Sen, A. B. and Bhattacharya, B. K. (1967). Cesticidal action of diphenyl sulphone 4:4′ diisothiocyanate. *Nature, Lond.* **214**, 708–709.

Kayser, H. J. S. (1980). Treatment of hydatid disease with mebendazole at Frere Hospital, East London: a preliminary report. *S. Afr. Med. J.* **58**, 560–563.

Keeling, J. E. D. (1968). The chemotherapy of cestode infections. *Adv. Chemother.* **3**, 109–152.

Kelly, J. D. and Bain, S. A. (1975). Critical test evaluation of micronized mebendazole against *Anoplocephala perfoliata* in the horse. *N.Z. Vet. J.* **23**, 229–232.

Kennedy, T. J. and Todd, A. C. (1975). Efficacy of fenbendazole against gastrointestinal parasites of sheep. *Am. J. Vet. Res.* **36**, 1465–1467.

Kern, P., Dietrich, M. and Volkmer, K. J. (1979). Chemotherapy of *Echinococcus* with mebendazole—clinical observations of 7 patients. *Tropenmed. Parasit.* **30**, 65–72.

Khalil, H. M. (1969). Treatment of cestode infections with Radeverm. *Trans. R. Soc. Trop. Med. Hyg.* **63**, 76–78.

Kirsch, R. (1975). Zur Wirksamkeit von Fenbendazol gegen adulte und immature Stadien von *Nematospiroides dubius* und *Heterakis spumosa. Zentralbl. Veterinaermed. Reihe B.* **22**, 441–447.

Köhler, P. and Bachmann, R. (1978). The effects of the antiparasitic drugs levamisole, thiabendazole, praziquantel, and chloroquine on mitochondrial electron transport in muscle tissue from *Ascaris suum. Mol. Pharm.* **14**, 155–163.

Köhler, P. and Bachmann, R. (1980). The possible mode of action of mebendazole in *Ascaris suum. In* "The Host Invader Inerplay", (H. Van den Bossche, ed.), pp. 727–730. Elsevier/North Holland Biomedical Press, Amsterdam.

Kruckenberg, S. M., Meyer, A. D. and Eastman, W. R. (1981). Preliminary studies on the effect of praziquantel against tapeworms. *Vet. Med. Small Anim. Clin.* **76**, 689–693.

Laclette, J. P., Guerra, G. and Zetina, C. (1980). Inhibition of tubulin polymerization by mebendazole. *Biochem. Biophys. Res. Commun.* **92**, 417–423.

Larkworthy, W. and Atiyeh, M. (1980). Mebendazole and hydatid disease. *Br. Med. J.* **280**, 1378.

Laws, G. F. (1967). Chemical ovicidal measures as applied to *Taenia hydatigena*, *Taenia ovis*, *Taenia pisiformis*, and *Echinococcus granulosus. Expl. Parasit.* **20**, 27–37.

Led, J. E., Yannarella, F. G., Manazza, J. A. and Denegri, G. M. (1979). Acción del albendazole sobre *Moniezia expansa* y *Thysanosoma actinoides* en lanares. *Gac. Vet.* **41**, 363–366.

Leopold, G., Ungethüm, W., Groll, E., Diekmann, H. W., Nowak, H. and Wegner, D. H. G. (1978). Clinical pharmacology in normal volunteers of praziquantel, a new drug against schistosomes and cestodes—example of a complex study covering both tolerance and pharmacokinetics. *Eur. J. Clin. Pharmacol.* **14**, 281–291.

Lloyd, S., Soulsby, E. J. L. and Theodorides, V. J. (1978). Effect of albendazole on the metacestodes of *Taenia saginata* in calves. *Experientia*, **34**, 723–724.

Loughran, C. F. and McCarey, A. G. (1980). Coincident pelvic and pulmonary hydatid disease in a young girl—the chest radiograph following treatment with mebendazole. *Brit. J. Radiol.* **53**, 1020–1021.

Lukacs, J., Tanaka, R. D., Kim, R. A. and MacInnis, A. J. (1980). Development of a cell-free protein-synthesizing system from *Schistosoma mansoni* and a comparison of the effects of hycanthone and praziquantel on this system. *J. Parasit.* **66**, 424–427.

MacNair, A. L. (1980). Mebendazole and hydatid disease. *Br. Med. J.* **280**, 1055.

Malan, F. S. (1980). Anthelmintic efficacy of fenbendazole against cestodes in sheep and cattle. *J. S. Afr. Vet. Assoc.* **51**, 25–26.

Malkin, M. F. and Camacho, R. M. (1972). The effect of thiabendazole on fumarate reductase from thiabendazole-sensitive and resistant *Haemonchus contortus*. *J. Parasit.* **58**, 845–846.

McBeath, D. G., Best, J. M. J. and Preston, N. K. (1977). Efficacy of fenbendazole against naturally acquired *M. expansa* infections in lambs. *Vet. Rec.* **101**, 408–409.

Mehlhorn, H., Becker, B., Andrews, P., Thomas, H. and Frenkel, J. K. (1981). *In vivo* and *in vitro* experiments on the effects of praziquantel on *Schistosoma mansoni*. *Arzneim-Forsch.* **31**, 544–545.

Metzger, H. and Düwel, D. (1973). Investigations of metabolism in the liver fluke (*Fasciola hepatica*) as an aid to the development of new anthelmintics. *Int. J. Biochem.* **4**, 133–143.

Middleton, K. R., Schaefer, F. W. III, and Saz, H. J. (1979). Chemotherapeutic effects of 4-isothiocyanato-4′-nitrodiphenylamine (C9333-Go/CGP 4540) on infections with *Nematospiroides dubius*, *Hymenolepis diminuta*, *Hymenolepis nana* and *Spirometra mansonoides*. *Experientia*, **35**, 243–244.

Miert, A. S. J. P. A. M. van and Groeneveld, H. W. (1969). Anthelmintics used for the treatment of fascioliasis as uncouplers of oxidative phosphorylation in warm blooded animals. *Eur. J. Pharm.* **8**, 385–388.

Miskovitz, P. F. and Javitt, N. B. (1980). Leukopenia associated with mebendazole therapy of hydatid disease. *Am. J. Trop. Med. Hyg.* **29**, 1356–1358.

Most, H., Yeoli, M., Hammond, J. and Scheinesson, G. P. (1971). Yomesan (niclosamide) therapy of *Hymenolepis nana* infections. *Am. J. Trop. Med. Hyg.* **20**, 206–208.

Mulhall, P. P. (1980). Treatment of a ruptured hydatid cyst of lung with mebendazole. *Br. J. Dis. Chest*, **74**, 306–308.

Muñst, G. J., Karlaganis, G. and Bircher, J. (1980). Plasma concentrations of mebendazole during treatment of echinococcosis. Preliminary results. *Eur. J. Clin. Pharm.* **17**, 375–378.

Murray-Lyon, I. M. and Reynolds, K. W. (1979). Complication of mebendazole treatment for hydatid disease. *Br. Med. J.* **2**, 1111–1112.

Nechay, B. R., Hillman, G. R. and Dotson, M. J. (1980). Properties and drug sensitivity of adenosine triphosphatases from *Schistosoma mansoni*. *J. Parasit.* **66**, 596–600.

Nolla Panades, R., Nolla Salas, M., Álvarez Lerma, F. and Torres Rodríguez, J. M. (1980). Treatment of microvesicular hydatidosis with mebendazole. *Rev. Clin. Esp.* **156**, 295–303.

Novak, M. (1977). Efficacy of a new cestocide, praziquantel, against larval *Mesocestoides corti* and *Taenia crassiceps* in mice. *J. Parasit.* **63**, 949–950.

Novak, M. and Evans, W. S. (1981). Mebendazole and the development of *Hymenolepis nana* in mice. *Int. J. Parasit.* **11**, 227–280.

Oğuz, T. (1976). The treatment of experimental cysticercosis (*Cysticercus tenuicollis*). *Ankara Univ. Vet. Fak. Derg.* **23**, 385–395.

Osborne, D. R. (1980a). Mebendazole and hydatid disease. *Br. Med. J.* **280**, 183.

Osborne, D. R. (1980b). Mebendazole and hydatid disease, Reply. *Br. Med. J.* **280**, 796.

Parnell, I. W. (1965). Some observations on taeniid ovicides: screening techniques, and the effects of some inorganic compounds. *J. Heminth.* **39**, 257–272.

Pawlowski, Z., Kozakiewicz, B. and Zatoriski, J. (1976a). The effect of mebendazole on hydatid cysts in pigs. *Vet. Parasit.* **2**, 299–302.

Pawlowski, Z., Kozakiewicz, B. and Wroblewski, H. (1976b). The effect of intraperitoneal inoculation of mebendazole on *Taenia saginata* cysticercosis in calves. *Vet. Parasit.* **2**, 303–306.

Pawlowski, Z., Kozakiewicz, B. and Wroblewski, H. (1978). The efficacy of mebendazole and praziquantel against *Taenia saginata* cysticercosis in cattle. *Vet. Sci. Commun.* **2**, 137–139.

Pax, R. A., Bennett, J. L. and Fetterer, R. H. (1978). A benzodiazepine derivative and praziquantel: Effects on musculature of *Schistosoma mansoni* and *Schistosoma japonicum*. *Naunyn-Schmiedeberg's Arch. Pharm.* **304**, 309–315.

Pax, R., Fetterer, R. and Bennett, J. L. (1979). Effects of fluoxetine and imipramine on male *Schistosoma mansoni*. *Comp. Biochem. Physiol.* **64C**, 123–127.

Paz, G. (1977). Tratamiento de teniasis saginata con praziquantel (Embay 8440). *Bol. Chil. Parasit.* **32**, 14–16.

Perera, D. R., Western, K. A. and Schultz, M. G. (1970). Niclosamide treatment of cestodiasis. *Am. J. Trop. Med. Hyg.* **19**, 610–612.

Pérez-Esandi, M. V., Colli, C. W. and Schantz, P. M. (1975). The ovicidal effect of selected chemicals against eggs of *Echinococcus granulosus*. *Bull. W.H.O.* **51**, 550–551.

Poole, J. B., Dooley, K. L. and Rollins, L. D. (1971). Efficacy of niclosamide for the removal of tapeworms (*Dipylidium caninum* and *Taenia pisiformis*) from dogs. *J. Am. Vet. Med. Assoc.* **159**, 78–80.

Prichard, R. K. (1970). Mode of action of the anthelminthic thiabendazole in *Haemonchus contortus*. *Nature (London)*, **228**, 684–685.

Prichard, R. K. (1973). The fumarate reductase reaction of *Haemonchus contortus* and the mode of action of some anthelmintics. *Int. J. Parasit.* **3**, 409–417.

Prichard, R. K. (1978). The metabolic profile of adult *Fasciola hepatica* obtained from rafoxanide-treated sheep. *Parasitology*, **76**, 277–288.

Putter, J. (1970). Zur biologisch-chemischen Wirkungsweise des Bundwurmmittels N; (2'-clor-4'-nitrophenyl)-5-chlor-salicylamid. 1. Mitteilung: Beeinflussung von Enzymsystemem. *Arzneim-Forsch.* **20**, 203–205.

Quilici, M., Dumon, H., Rampal, M. and Alimi, J. C. (1979). Hydatidose: traitement préopératoire par fluoromebendazole. *Nouv. Presse Med.* **8**, 524.

Quinlan, R. A., Roobol, A., Pogson, C. I. and Gull, K. (1981). A correlation between *in vivo* and *in vitro* effects of the microtubule inhibitors colchicine, parabendazole and nocodazole on myxamoebae of *Physarum polycephalum*. *J. Gen. Microbiol.* **122**, 1–6.

Rahman, M. S. and Bryant, C. (1977). Studies of regulatory metabolism in *Moniezia expansa*: Effects of cambendazole and mebendazole. *Int. J. Parasit.* **7**, 403–409.

Rahman, M. S., Cornish, R. A., Chevis, R. A. F. and Bryant, C. (1977). Metabolic changes in some helminths from sheep treated with mebendazole. *N.Z. Vet. J.* **25**, 79–83.

Reisin, I. L., Rabito, C. A., Rotunno, C. A. and Cereijido, M. (1977). The permeability of the membranes of experimental secondary cysts of *Echinococcus granulosus* to [^{14}C] mebendazole. *Int. J. Parasit.* **7**, 189–194.

Reuss, U. (1979). Behandlungsversuche des Magendarm und Bandwurmbefalls der Schafe mit dem Breitbandanthelminthikum "Synanthic" unter tierärzthichen Praxisbedingungen. *Tieraerztl. Umsch.* **34**, 836–842.

Rew, R. S. (1978). Mode of action of common anthelmintics. *J. Vet. Pharmacol. Ther.* **1**, 183–198.

Richards, R. J. and Somerville, J. M. (1980). Field trials with nitroscanate against cestodes and nematodes in dogs. *Vet. Rec.* **106**, 332–335.

Rim, H. J., Park, C. Y., Lee, J. S., Joo, K. H. and Lyn, K. S. (1978). Therapeutic effects of praziquantel (Embay 8440) against *Hymenolepis nana* infection. *Kor. J. Parasit.* **16**, 82–87.

Roberson, E. L. (1976). Comparative effects of uredofos, niclosamide, and bunamidine hydrochloride against tapeworm infections in dogs. *Am. J. Vet. Res.* **37**, 1483–1484.

Roberson, E. L. (1977). Anticestodal and antitrematodal drugs. *In* "Veterinary Pharmacology and Therapeutics", (L. M. Jones, ed.), pp. 1052–1078. Iowa State University Press, Iowa.

Roberson, E. L. and Ager, A. L. (1976). Uredofos: Anthelmintic activity against nematodes and cestodes in dogs with naturally occurring infections. *Am. J. Vet. Res.* **37**, 1479–1482.

Roberson, E. L. and Burke, T. M. (1980). Evaluation of granulated fenbendazole (22.2%) against induced and naturally occurring helminth infections in cats. *Am. J. Vet. Res.* **41**, 1499–1502.

Romanowski, R. D., Rhoads, M. L., Colglazier, M. L. and Kates, R. C. (1975). Effect of cambendazole, thiabendazole and levamisole on fumarate reductase in cambendazole-resistant and -sensitive strains of *Haemonchus contortus*. *J. Parasit.* **61**, 777–778.

Rommel, M., Grelck, H. and Hörchner, F. (1976). Zur Wirksamkeit von Praziquantel gegen Bandwürmer in experimentell infizierten Hunden und Katzen. *Berl. Münch. Tierärztl. Wochenschr.* **89**, 255–257.

Ronald, N. C. and Wagner, J. E. (1975). Treatment of *Hymenolepis nana* in hamsters with Yomesan[R] (niclosamide). *Lab. Anim. Sci.* **25**, 219–220.

Round, M. C., Simpson, D. J., Haselden, C. S., Glendinning, E. S. A. and Baskerville, R. E. (1974). Horse strongyles' tolerance to anthelmintics. *Vet. Rec.* **95**, 517–518.

Sakamoto, T. (1973). Studies on echinococcosis. XXV. Anthelmintic action of drugs on larval *Echinococcus multilocularis in vitro. Jap. J. Vet. Res.* **21**, 73–91.

Sakamoto, T. (1979). Relationships between anthelmintic effects of drugs against *Echinococcus multilocularis in vitro* and *in vivo. Mem. Fac. Agric. Kagoshima Univ.* **15**, 115–123.

Sakamoto, T. and Gemmell, M. A. (1975). Studies on effects of drugs upon protoscoleces of *Echinococcus granulosus in vitro.* I. Scolicidal effect of salicylanilide and bisphenol derivatives against *Echinococcus granulosus in vitro. Jap. J. Vet. Res.* **23**, 81–94.

Saz, H. J. (1972). Effects of anthelmintics on P^{32} esterification in helminth metabolism. *In* "The Comparative Biochemistry of Parasites", (H. Van den Bossche, ed.), pp. 445–454. Academic Press, New York and London.

Saz, H. J. and Lescure, O. L. (1968). Effects of anticestodal agents on mitochondria from the nematode, *Ascaris lumbricoides. Mol. Pharmacol.* **4**, 407–410.

Saz, H. J., Berta, J. and Kowalski, J. (1972). Transhydrogenase and anaerobic phosphorylation in *Hymenolepis diminuta* mitochondria. *Comp. Biochem. Physiol.* **43B**, 725–732.

Schacht, M., Düwel, D. and Kirsch, R. (1971). Über die Beeinflussung des Stoffwechsels von *Hymenolepis diminuta* durch Resorantel. *Z. ParasitKde*, **37**, 278–287.

Schantz, P. M. and Prezioso, U. (1976). Efficacy of divided doses of fospirate against immature *Echinococcus granulosus* infections in dogs. *Am. J. Vet. Res.* **37**, 619–620.

Schantz, P. M., Prezioso, U. and Marchevsky, N. (1976). Efficacy of divided doses of GS-23654 against immature *Echinococcus granulosus* infections in dogs. *Am. J. Vet. Res.* **37**, 621–622.

Schantz, P. M., van den Bossche, H. and Eckert, J. (1982). Chemotherapy for larval echinococcosis in animals and humans: report of a workshop. *Z. ParasitKde*, **67**, 5–26.

Scheibel, L. W., Saz, H. J. and Bueding, E. (1968). The anaerobic incorporation of ^{32}P into adenosine triphosphate by *Hymenolepis diminuta. J. Biol. Chem.* **243**, 2229–2235.

Schenone, H. (1980). Praziquantel in the treatment of *Hymenolepis nana* infection in children. *Am. J. Trop. Med. Hyg.* **29**, 320–321.

Schenone, H., Galdames, M., Rivadeneira, A., Morales, E., Hoffman, M. T., Asalgado, N., Meneses, G., Mora, M. V. and Cabrera, G. (1977). Tratamiento de las infecciones por *Hymenolepis nana* en niños con una dosis oral única de praziquantel (Embay 8440). *Bol. Chil. Parasit.* **32**, 11–13.

Sen, A. B. and Hawking, F. (1960). Screening of cesticidal compounds on a tapeworm *Hymenolepis nana in vitro. Brit. J. Pharmacol.* **15**, 436–439.

Sen, H. G. and Deb, B. N. (1981). Anthelmintic efficacy of amoscanate (C 9333-

Go/CGP 4540) against various infections in rodents, dogs and monkeys. *Am. J. Trop. Med. Hyg.* **30**, 992–998.

Seth, D. and Lovekar, C. D. (1972). Cesticidal effect of emetine on *H. nana* infection in mice. *Indian J. Med. Res.* **60**, 1251–1253.

Sharma, S., Dubey, S. K. and Iyer, R. N. (1980). Cestode infestation chemotherapy. *Prog. Drug Res.* **24**, 217–266.

Shearer, G. C. and Gemmell, M. A. (1969). The efficiency of bunamidine hydroxynaphthoate against *Echinococcus granulosus* in dogs. *Res. Vet. Sci.* **10**, 296–299.

Sheir-Neiss, G., Lai, M. H. and Morris, N. R. (1978). Identification of a gene for β-tubulin in *Aspergillus nidulans*. *Cell*, **15**, 639–647.

Singh, H., Sharma, S., Singh, A. K., Iyer, R. N., Govil, P., Katiyar, J. C. and Sen, A. B. (1976). 5-Chloro-3′-nitro-4′-cyclohexylaminosalicylanilide: a new cestodicidal agent. *Indian J. Exp. Biol.* **14**, 332–333.

Slocombe, J. O. (1979). The prevalence and treatment of tapeworms in horses. *Can. Vet. J.* **20**, 136–140.

Stampa, S. (1967). A contribution towards the influence of tapeworms on liveweights of lambs. *Vet. Med. Rev., Leverkusen*, **1**, 81–85.

Standen, O. D. (1963). Chemotherapy of helminth infections. *In* "Experimental Chemotherapy", (R. J. Schnitzer and F. Hawking, eds), Vol. 1, pp. 701–892. Academic Press, New York and London.

Stevenson, P., Holmes, P. W. and Muturi, J. M. (1981). Effect of albendazole on *Taenia saginata* cysticerci in naturally infected cattle. *Vet. Rec.* **109**, 82.

Szanto, J., Lillis, W. G., Brown, W. E., Sutphin, C. F. and Maplesden, D. C. (1979). Critical evaluation of taeniacidal antibiotic S15-1 (SQ 21,704) for removal of natural tapeworm infections in dogs and cats. *Am. J. Vet. Res.* **40**, 673–675.

Taffs, L. F. (1975). Continuous feed medication with thiabendazole for the removal of *Hymenolepis nana*, *Syphacia obvelata* and *Aspiculuris tetraptera* in naturally infected mice. *J. Helminth.* **49**, 173–177.

Taffs, L. F. (1976). Further studies on the efficacy of thiabendazole given in the diet of mice infected with *H. nana*, *S. obvelata* and *A. tetraptera*. *Vet. Rec.* **99**, 143–144.

Tanowitz, H. B. and Wittner, M. (1973). Paromomycin in the treatment of *Diphyllobothrium latum* infection. *J. Trop. Med. Hyg.* **76**, 151–152.

Telléz-Girón, E., Ramos, M. C. and Montante, M. (1981). Effect of flubendazole on *Cysticercus cellulosae* in pigs. *Am. J. Trop. Med. Hyg.* **30**, 135–138.

Tetzlaff, R. D. and Weir, W. D. (1978). Anthelmintic control of concurrent *Hymenolepis nana* and *Syphacia obvelata* infections in the mouse with uredofos. *Lab. Anim. Sci.* **28**, 287–289.

Thakur, A. S., Prezioso, U. and Marchevsky, N. (1978). The efficacy of droncit against *Echinococcus granulosus* infection in dogs. *Am. J. Vet. Res.* **39**, 859–860.

Thakur, A. S., Prezioso, U. and Marchevsky, N. (1979). *Echinococcus granulosus*: ovicidal activity of praziquantel and bunamidine hydrochloride. *Expl. Parasit.* **47**, 131–133.

Theodorides, V. J. (1976). Anthelmintics: from laboratory animals to the target species. *In* "Chemotherapy of Infectious Disease", (H. H. Gadebusch, ed.), pp. 71–96. CRC Press, Cleveland.

Theodorides, V. J., Nawalinski, T. and Chang, J. (1976). Efficacy of albendazole against *Haemonchus*, *Nematodirus*, *Dictyocaulus*, and *Moniezia* of sheep. *Am. J. Vet. Res.* **37**, 1515–1516.

Thienpont, D., Vanparijs, O. and Hermans, L. (1974). Anthelmintic activity of mebendazole against *Cysticercus fasciolaris*. *J. Parasit.* **60**, 1052–1053.

Thienpont, D., Vanparijs, O., Niemegeers, C. and Marsboom, R. (1978). Biological and pharmacological properties of flubendazole. *Arzneim-Forsch.* **28**, 605–612.

Thomas, H. and Andrews, P. (1977). Praziquantel—a new cestocide. *Pestic. Sci.* **8**, 556–560.

Thomas, H. and Gönnert, R. (1977). The efficacy of praziquantel against cestodes in animals. *Z. ParasitKde* **52**, 117–127.

Thomas, H. and Gönnert, R. (1978a). Zur Wirksamkeit von Praziquantel bei der experimentellen Cysticercose und Hydatidose. *Z. ParasitKde*, **55**, 165–179.

Thomas, H. and Gönnert, R. (1978b). The efficacy of praziquantel against cestodes in cats, dogs and sheep. *Res. Vet. Sci.* **24**, 20–25.

Todd, K. S. (1978). Albendazole in the treatment of *Mesocestoides corti* (Cestoda) infections in dogs. *Vet. Med. Small Anim. Clin.* **73**, 453–454.

Todd, K. S., Howland, T. P. and Woerpel, R. W. (1978). Activity of uredofos, niclosamide, bunamidine hydrochloride and arecoline hydrobromide against *Mesocestoides corti* in experimentally infected dogs. *Am. J. Vet. Res.* **39**, 315–316.

Townsend, R. B., Kelly, J. D., James, R. and Weston, I. (1977). The anthelmintic efficacy of fenbendazole in the control of *Moniezia expansa* and *Trichuris ovis* in sheep. *Res. Vet. Sci.* **23**, 385–386.

Trejos, A., Szyfres, B. and Marchevsky, N. (1975). Comparative value of arecoline hydrobromide and bunamidine hydrochloride for treatment of *Echinococcus granulosus* in dogs. *Res. Vet. Sci.* **19**, 212–213.

Vakil, B. J., Dalal, N. J. and Enjetti, E. (1975). Clinical trials with mebendazole a new broad spectrum anthelmintic. *J. Trop. Med. Hyg.* **78**, 154–158.

Van den Bossche, H. (1972a). Biochemical effects of the anthelmintic drug mebendazole. *In* "The Comparative Biochemistry of Parasites", (H. Van den Bossche, ed.), pp. 139–157. Academic Press, New York and London.

Van den Bossche, H. (1972b). Studies on the phosphorylation in *Ascaris* mitochondria. *In* "The Comparative Biochemistry of Parasites", (H. Van den Bossche, ed.), pp. 455–468. Academic Press, New York and London.

Van den Bossche, H. (1976). The molecular basis of anthelmintic action. *In* "Biochemistry of Parasites and Host-Parasite Relationships", (H. Van den Bossche, ed.), pp. 553–572. Elsevier/North Holland Biomedical Press, Amsterdam.

Van den Bossche, H. (1978). Chemotherapy of parasitic infections. *Nature (London)*, **273**, 626–630.

Van den Bossche, H. (1980a). Chemotherapy of hymenolepiasis. *In* "Biology of the Tapeworm *Hymenolepis diminuta*", (H. P. Arai, ed.), pp. 639–693. Academic Press, New York and London.

Van den Bossche, H. (1980b). Peculiar targets in anthelmintic chemotherapy. *Biochem. Pharm.* **29**, 1981–1990.

Van den Bossche, H. (1981). A look at the mode of action of some old and new antifilarial compounds. *Belgian Ann. Trop. Med.* **61**, 287–296.

Van den Bossche, H. and De Nollin, S. (1973). Effects of mebendazole on the absorption of low molecular weight nutrients by *Ascaris suum*. *Int. J. Parasit.* **3**, 401–407.

Van den Bossche, H., Verhoeven, H., Vanparijs, O., Lauwers, H. and Thienpont, D. (1979). Closantel, a new antiparasitic hydrogen ionophore. *Arch. Int. Physiol. Biochem.* **87**, 851–852.

Van Schalkwyk, P. C., Geyser, T. L., Recio, M. and Erasmus, F. P. G. (1979). The anthelmintic efficacy of albendazole against gastrointestinal roundworms, tapeworms, lungworms and liverflukes in sheep. *J. S. Afr. Vet. Assoc.* **50**, 31–35.

Verheyen, A., Borgers, M., Vanparijs, O. and Thienpont, D. (1976). The effects of mebendazole on the ultrastructure of cestodes. *In* "Biochemistry of Parasites and Host-Parasite Relationships", (H. Van den Bossche, ed.), pp. 605–618. Elsevier/North Holland Biomedical Press, Amsterdam.

Verheyen, A., Vanparijs, O., Borgers, M. and Thienpont, D. (1978). Scanning electron microscopic observations of *Cysticercus fasciolaris* (= *Taenia taeniaeformis*) after treatment of mice with mebendazole. *J. Parasit.* **64**, 411–425.

Verheyen, A., Vanparijs, O., Lauwers, H. and Thienpont, D. (1980). The influence of closantel administration to sheep on the ultrastructure of the adult liver fluke *Fasciola hepatica* L. *In* "The Host Invader Interplay", (H. Van den Bossche, ed.), pp. 705–708. Elsevier/North Holland Biomedical Press, Amsterdam.

Verster, A. and Marincowitz, G. (1980). The treatment of *Stilesia hepatica* infestation. *J. S. Afr. Med. Assoc.* **51**, 249–250.

Verster, A., Tustin, R. C. and Reinecke, R. K. (1978). Attempt to treat the larval stage of *Taenia multiceps* and a resume of its neural and extraneural distribution in sheep. *Onderstepoort J. Vet. Res.* **45**, 257–259.

Walley, J. K. (1966). Oxyclozanide (3,3′,5,5′,6-pentachloro-2,2′-dihydroxybenzanilide—"Zanil") in the treatment of the liver fluke *Fasciola hepatica* in sheep and cattle. *Vet. Rec.* **78**, 267–276.

Walther, M. and Grossklaus, D. (1979). Zur Wirksamkeit von Praziquantel bei der experimentallen Saginata-Zystizerkose des Rindes. *Zbl. Veterinärmed.* **26**, 828–834.

Walther, M. and Koske, J. K. (1979). The efficacy of praziquantel against *Taenia saginata* cysticercosis in naturally infected calves. *Tropenmed. Parasit.* **30**, 401–403.

Watts, S. D. M. (1981). Colchicine binding in the rat tapeworm, *Hymenolepis diminuta*. *Biochim. Biophys. Acta*, **667**, 59–60.

Werczberger, A., Golhman, J., Wertheim, G., Gunders, A. E. and Chowers, I. (1979). Disseminated echinococcosis with repeated anaphylactic shock—treated with mebendazole. *Chest*, **76**, 482–484.

Wescott, R. B. (1967). Efficacy of niclosamide in the treatment of *Taenia taeniaeformis* infections in cats. *Am. J. Vet. Res.* **28**, 1475–1477.

Williams, J. F., Colli, C. W., Leid, R. W. and MacArthur, R. (1973). Effects of bunamidine hydrochloride on the infectivity of taenid ova. *J. Parasit.* **59**, 1141–1144.

Wilson, J. F., Davidson, M. and Rausch, R. L. (1978). Clinical trial of mebendazole in the treatment of alveolar hydatid disease. *Am. J. Resp. Dis.* **118**, 747–757.

Witassek, F., Eckert, J. and Bircher, J. (1980). Monitoring of mebendazole plasma concentrations in the treatment of alveolar echinococcosis. *Experientia*, **36**, 715.

Wittner, M. and Tanowitz, H. (1971). Paromomycin therapy of human cestodiasis with special reference to hymenolepiasis. *Am. J. Trop. Med. Hyg.* **20**, 433–435.

Wollweber, H., Niemers, E., Flucke, W., Andrews, P., Schulz, H.-P. and Thomas, H. (1979). Amidantel, a potent anthelminthic from a new chemical class. *Arzneim-Forsch.* **29**, 31–32.

Yorke, R. E. and Turton, J. A. (1974). Effects of fasciolicidal and anti-cestode agents on the respiration of isolated *Hymenolepis diminuta* mitochondria. *Z. ParasitKde*, **45**, 1–10.

Index